Walker's

Primates
of the World

D0931757

Walker's
Primates
of the World

Ronald M. Nowak

Introduction by
Russell A. Mittermeier,
Anthony B. Rylands, and
William R. Konstant

The Johns Hopkins University Press
Baltimore and London

Portions of this book have been adapted from *Walker's Mammals of the World,* 6th edition,
by Ronald M. Nowak, © 1999 by the Johns Hopkins University Press

The Johns Hopkins University Press
2715 North Charles Street
Baltimore, Maryland 21218-4363
www.press.jhu.edu

A catalog record for this book is available from the British Library.

Library of Congress Cataloging-in-Publication data

Nowak, Ronald M.
    Walker's primates of the world / Ronald M. Nowak ; introduction by
Russell A. Mittermeier, Anthony B. Rylands, and William R. Konstant.
        p.   cm.
    "Portions of this book have been adopted from Walker's mammals of
the world, 6th ed., by Ronald M. Nowak, c1999"—T.p. verso.
    Includes bibliographical references (p.   ).
    ISBN 0-8018-6251-5 (pbk. :   alk. paper)
        1. Primates.   2. Primates Classification.   I. Walker, Ernest P.
(Ernest Pillsbury), 1891–   Walker's mammals of the world.   II. Title.
QL737.P9N66   1999
599.8—dc21
99-28958
CIP

Contents

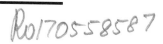

Walker's

Primates
of the World

Primates of the World:
An Introduction

Russell A. Mittermeier, Anthony B. Rylands,
and William R. Konstant

Primates are one of the more diverse groups of mammals. They are similar in number of species to the carnivores and the artiodactyls (even-toed hoofed mammals) and surpassed only by the incredibly diverse bats, insectivores, and rodents. In size range, they are similar to the order Carnivora. Varying in size by several orders of magnitude, they range from the tiniest of the mouse lemurs (*Microcebus myoxinus*) at a mere 24–38 grams to the gargantuan adult male mountain gorilla (*Gorilla gorilla beringei*), which can tip the scales at more than 200 kg. The largest primate that ever lived was a Pleistocene ape recorded from China and Viet Nam, *Gigantopithecus blacki,* believed to have weighed around 300 kg (Fleagle 1988). Seventy million years of evolving primate anatomy has been influenced significantly by an arboreal lifestyle and has resulted in such distinctive characteristics as stereoscopic vision, a relatively large brain, grasping hands and feet (having nails rather than claws), and superior levels of dexterity and muscular coordination. Primates are largely tropical creatures. Natural populations occur in the neotropical region of South America, through Central America as far north as southern Mexico, in northwestern and throughout sub-Saharan Africa, in the southern part of the Arabian peninsula, on the island of Madagascar, throughout the Indian subcontinent and southern China, in northeastern China and Japan, throughout Southeast Asia and Sundaland, and into Wallacea. Primates are absent from New Guinea, Australia, and the oceanic islands. As a group, primates eat a wide range of foods, including fruits, leaves, shoots, flowers, lichens, fungi, nectar, barks, gums, invertebrates, and vertebrates. Primate reproductive patterns and social units are also diverse, ranging from monogamous, territorial families to large, gregarious, multimale troops and to complex fission-fusion communities in which associations change frequently based on the availability of food and sexually receptive females. Despite their evolutionary success, however, primates are seriously threatened by a number of human activities throughout the tropical regions of the world, a situation that has placed more than 100 species and subspecies at great risk.

Taxonomy, Distribution, and Patterns of Diversity

The order Primates includes 15 families, 77 genera, and approximately 280 species that lived in historical times (the last 5,000 years) (Mittermeier and Konstant 1996/1997). Two of the families and 11 genera are extinct. The species comprise about 626 taxa, that is, monotypic species plus the number of subspecies within polytypic species. Groves (1993) and Rowe (1996, illustrated) provided listings of all modern primate species. Two suborders are recognized: the Strep-

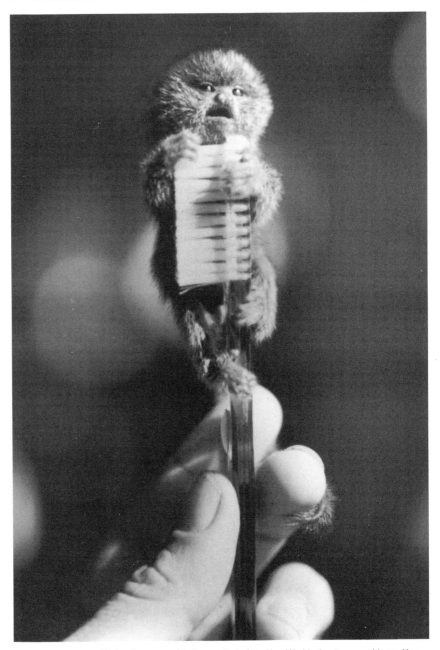

The pygmy marmoset (*Cebuella pygmaea*) is the smallest of the New World primates, as evidenced by this infant perched atop a toothbrush. Photo by Maria de Soive.

sirhini (prosimian or lower primates), with 8 families, and the Haplorhini (higher primates), with 7 (Table 1). As their name implies, living prosimians retain a suite of anatomical characteristics (largely cranial and dental) that were common to ancestral primates; thus they are generally regarded as being more primitive than the anthropoids. The typical prosimian skull, for example, has large orbits, a reduced braincase, an elongated snout, and more highly developed olfactory and auditory regions than the skull of a similar-sized monkey or in proportion to the larger skull of an anthropoid ape.

TABLE 1. PRIMATE DIVERSITY BY FAMILY

Family	Common name(s)	Distribution	Genera	Species	Taxa
Suborder Strepsirhini					
Lorisidae	Angwantibos, lorises, pottos, and galagos	Africa, Asia	9	18	44
Cheirogaleidae	Dwarf, mouse, and fork-marked lemurs	Madagascar	5	8	12
Lemuridae	Ring-tailed, gentle, brown, black, crowned, red-bellied, mongoose, and ruffed lemurs	Madagascar	5	11	20
Megaladapidae	Sportive lemurs and koala lemurs	Madagascar	2	10	10
Indriidae	Indri, avahi, and sifakas	Madagascar	3	6	12
Palaeopropithecidae	Sloth lemurs	Madagascar	4	5	5
Archaeolemuridae	Baboon lemurs	Madagascar	2	3	3
Daubentoniidae	Aye-aye	Madagascar	1	2	2
			31	63	108
Suborder Haplorhini					
Tarsiidae	Tarsiers	Asia	1	5	5
Cebidae	Owl, titi, squirrel, capuchin, saki, bearded saki, howling, spider and woolly monkeys, uakaris, and muriquis	Neotropics	14	65	150
Callitrichidae	Marmosets, tamarins, lion tamarins, and Goeldi's monkey	Neotropics	5	35	58
Cercopithecidae	Macaques, baboons, drills, mangabeys, patas, guenons, grivets, vervets, talapoins, swamp monkey, colobus, langurs, snub-nosed monkeys, and proboscis monkeys	Africa, Asia	21	96	280
Hylobatidae	Lesser apes: gibbons and siamang	Asia	1	11	26
Pongidae	Great apes: chimpanzees, gorillas, and orangutans	Africa, Asia	3	6	9
Hominidae	Humans	Global	1	1	4
			46	219	534
		Total	77	282	642

Note: The families Palaeopropithecidae and Archaeolemuridae, one genus and one species of the Lemuridae, one genus and three species of the Megaladapidae, one species of the Daubentoniidae, and three genera and three species of the Cebidae became extinct about 500 to 1000 years ago.

Although considered prosimians in the past, the tarsiers (infraorder Tarsii or Tarsiiformes) are the remnants of a quite separate radiation from that leading to the lemurs and lorises. Extinct "tarsierlike" primates of the family Omomyidae include the early Eocene *Teilhardina* found in deposits in Wyoming and Belgium, indicating that tarsiiforms separated from the lemurs and lorises some 65 to 60 million years ago, representing as such the first main division among the primates surviving today (Krishtalka and Schwartz 1978; Schwartz 1978). The tarsiers sometimes are included with the lower primates in a suborder designated Prosimii, and the remaining primates are then placed in a suborder called An-

The night-hunting tarsier (*Tarsius syrichta*), noted for its huge eyes and acrobatic leaps, is sometimes considered a prosimian, like the lemurs and lorises, but probably is more closely related to the monkeys and apes. Photo by David Haring, Duke University Primate Center.

thropoidea. Such a subordinal division would have the advantage of all early primates being classified as prosimians, but much evidence suggests that modern tarsiers are more closely related to the monkeys and apes than they are to the lemurs and lorises. That evidence requires a division of the primates, as given above, into the suborders Strepsirhini (lemurs and lorises) and Haplorhini (tarsiers, monkeys, and apes) (see, for example, Groves 1998; Martin 1978, 1979, 1995). Groves (1998, p. 13) wrote: "The term prosimian is no longer appropriate in a formal taxonomic sense, but it is still acceptable as long as one realizes that it belongs to the realm of Folk Taxonomy, designating a small brained primate, i. e. one that is not a monkey or ape." The living Strepsirhini all have a dental toothcomb (procumbent lower incisors); a claw, the "grooming claw," on the second digit of their feet; a postorbital bar (no postorbital closure); a well-developed rhinarium (a moist nose); and a *tapetum lucidum* in the eye, a reflective layer behind the retina that aids nocturnal vision. Archaic fossil prosimians (infraorder Plesiadapiformes) are believed to have originated in the Northern Hemisphere (North America and Europe) over 65 million years ago. The first modern primates appeared in the early Eocene and involved a diverse group in terms of dietary adaptations in about 40 genera, originating in Africa or possibly South Asia some 56 million years ago. It is believed that at this time primates also divided into two major lineages. One led to the anthropoids and the other to the modern prosimians (infraorders Lemuriformes and Lorisiformes).

Nearly three-quarters of the strepsirhines (and all of the tarsiers) are nocturnal, resting during the day in nests or tree hollows. Strepsirhine adaptations for a nocturnal niche include relatively large eyes; sensitive nocturnal vision (*tapetum lucidum*); large, independently movable ears; elaborate tactile hairs; and a well-developed sense of smell (rhinarium, along with relatively large olfactory bulbs in comparison to diurnal primates) (Bearder 1987; Martin 1979). Social relations and communication systems, constrained by the dark, are very different from those typical of the diurnal species (Bearder 1987). Some (or all) of the true lemurs (*Eulemur*), for example the mongoose lemur (*E. mongoz*) and the black lemur (*E. macaco*), may be active during the day and the night. This pattern is referred to as cathemerality (Tattersall 1988) and is believed to be determined by a number of factors, including thermoregulation (avoiding especially cold nights) and predation (avoiding daytime foraging during seasons when leaf cover is reduced in deciduous forests) (Colquhoun 1998; Curtis 1997).

Living prosimians occur only in the Old World tropics. Of the seven extant families, five (Cheirogaleidae, Lemuridae, Megaladapidae, Indriidae, and Daubentoniidae) occur naturally only in Madagascar, where they are represented by 14 genera, 32 species, and 51 taxa (Mittermeier et al. 1994). The family Cheirogaleidae includes five genera (*Microcebus, Mirza, Allocebus, Cheirogaleus,* and *Phaner*) and a total of nine small, nocturnal lemur species that move about quadrupedally and sleep during the day in tree holes or nests made from dead leaves. At least four species of mouse lemurs (*Microcebus*) range widely throughout Madagascar. One of these, *Microcebus ravelobensis*, sympatric with *Microcebus murinus*, was described only in 1998 (too late for coverage in the main text), and Malagasy primatologists have recently discovered a number of other new species (Zimmerman et al. 1998). *Allocebus* is a monotypic genus represented by the tiny hairy-eared dwarf lemur (*Allocebus trichotis*), once believed to be extinct but rediscovered in 1989 in the rainforests of northeast Madagascar (Meier and Albignac 1991). The genus *Mirza* is also monotypic; Coquerel's dwarf lemur (*Mirza coquereli*) is found in two areas of Madagascar's deciduous western forests and, at approximately 300 grams, weighs more than three times as much as the mouse lemurs with which it was once grouped. The two species of dwarf lemurs (*Cheirogaleus*) range from about 150 to 600 grams, weights that vary with the seasonal storage of fat reserves in their tails and periods of torpor during the southern winter. The fork-marked lemur, genus *Phaner*, contains a single species with four subspecies known from scattered populations around the island of Madagascar. It vies with the greater dwarf lemur (*Cheirogaleus major*) for being the largest cheirogaleid and is easily the most vocal.

The family Megaladapidae contains the single living genus *Lepilemur* and seven species of medium-sized, nocturnal lemurs widely distributed throughout Madagascar. However, many local variants exist, and this genus may eventually yield new taxa as it becomes better known. A much larger genus, *Megaladapis*, is now extinct.

The living genera *Lemur, Eulemur, Varecia,* and *Hapalemur,* together with the extinct *Pachylemur,* make up the widespread family Lemuridae, or "true" lemurs. All are medium-sized and largely diurnal. The monotypic ring-tailed lemur (*Lemur catta*), with its characteristic black-and-white striped tail, is perhaps the best known and was recently split off from the five species of *Eulemur.* The gentle or bamboo lemurs (*Hapalemur*) contain three species, all of which display a characteristic vertical resting posture, and one of which (*Hapalemur aureus*) was discovered only in 1986 (Meier et al. 1987). The large, quadrupedal ruffed lemurs (*Varecia*) are represented by a single species and two subspecies that inhabit Madagascar's eastern rainforests.

Largest of all Madagascar's living lemurs are the indri (*Indri*) and the sifakas (*Propithecus*) of the family Indriidae, which also includes the smaller woolly lemurs (*Avahi*). The first two are diurnal, the last nocturnal, but all have greatly elongated legs that assist in a form of locomotion known as vertical clinging and leaping. *Indri* is a monotypic genus confined to the eastern rainforests, *Propithecus* is represented by three species and nine taxa distributed around Madagascar's periphery, and *Avahi* consists of two distinctive subspecies (or two full species), one eastern and one western. Closely related to the Indriidae are two extinct families of much larger animals, the Palaepropithecidae (sloth lemurs) with four genera and five species, and the Archaeolemuridae (baboon lemurs) with two genera and three species.

The most distinctive and bizarre of the lemurs is the aye-aye (*Daubentonia madagascariensis*), the sole living representative of the family Daubentoniidae. It has a unique dentition among the primates, characterized by ever-growing incisors, and an unusual, skeletonlike middle finger, both of which are feeding specializations (Erickson 1995). Once believed to be near extinction, *D. madagascariensis* has recently been found to be widely distributed in Madagascar, though nowhere does it appear to occur in large numbers (Sterling 1994). A larger species, *D. robusta*, is extinct.

The nine genera, 18 species, and 44 taxa of the family Lorisidae are divided into two subfamilies—Lorisinae (lorises) and Galaginae (galagos)—distributed throughout mainland sub-Saharan Africa, India, Southeast Asia, and the East Indies. All are small, nocturnal primates with a common dental formula (2/2, 1/1, 3/3, 3/3 = 36). In Africa there are three lorisid genera, *Pseudopotto*, *Arctocebus*, and *Perodicticus*, containing 3 species, along with at least a dozen species of galagos of the genera *Galago*, *Galagoides*, *Euoticus*, and *Otolemur* (Oates 1996). In India and Asia, lorisids are represented by two genera (*Loris* and *Nycticebus*), which contain 3 species and 11 taxa overall (Eudey 1996/1997), although Groves (1998) concluded that they are taxonomically more diverse than has been recognized to date.

The small, nocturnal tarsiers (Tarsiidae) of Indonesia, Malaysian Borneo, and the Philippines possess certain anthropoid traits in addition to some primitive features shared with the lemurs, lorises, and galagos. For example, neither anthropoids nor tarsiers have the *tapetum lucidum* found in strepsirhines. Tarsiers, like anthropoids, have a fovea, which promotes visual acuity. They also have a dry nose, lacking the damp nasal skin (rhinarium) characteristic of the lemurs and lorises as well as other mammals. In addition, tarsier cranial anatomy of the auditory and olfactory regions (small bulbs) is more similar to that of monkeys than it is to other prosimians. These features indicate that tarsiers (along with night monkeys, *Aotus*) are secondarily nocturnal (Martin 1978, 1979). The most distinctive features of the five *Tarsius* species, however, are the elongated tarsal bones of their legs, which provide great leaping ability; their huge, fixed eyes (each larger than their brain, and in a nearly complete bony socket as in the anthropoids); and their large hands and feet, with long fingers and toes, which aid in support on vertical stems as well as in locating and capturing insect prey.

Compared with the prosimians, the anthropoid primates are much more diverse, having almost twice the number of genera and more than four times the number of species and taxa overall. Of the six anthropoid families, two (Callitrichidae and Cebidae) occur only in the New World tropics, two (Cercopithecidae and Pongidae) occur throughout much of Africa and Asia, one (Hylobatidae) is restricted to Asia, and the last (Hominidae), represented by our own species, is global in its distribution.

New World primates are grouped in the infraorder Platyrrhini, a name that describes their flat, outward facing nostrils. They comprise 19 genera (including 3 that may have occurred in the West Indies up until the time of Columbus but

The hairy-eared dwarf lemur (*Allocebus trichotis*) was believed extinct. It was rediscovered in 1989 in the rainforests of northeastern Madagascar by the photographer (Bernhard Meier).

are now extinct), about 100 species, and 205 taxa; the taxonomy of many genera is still controversial (Rylands et al. 1996/1997; van Roosmalen et al. 1998). Over the last decade, a phylogenetic approach to the systematics of these primates has identified three major groups, or clades, which some authorities consider distinct at the family level: Cebidae (the marmosets and tamarins along with the squirrel monkeys and capuchin monkeys), Atelidae (the howling monkeys, spider monkeys, woolly monkeys, and muriquis), and the Pitheciidae (the uakaris, sakis, and titi monkeys) (see Schneider and Rosenberger 1996). Both morphological and molecular genetic studies have agreed with this classification except in the case of the night monkey, *Aotus,* which although still enigmatic, is increasingly seen as a very early platyrrhine offshoot and perhaps even worthy of a separate family, the Aotidae. Still other suggested classifications are discussed in the main text. For convenience, in the following, however, we maintain the traditional division of these genera in two families, Callitrichidae (marmosets and tamarins) and Cebidae (the remaining genera).

Five genera of marmosets and tamarins compose the family Callitrichidae. All are small (less than 1 kg), have clawlike nails on all digits but their big toes, and lack the prehensile tail characteristic of some cebids (Hershkovitz 1977). The two subspecies of the pygmy marmoset (*Cebuella pygmaea*) are the smallest monkeys (adults weigh about 125 grams), and though they are quite distinct in some aspects of their appearance and behavior from other callitrichids, a number of authors now regard them as congeneric with the "true" marmosets (*Callithrix*), a result of both morphological and genetic studies (Rosenberger 1981; Tagliaro et al. 1997). Goeldi's monkey, the monotypic genus *Callimico,* is restricted to the upper Amazon basin. For many years it had been believed to be the closest living relative of the family's common ancestor because it retains certain primitive characteristics (a third molar—other callitrichids have two—and giving birth to single young—other callitrichids have twins). However, immunological and genetic studies are changing our thinking in this respect, placing *Callimico* as more closely related to *Callithrix* than *Callithrix* is to the tamarins (*Saguinus*) (Cronin and Sarich 1978; Pastorini et al. 1998). The genera *Callithrix* and *Saguinus* are the most geographically widespread and speciose of the callitrichids, represented by 18 and 12 species, respectively, that range from Central America south through Amazonia and into both dry and wet forest formations of central and eastern South America (Ferrari 1993). The four species of lion tamarins (*Leontopithecus*) are endemic to Brazil's Atlantic forest, where their few small, isolated populations are critically endangered.

Physically and taxonomically, monkeys of the family Cebidae are much more diverse than the callitrichids. There are 14 (including 3 extinct) genera, about 65 species, and 149 taxa ranging in size from 1 kg (*Saimiri*) to 15 kg (*Brachyteles*). A number of species have prehensile tails. There is one unusual nocturnal genus, *Aotus*, the night or owl monkey. Because of taxonomic reassessment, the number of night monkey species has grown from 1 to 10 in the last decade. However, *Aotus* remains one of the least known of the New World monkey genera due to its nocturnal habits, though in the chaco habitat in the south of its range it is active during the day as well (cathemeral). The other living small- to medium-sized cebids include 5 species (12 taxa) of squirrel monkeys (*Saimiri*), 6 species (33 taxa) of capuchins (*Cebus*), 13 species (26 taxa) of titi monkeys (*Callicebus*), 5 species (7 taxa) of saki monkeys (*Pithecia*), 2 species (4 taxa) of bearded sakis (*Chiropotes*), and 2 species (6 taxa) of uakaris (*Cacajao*) (see Rylands et al. 1995). Distinctive characteristics of these monkeys include the prehensile tail of the capuchins; the monogamous family units of the titis and the incredibly large troops of the squirrel monkeys; and the specialized incisors and canines of the sakis, bearded sakis, and uakaris, which allow them to eat hard fruits and nuts (Ayres 1989).

The four remaining living genera of New World primates are all large-bodied and prehensile-tailed monkeys of the subfamily Atelinae (Rosenberger and Strier 1989). Nine species and about 23 taxa of howling monkeys (*Alouatta*) range widely throughout Central and South America and are easily recognized by their relatively slow, quadrupedal locomotion, sexual dimorphism (males larger than females), and loud territorial vocalizations, amplified by a large, hollow resonating chamber formed by the hyoid bone, from which their common name is derived. The spider monkeys (*Ateles*), whose distribution is similar to that of howlers, though not extending south of the Amazon forests, include six species and 17 taxa. They lack thumbs entirely and are named for their long limbs and brachiating locomotion. Woolly monkeys (*Lagothrix*) are similar in size, but bulkier. One species, *Lagothrix lagotricha* (with four subspecies), is widespread throughout western Amazonia, while the other, *Lagothrix flavicauda*, is restricted to the Peruvian Andes (Mittermeier et al. 1977; Ramirez 1988). The largest of the neotropical primates are the two distinct subspecies or two full species of muriqui (*Brachyteles*), which, like the lion tamarins, are restricted to Brazil's Atlantic forest and are critically endangered (Strier 1992).

The Old World monkeys (Cercopithecidae) belong to the infraorder Catarrhini, a name that describes their narrow, downward-facing nostrils. There are 21 genera and approximately 100 species, all of which are diurnal, have longer hind limbs than forelimbs, and have flattened nails on their digits. None have the prehensile tails found in some New World monkeys. Most species are at least partly terrestrial and possess prominent ischial callosities (bare calloused patches on their buttocks). Catarrhine monkeys are divided into two subfamilies, Cercopithecinae and Colobinae, based largely on dietary adaptations. The cercopithecines have molars with rounded, low cusps, a deep jaw, and orbits that are close together; they are primarily frugivorous and typically possess cheek pouches in which they store food; the tail is short; the arms and legs are of similar size (for quadrupedal locomotion); and they have well-developed thumbs. The

Milne-Edwards's sportive lemur (*Lepilemur edwardsi*) is a nocturnal primate found in the deciduous forests of western Madagascar. Leaves appear to be the mainstay of its diet. Photo by Russell A. Mittermeier.

The aye-aye (*Daubentonia madagascariensis*) is the most bizarre of the living lemurs. Once thought to be nearly extinct, it is now known to occur in small numbers throughout much of Madagascar. Photo by David Haring, Duke University Primate Center.

colobines, by comparison, have high-cusped, shearing molars, specialized multi-chambered stomachs for fermenting fibrous leaves, widely spaced orbits, long legs (for leaping), short or absent thumbs, and a very long tail. Colobines lack cheek pouches.

Cercopithecine primates include the macaques (*Macaca*), predominantly Asian species save for the Barbary macaque (*Macaca sylvanus*), which is found in northern Africa and on Gibraltar (Lindburg 1980). *Macaca* has 20 species and 42 taxa overall (Eudey 1996/1997). The baboons (*Mandrillus, Papio* and *Theropithecus*) are the largest of the cercopithecines, with at least 1 species, the hamadryas baboon (*Papio hamadryas*) approaching 40 kg. This is also the only one of 8 baboon species whose distribution extends beyond continental Africa to include a thin strip along the southwestern coast of the Arabian peninsula. Seven additional genera and at least 33 species of cercopithecines are distributed throughout much of central Africa, including 5 species of mangabeys (*Cercocebus* and *Lophocebus*), 24 species of guenons (*Chlorocebus* and *Cercopithecus*), Allen's swamp monkey (*Allenopithecus nigroviridis*), the highly terrestrial patas monkey (*Erythrocebus patas*), and the smallest members of the group, the 2 species of talapoin monkey (*Miopithecus*), which barely weigh more than a kilogram (Oates 1996). Taxonomically, the most problematic of the cercopithecines are the guenons, whose name is derived from a French word for "fright," which describes a characteristic grimace displayed by many members of this group.

The Old World colobines are also called leaf-monkeys because of their predominantly specialist folivorous diets. This subfamily is represented by two major evolutionary radiations, one in Africa and the other in Asia. African colobines are commonly called colobus monkeys. Oates (1996) recognized two genera (*Procolobus* and *Colobus*), 7 species, and 19 subspecies or populations. Other authors recognize a third genus (*Piliocolobus*) and as many as 26 or 28 taxa (Kingdon

1997; Rowe 1996). Our main text lists three genera and 11 species, but a taxonomic revision of this group is clearly needed.

Asian leaf-eating monkeys are commonly referred to as langurs, and seven genera and as many as 30 (the main text lists 35) species and 89 taxa are recognized (Eudey 1996/1997). Ten species and 31 taxa of the genus *Presbytis*, the young of which sport a characteristic white natal coat, are found on Peninsular Malaysia and in Indonesia. Young of the genus *Trachypithecus*, by contrast, are covered with bright yellow or orange hair at birth. Eudey (1996/1997) recognizes 12 *Trachypithecus* species and 36 taxa from southern India, Sri Lanka, Southeast Asia, and Indonesia. The remaining four genera are much smaller in number of taxa and less widespread geographically. Eleven subspecies of the Hanuman langur (*Semnopithecus entellus*) are distributed throughout the Indian subcontinent and north into the foothills of the Himalayas. The monotypic proboscis monkey (*Nasalis larvatus*) is endemic to the island of Borneo; two subspecies of the snub-nosed langur (*Simias concolor*) are endemic to Indonesia's Mentawai Islands; and two or three subspecies of Douc langur (*Pygathrix nemaeus*) occur in Viet Nam, Laos, and Cambodia (Nadler 1997). The genus *Rhinopithecus* includes 4 highly endangered species, 1 (*Rhinopithecus avunculus*) from Viet Nam and 3 (5 taxa) from temperate forest habitats in southwestern China.

Gibbons and siamangs (*Hylobates* and *Symphalangus*) are commonly referred to as the lesser apes. Although they are similar in size to the Southeast Asian leaf monkeys whose habitats they share, they are readily distinguished from them by their lack of a tail and brachiation as their mode of locomotion. Eleven hylobatid species and 26 taxa are presently recognized (Eudey 1996/ 1997), although the taxonomy of this family is still under debate. All are monogamous, highly territorial, and vocal. Their vocalizations may be a useful tool for sorting out gibbon taxonomy (Leighton 1987).

The great apes of the family Pongidae include the largest of the living primates. As a group they also have larger brains relative to other primates and are sexually dimorphic, males being larger than females. Social behavior within this small taxonomic group, however, varies significantly from one species to the other. At least two species (four taxa) of chimpanzees (*Pan*) live in the tropical forests of West and Central Africa, as do the three or four subspecies of gorilla (*Gorilla gorilla*), which, averaging 117 kg in weight, are the world's largest living primate. Both chimpanzees and gorillas live in large, sometimes complex social groups. The orangutan (*Pongo*), represented by two distinct forms on the islands of Borneo and Sumatra, is a much more solitary creature, with males and females typically coming together only to mate.

Looking at global primate diversity from a regional perspective, we see that it is by no means evenly distributed. For example, the remaining tropical forests of Madagascar are dwarfed by the extensive tropical forests of Africa, Asia, and Central and South America, yet Madagascar has by far the densest concentration of primate diversity anywhere on Earth within an area of less than 600,000 km². Although usually considered part of the Afrotropical zoogeographical region, Madagascar is in many respects a zoogeographical region in itself, and especially so with respect to its unique mammal, and especially primate, fauna. All of Madagascar's primates are endemic, with *Eulemur mongoz* and *E. fulvus* also occurring on the nearby Comoro Islands, where they were almost certainly introduced by people (Mittermeier et al. 1994; Tattersall 1982; Tattersall and Sussman 1998).

With the exception of the Barbary macaque (*Macaca sylvanus*), all primates of Africa and nearby islands (e.g., Zanzibar, Bioko) occur within the Ethiopian

The indri vies with the diademed sifaka for the title of the largest living lemur. It is restricted to the eastern rain forest region of Madagascar. Photo by Russell A. Mittermeier.

zoogeographical region. They include 22 genera, about 70 species, and 190 to 200 taxa overall.

Within the Neotropics, primates occur from southern Mexico, through Central America and northern South America, to as far as southern Brazil, northern Argentina, and Paraguay (but not Chile). Native primate populations also occur on Trinidad; there are several introduced populations of African primates on the islands of St. Kitts, Nevis, Barbados, and Grenada; both New World and Old World species have been introduced to the island of Puerto Rico; and at least 1 squirrel monkey species and 1 macaque species occur as free-living populations in southern Florida. Sixteen living genera and 203 taxa are found within the Neotropics, figures comparable to those for Africa. However, the approximately 100 primate species of the Neotropics are the most for any major region, account for 36 percent of global primate diversity, and are roughly equivalent to the number of species inhabiting Africa and Madagascar combined.

Asian primates are found mainly in the Oriental zoogeographical region, as well as in the southeastern portion of the Palearctic and in Wallacea (the transition zone between the Oriental and Australian regions). In south Asia primates are widely distributed from the Indian subcontinent and the island of Sri Lanka throughout Southeast Asia as far as the Philippines and the Indonesian islands of Halmahera and Sulawesi, and in central and north Asia from Afghanistan through southern China (including the islands of Hainan and Taiwan) to Japan. In contrast to Africa (excluding Madagascar) and the Neotropics, where primates are basically continental, Asian primates are found in large numbers on islands as well. The region is home to 13 genera and about 70 species and 176 taxa, a slightly lower level of generic diversity compared to the Neotropics and Africa, but a level of species diversity roughly equivalent to that of Africa.

Ninety-two of the world's 192 sovereign nations have wild primate popula-

TABLE 2. WORLD'S TOP SEVEN COUNTRIES
FOR PRIMATE SPECIES DIVERSITY
(>30 SPECIES)

Country	Species	Genera
Brazil	77	16
Democratic Republic of the Congo	37	18
Indonesia	36	9
Madagascar	33	14
Cameroon	32	18
Peru	32	12
Colombia	31	12

tions; the top seven countries for primate species diversity are listed in Table 2. Brazil, with 77 species, is far and away the leader and accounts for slightly more than three-fourths of all neotropical primate species. Together, the top four countries (Brazil, Democratic Republic of the Congo, Indonesia, and Madagascar), which also represent the world's four major primate habitat regions, account for 182 species, or two-thirds of all living primates.

Furthermore, three of the top four countries for living primate diversity—Madagascar, Brazil, and Indonesia—also head the list of the world's top countries for primate endemism (Table 3). Madagascar is by far the international superstar with 33 unique species, 14 unique genera, and five unique families, all three representing 100 percent levels of endemism. Primate faunas for the next two countries on the list, Brazil and Indonesia, are roughly 50 percent endemic. Although Brazil has more endemic species (38) than Madagascar, it covers an area almost 15 times as large, and only 2 (*Leontopithecus, Brachyteles*) of its 16 primate genera (12.5 percent) are endemic (one-seventh the number for Madagascar). Indonesia is a distant third on the list with 19 endemic species and 1 endemic genus (*Simias*), after which the numbers of endemic species fall off precipitously, and no other country can claim an endemic primate genus.

The key point here is that, within the broad geographic regions that provide critical habitat for wild populations of nonhuman primates, there are a handful of countries that harbor a disproportionately large share of the world's primate fauna. The four "megadiversity" countries—Brazil, Democratic Republic of Congo, Indonesia, and Madagascar—must therefore rank among the highest global priorities for conserving primates. Furthermore, if we extend the analysis to consider subspecies as well, we find that other countries will rise toward the

TABLE 3. WORLD'S TOP SIX COUNTRIES FOR PRIMATE ENDEMISM

Country	Endemic species	Percent	Endemic genera	Percent
Madagascar	33	100	14	100
Brazil	39	51	2	12.5
Indonesia	19	53	1	11.1
Colombia	4	13	0	0
Peru	3	9	0	0
Democratic Republic of the Congo	3	8	0	0

The ring-tailed lemur (*Lemur catta*) is the most terrestrial of Madagascar's living primates, although this photo depicts one in the terminal branches of a flowering sisal plant in the southern spiny desert. Photo by Russell A. Mittermeier.

top of the priority list. Take India, for example. With only 15 primate species and 3 endemics, it falls well below many other countries on the previous lists. However, of India's 34–35 primate taxa, 22 are endemic, and the additional 19 endemic subspecies represent important wild populations that should be considered when establishing conservation priority rankings.

The slender loris (*Loris tardigradus*) is a nocturnal prosimian native to southern India and Sri Lanka. Its diet consists mostly of insects. Photo by Russell A. Mittermeier.

Primate Habitats and Communities

Primates are typically animals of the tropical rainforest, largely confined to 23°5' north and south of the equator, but extending as far as 30° S in South America (the Atlantic forest and Chaco) and 30° N in Southeast Asia. Tropical rainforest is composed of an enormous diversity of plants and life forms (canopy trees, including stranglers, understory trees, shrubs, lianas, understory herbs, and vascular epiphytes, parasites, and saprophytes), and the morphological characteristics of primates reflect their adaptations in terms of moving about in, and obtaining food from, this extremely complex, three-dimensional, species-rich, varied, and dynamic environment. These forests not only vary in terms of species composition, but also comprise complex mosaics of different forest types depending on such factors as altitude, rainfall and temperature, soil types, and flooding. The rainforest of the Amazon basin, for example, can be divided into upper and lower montane forest, lowland forest, swamp forests of various types (blackwater inundated forest or *igapó*, white-water inundated forest or *várzea*, *Mauritia* palm swamp), heath forests (including tall forest on sandy soils, or *campinarana*, and scrub on sandy soils, or *campina*), gallery forests, and dry, semideciduous forests with marked seasonality in rainfall.

In South America, the lowland forests of the Amazon basin have the richest communities, but they vary from as few as 3 sympatric species in the tidal gallery forests of the lower Amazon, to 5 in the *várzeas* of the lower Rio Japurá, to 7 in eastern Amazonia, and as many as 14 in lowland forests of the upper Amazon along the Rio Juruá (Rylands 1987; Ferrari and Lopes 1996; Peres 1997*a*). In the Brazilian Atlantic forest the maximum number of primates found in any one locality is 6 (Rylands et al. 1996). In Madagascar, 11–12 sympatric species occur in the eastern lowland evergreen rainforests (annual rainfall between 2,000 and 3,000 mm and no dry season), in, for example, the Ranomofana National Park (Mittermeier et al. 1994; Wright 1996*a*). As pointed out by Ganzhorn et al. (1997), however, this is the situation today whereas, in the recent past, communities may have been much richer. At Ankarana in northern Madagascar, for example, there are 10 extant lemurs, but a further 6 species have been found as subfossils, 4 of them extinct and 2 (*Hapalemur simus* and *Indri indri*) still surviving elsewhere (Wilson et al. 1995). A full complement, in this case would involve a community of 16 primates, as rich as any recorded to date. In Africa, the richest communities have been recorded for the rainforests of Gabon, for example, the Makokou region, M'passa plateau with 16 species, 5 nocturnal and 11 diurnal (Charles-Dominique 1977; Gautier-Hion 1978; Gautier-Hion and Gautier 1974). The swamp forest of the Liboy River, also in the Makokou region totals 13 species (Quris 1976). In Asia, the numbers of sympatric species tend to be lower. The richest communities have been recorded at the Kutai Nature Reserve, Borneo, with 10 species (Rodman 1978; Waser 1987), and the Krau Game Reserve, Malaysia (Chivers 1980), and the Gunung Leuser National Park, Sumatra (Van Schaik and Mirmanto 1985), each with 7 species.

Whereas the composition of primate communities reflect strongly the biogeographic distributions of distinct clades of primates, a number of authors have looked for patterns worldwide to examine to what extent and how the component species fill similar niches or are convergent. Analyses have been made both within (for example, Eisenberg [1979] and Peres [1997*a*, 1997*b*], for the Neotropics; Ganzhorn [1998], Kappeler and Ganzhorn [1993], and Tattersall [1982], for Madagascar; Oates et al. [1990] for Africa; and Brandon-Jones [1996] for Asia) and between geographical regions (for example, Bourlière [1979, 1985], Kappel-

Goeldi's monkey (*Callimico goeldii*) was once thought to be the sole member of the family Callimiconidae, but it is now lumped with marmosets and tamarins in the family Callitrichidae. Photo by Russell A. Mittermeier.

er and Heymann [1996], Terborgh and Van Schaik [1987], and Waser [1987]). Fleagle and Reed (1996) compared primate communities in eight localities, two in each of the continents and Madagascar, examining 10 variables for the 70 species included regarding diet, locomotor behavior, activity cycles, and body weight. Their study indicated the main adaptive niches occupied by primates in each of the major habitat regions. They found that the Malagasy primates today are characterized by two distinct ecological clusters—small- to medium-sized folivore leapers, and small faunivorous-frugivorous arboreal quadrupeds—and, as also pointed out by Tattersall (1982) and Terborgh and Van Schaik (1987), that Madagascar stands out in its abundance of folivores and frugivore-insectivores though, as mentioned above, communities may be considered to be incomplete when taking into account recent extinctions, especially of the large suspensory and terrestrial species (Mittermeier et al. 1994). Madagascar harbors more nocturnal species, it is home to all but 1 of the species that show irregular bursts of activity, and its lemurs live in significantly smaller groups than other primates (the majority are solitary or pair living) (Kappeler and Heymann 1996). Distinctive of the African primate communities are numerous sympatric cercopithecines, notably among the guenons (*Cercopithecus*), with 4 to 6 species occurring together and in many areas forming polyspecific associations (Gautier-Hion et al. 1988). Asian communities are marked by a reduced number of species even though the total number of species is on a par with Africa. The reasons for this are believed to lie in the complex zoogeographical history of Southeast Asia, involving numerous isolated islands and peninsulas (see Brandon-Jones 1996). Speciation has resulted more in large numbers of ecologically similar, allopatric, congeneric sister species, than in ecologically divergent species that would contribute to richer communities, in a situation where many islands restrict species-packing due to their small area. Terborgh and Van Schaik (1987) argued that the reduced numbers of species in the communities is related to a highly irregular fruit production when compared, for example, to Africa or the Neotropics. Likewise, Caldecott (1986) pointed to the abundance of dipterocarps that have inedible fruits, though these arguments are contested by Fleagle and Reed (1996), who insisted that low numbers of species and low overall ecological diversity do not indicate a narrower resource base, and cited Oates and Davies (1994), who reported that the guts of Asian colobines allow them to occupy broader niches than other primate species. The most unique aspect of the Asian primate communities is the number of suspensory species: gibbons, siamang, and orangutan. Asia is the only region with strictly faunivorous primates (the tarsiers), but it lacks primarily gumivorous species (Kappeler and Heymann 1996). The Neotropics, by comparison to other regions, show a low adaptive diversity, with few folivores (Kappeler and Heymann 1996), only one nocturnal species (*Aotus*) (Wright 1981, 1996b), no solitary, and no terrestrial species. The small, nocturnal, insectivore-frugivore niches are occupied by didelphid marsupials. Size diversity is also low, and more comparable to the extant prosimians than to the anthropoid communities of Asia and Africa (Terborgh and Van Schaik 1987). Terborgh and Van Schaik (1987) argued that the small size and only limited folivory in the platyrrhines is due to strong seasonality and the overlap in fruiting and leafing cycles, resulting in accentuated periods of food shortage. The recent finding of two large late Pleistocene atelines of more than 20 kg (Cartelle and Hartwig 1996; Hartwig and Cartelle 1996), which Heymann (1998) postulated may even have been terrestrial, indicates a similar situation to that in Madagascar, with a broader range size and more diverse communities in the past (Fleagle and Reed 1996).

Wright (1997) provided a detailed comparison of the strepsirhine radiation with that of the platyrrhines in terms of their zoogeographical history, some morphological aspects, dietary adaptations, locomotor behavior, reproduction and behavioral ecology, and social organization. Smith and Ganzhorn (1996) also made a most elegant comparison of the arboreal mammal communities in very similar habitats in Australia (possums and gliders) and Madagascar (lemurs). Although they found little convergence in terms of community structures, there are strong similarities in dietary adaptations at the species and niche level. This analysis highlighted the importance of founding effects, competition from other vertebrate taxa, coevolution between the species and their food plants, and differences and similarities in biophysical environments as determinants of mammalian radiation and convergence.

Regional differences within forest types are notable in the extent and seasonality of rainfall, resource productivity related to soil types, and natural disturbance regimes. As such the diversity and the temporal (seasonal) and spatial abundance, quality, and distribution of primate foods are key aspects responsible for the richness and structure of primate communities (Bourliére 1979, 1985). Rainfall is believed to be a good indicator of resource productivity for primates and hence of the number of species in a given community. Reed and Fleagle (1995) analyzed species diversity at more than 70 sites worldwide and showed that higher numbers of species occur in areas with higher mean annual rainfall for South America, Africa, and Madagascar, but not for Asia, possibly, as mentioned above, because of the number of islands, which force an upper limit on potential primate diversity (size effect) independent of rainfall. The richest communities are found in areas of high annual rainfall and low seasonality in the Neotropics (Eisenberg 1979; Pastor-Nieto and Williamson 1998). Ganzhorn et al. (1997) examined tree species diversity, structural complexity, and the extent of disturbance in forests

The white uacari (*Cacajao calvus calvus*) inhabits flooded, white-water forests of western Amazonia. Its strong jaws and specialized teeth are adapted for eating hard nuts and seeds. Photo by Russell A. Mittermeier.

as factors determining lemur species richness in Madagascar. The number of species increased with increasing structural diversity and the number of tree species, with floristic diversity overriding the effects of forest structure. At one undisturbed, floristically very rich site, however, the number of lemur species dropped, which may have been due to the resulting rarity of key resources. Slight habitat disturbance and increasing fruit and leaf production also contributed to increasing lemur diversity. Overall, the presence or absence of a species in a particular forest within its geographic range depends on the availability of key re-

The night monkeys (*Aotus* spp.) are the only nocturnal Neotropical primates. The taxonomy of this genus is disputed and requires further examination. This individual is from Panama. Photo by Russell A. Mittermeier.

sources (and possibly competition). To understand why a certain species is absent or occurs in low densities in some areas, while occurring at high densities in others, demands a detailed and long-term knowledge of its feeding ecology for a full comprehension of the limiting factors. Hunting is a confounding variable but can be accounted for in such analyses. Peres (1997*b*), for example, found that variation in density and abundance of the folivorous howling monkeys (*Alouatta*) in 106 localities, while affected by hunting, was related to the structural heterogeneity of the canopy and, to a lesser extent, to rainfall seasonality, total rainfall, and latitude. The best single predictor of density was the distance to white-water rivers, which covaried with forest heterogeneity and soil fertility. He concluded that howlers face strong dietary constraints related to the quality, phenology, and productivity of digestible foliage—undoubtedly also true for other folivorous primates.

Polyspecific associations involving mixed-species groups occur in tamarins (*Saguinus fuscicollis* associate with mustached tamarin species, *S. mystax*, *S. labiatus*, or *S. imperator*) (Buchanan-Smith 1990; Heymann 1990; Pook and Pook 1982); between squirrel monkeys (*Saimiri*) and capuchin monkeys (*Cebus*) (Fleagle et al. 1981; Podolsky 1990) in the Neotropics; and, most notably, between a number of species of guenons in West Africa (Gautier-Hion 1988; Gautier-Hion and Gautier 1974; Gautier-Hion et al. 1983). For example, at Makokou, Gabon, the putty-nosed monkey (*Cercopithecus nictitans*), the crowned monkey (*C. pogonias*), and the mustached monkey (*C. cephus*) associate regularly and for long periods. *C. nictitans* and *C. pogonias* groups can spend up to 97 percent of their day together (Gautier-Hion and Gautier 1974). At Makokou, guenons also associate with the gray-cheeked mangabeys (*Lophocebus albigena*) and talapoins (*Miopithecus*) (Waser 1980). There is much discussion as to the behavioral mechanisms involved in these associations, as well as the reasons they exist in functional adaptive terms. They evidently vary between the species involved, but hinge mainly on aspects concerning protection from predators and increased foraging efficiency (Gautier-Hion et al. 1983; Peres 1992*a*, 1992*b*, 1993; Struhsaker 1981; Waser 1987).

Primate communities in more seasonal habitats have fewer species, going from semideciduous to deciduous forest, savannah woodland, and wooded grassland, to semidesert scrub and deserts. In South America, only titi monkeys (*Callicebus*), marmosets (*Callithrix*), howlers (*Alouatta*), and capuchin monkeys (*Cebus*) occur in savanna regions, where they occupy deciduous and semideciduous forest patches and gallery forests. In the chaco of Bolivia, Paraguay, and Argentina there are also night monkeys (*Aotus*). In Africa, the chief savanna species include, in woodland savanna, vervets and grivets (*Chlorocebus* and *Cercopithecus*), patas monkeys (*Erythrocebus*), and baboons (*Papio* and *Theropithecus*), which extend into open montane grassland, desert scrub, and even desert. Some langur species, for example, the Hanuman langur (*Semnopithecus entellus*) and the Nilgiri langur (*Trachypithecus johnii*), and some macaques, *Macaca mulatta*, for example, occur in open woodland savanna in Asia. In Madagascar, the mouse lemur (*Microcebus murinus*), the white-footed sportive lemur (*Lepilemur leucopus*), the ring-tailed lemur (*Lemur catta*), and Verreaux's sifaka (*Propithecus v. verreauxi*) occur in the dry, highly seasonal, thorn scrub in the south.

The numbers of species in primate communities also tend to drop with increasing altitude, associated with lower temperatures, changes in soil types, and the structure and species' composition of the forest types. A number of species, however, are well adapted to upper montane forest and very cold temperatures. In South America, the yellow-tailed woolly monkey (*Lagothrix flavicauda*) occurs in the Andes, from 700 meters to as high as 3,000 meters, where temperatures can reach 4°C (Mittermeier et al. 1977). The Hanuman langur (*Semnopithecus entellus*) has been recorded at altitudes above 4,000 meters, with temperatures reaching −3°C in alpine scrub in the Himalayas (Oppenheimer 1977). The Japanese macaque (*Macaca fuscata*) lives at the northernmost latitude of any

The monk saki (*Pithecia monachus*) inhabits the western Amazon Basin, where it lives in monogamous social groups. Photo by Russell A. Mittermeier.

of the nonhuman primates, in subtropical forest but also subalpine deciduous broadleaf forests up to 1,500 meters, which are snow covered during the winter months. The champions in terms of coping with extremely cold climates, however, are the snub-nosed monkeys. The Sichuan golden snub-nosed monkey (*Rhinopithecus roxellana*) lives in mixed bamboo, conifer, and deciduous forest up to 3,150 meters, and the Yunnan snub-nosed monkey (*R. bieti*) lives in temperate montane forest in China, from 3,000 to 4,500 meters, with snow and temperatures averaging below zero for several months of the year (Ren et al. 1996/1997). Generally only the larger primates can habitually occupy these extremely cold habitats, but some of the galagos, for example, can also endure very cold temperatures. In South Africa, *Galago senegalensis* may experience nighttime temperatures as low as −6.5°C, and minimum temperatures can remain around the freezing point for four weeks or more (Bearder and Martin 1980). During this time they survive on gums, and many curtail their activity, returning to their sleeping sites around midnight to escape the cold.

Mouse lemurs (*Microcebus*) and dwarf lemurs (*Cheirogaleus* and *Allocebus*) hibernate or pass through periods of lethargy at times of extreme food shortage (winter dry season), using fat reserves stored in the tail. *Allocebus* also accumulates fat all over its body (Meier and Albignac 1991). *Cheirogaleus* may hibernate for six to eight months of each year, losing much weight during this time. An adult *C. medius,* for example, can drop from 217 grams to 142 grams (Hladik 1988; Hladik et al. 1980).

Numerous factors define the ecological separation of primates that are sympatric, but the principal ones are body size; diet and foraging behavior; activity patterns; and the forest strata used for locomotion, sleeping, and foraging. Examples of in-depth, long-term, comparative ecological-behavioral studies demonstrating niche divergence allowing for sympatry in species-rich communities include those carried out at Makokou, Gabon (Charles-Dominique 1977; Gautier-Hion 1978; Gautier-Hion and Gautier 1974), the Kibale Forest, Uganda (Gebo and Chapman 1995; Struhsaker 1975, 1978, 1997; Struhsaker and Leland 1979), the Krau Game Reserve, Peninsular Malaysia (Chivers 1980; MacKinnon and MacKinnon 1980), the island of Barro Colorado, Panama (Leigh et al. 1982), Cocha Cashu in the Manu National Park in Peru (Terborgh 1983, 1990), and the Raleighvallen-Voltzberg Nature Reserve in central Suriname (Fleagle and Mittermeier 1980; Fleagle et al. 1981; Mittermeier and van Roosmalen 1981; van Roosmalen, 1985).

Locomotion and Posture

Living in three-dimensional habitats means that primates show a large number of anatomical specializations for locomotion. A combination of different forms of posture and locomotion—walking, running, bounding, climbing, vertical clinging, leaping, brachiation, and hanging, for example—can be used by any particular species at some time, but most show definite tendencies to predominate in one form or another depending on such factors as their size, the typical substrates they use, forest types, diets (leaf-eaters, for example, have different locomotory requirements from those that hunt animal prey [see Fleagle 1984]), and the height of the forest in which they spend most of their time (see, for example, Charles-Dominique [1977], who studied sympatric nocturnal prosimians in Gabon, and Fleagle and Mittermeier [1980], who compared locomotion in seven primates in Suriname). Mittermeier (1978*b*) also compared lo-

The Tonkean macaque (*Macaca tonkeana*) is a little-known species found only on the Indonesian island of Sulawesi. Photo by Russell A. Mittermeier.

comotion in *Ateles* using two different species in two different parts of the Neotropics. A number of species are highly specialized, but in most there is a tendency for just one or two forms of locomotion to predominate, with their characteristic, associated anatomical adaptations, in five major categories as described by Fleagle (1988, 1992): arboreal quadrupedalism, terrestrial quadrupedalism, leaping, hanging, and bipedalism.

Arboreal quadrupedalism (walking, running, and bounding) is the most common form of locomotion and probably characteristic of the earliest primates. It is typical, for example, of the guenons, macaques, colobines, capuchin monkeys, marmosets and tamarins, and bearded sakis. Arboreal quadrupedalism is associated with such anatomical features as a long tail (helping balance); relatively short limbs of equal length (often flexed while walking which, combined, lower the center of gravity); a narrow thorax; narrow, laterally placed scapulae; grasping feet; a long olecranon process (leverage for the triceps when the elbow is flexed); deep, robust ulna (suggesting an important role in support); and relatively long digits, especially the big toe, for grasping (not as long as suspensory species, but longer than in terrestrial). Whereas the legs and feet are important for stability and propulsion, the arms and hands are used more for "steering" on supports, which are often narrow and unstable.

Terrestrial quadrupedalism is typical of baboons, some macaques, and the patas monkey, which with their long legs may even be considered cursorial. In these species the tail tends to be reduced in size; the limbs are of similar length and are designed more for speed and simple fore-aft movements than for power or for relatively complex, rotational movements at the joints (when compared to the arboreal quadrupeds on uneven, rather complex substrates). The carpal and tarsal bones are robust for weight bearing, and the digits are relatively short.

Some primates may walk on their hind legs for a short period, but bipedalism as such is restricted to humans. In bipedalism the spine bears the weight of the body and shows a strong dual curvature, dorsally convex (kyphosis) in the thoracic region, and ventrally convex (lordosis) in the lumbar region (kyphosis extends the entire length of the spine in most other primates). The size of each vertebra increases from the cervical to the lumbar (each having progressively more weight to bear). The legs are long, the head of the femur is large, and the ischium and ilium are short. The feet act as levers, the big toe is adducted, and prehensility is lost.

Many primates are excellent leapers, and adaptations for this form of locomotion have evolved independently in a number of groups, for example, the prosimians, especially the Indriidae, the tarsiers and the anthropoids, especially the pygmy marmoset (*Cebuella*), and the saki monkeys (*Pithecia*) (Kinzey et al. 1975; Walker 1996). The hind limbs and the lumbar region are long for springing, and there are special and varied anatomical adaptations in the pelvic bones for propulsion (a long ischium, for example). The forelimbs tend to be shorter than the hind limbs. Suspension, hanging below arboreal supports, involves a range of combinations of the arms and legs and, in some of the New World monkeys, the prehensile tail. Suspensory species include the atelines (Rosenberger and Strier 1989), the orangutan, and such brachiating (hand-over-hand suspensory locomotion) species as the gibbons (Fleagle 1980). Both arms and legs are long, the trunk is short, the thorax is deep, with a broad fused sternum, and the lumbar region is short to reduce the bending of the trunk during hanging and reaching out. A deep narrow scapula is placed dorsally rather than laterally, which facilitates the movements of the forelimbs in different directions. The forelimbs and the hind limbs and pelvis are also adapted for flexibility in movement, and the digits are long.

Diets and Feeding Behavior

There are three main sources of foods for the primates: small animal prey, mostly insects and other arthropods and sometimes small vertebrates and birds' eggs, providing mainly lipids, proteins and essential amino acids (insectivory or faunivory); fruits (carbohydrates) and seeds (lipids and carbohydrates) (frugivory); and leaves and flowers (structural carbohydrates, such as cellulose and hemicellulose, and proteins) (folivory).

All primates include fruits, seeds, or leaves in their diet, except for the tarsiers (*Tarsius*), which are entirely faunivorous. The galagos and the small New World primates (marmosets, tamarins, and squirrel monkeys) are all highly insectivorous, but supplement their diets with fruits or gums (see below). Likewise, small cheirogaleid lemurs from Madagascar are insect foragers, but eat also fruits, flowers, nectar, and leaf buds, besides larval secretions, birds' eggs, and small vertebrates such as frogs and chameleons (Mittermeier et al. 1994). In terms of specialized insectivory, two primate species stand out prominently from the rest. The aye-aye (*Daubentonia madagascariensis*) of Madagascar has evolved three physical features—two that are unique among primates—that aid in securing the larvae of ambycid beetles hidden beneath tree bark (Erickson 1995). First, this relatively large nocturnal lemur uses its huge batlike ears to listen for even the slightest movements as the larvae shift about in their chambers, while tapping the branch. Once it locates its prey, the aye-aye uses its continuously growing incisors to gnaw through the bark and into the wood, opening a small hole to expose the larvae. Then it probes the hole with a long skeletal middle finger, essentially macerating the beetle grub and drawing it to the surface in a thick "soup" to be licked from the finger. The aye-aye's chisel-like incisors also allow it to break open coconuts and the hard nuts of the *ramy* tree as additional sources of protein and energy. Across the Mozambique Channel in Africa, chimpanzees have learned to fashion simple tools and use them to fish termites from their nests. Jane Goodall (1996) discovered this behavior in 1960 when she saw several chimps break off long blades of grass, peel them to a desired shape, and then insert them in the holes of earthen termite mounds. Within seconds, the chimps would withdraw the grass stem to which termites were attached. The tiny morsels were then carefully plucked off with the lips and the stem reinserted for another try. Termite-fishing was the first evidence of tool-making and tool use in a nonhuman primate.

Fruits are poor in protein, and the smaller primates supplement their diets with animal prey, whereas the larger ones also eat leaves (Chivers and Hladik 1980; Ford and Davis 1992; Kay 1984). In faunivores, the gut is dominated by the small intestine (for absorption), the stomach is simple, and colon is short with a vestigial or absent cecum. Folivores, on the other hand, have either a large and/ or complex stomach (for example, the langurs and colobines) or a large cecum/ colon (for example, the howling monkeys, gorillas, and the folivorous sportive lemur, *Lepilemur*), which acts as a chamber for microbial fermentation necessary for the digestion of celluloses and lignin. The problems of digesting plant fiber result in selectivity on the part of folivorous primates for immature leaves. However, many plants also protect their leaves through chemical defenses, including compounds that reduce digestibility (tannins and enzymes that bind to proteins) or are straight toxins, including alkaloids and phenols. These are dealt with in mammals in the liver or kidneys or by means of the gut microorganisms; they also result in selective feeding (Glander 1982).

Frugivores have guts that are intermediate in proportions to those of faunivores and folivores. The stomach and large intestine tend to be larger than in the small faunivores, due to the need to digest large quantities of fruits, and also in the larger species that supplement their diet with leaves (Chivers 1992; Chivers and Hladik 1980). Plant exudates, mainly gums, but also latex and nectar, assume importance for some of the primates (reviewed by Nash 1986). Gums are composed mainly of complex polysaccharides (polymerized pentose sugars) and require the action of symbiotic bacteria for their digestion in an enlarged cecum (Bearder and Martin 1980; Coimbra-Filho et al. 1980; Ferrari and Martins 1992). Gums are usually produced following damage, either by wind or by insects (some beetle and moth larvae). Notable among the prosimians are the fork-marked lemur (*Phaner furcifer*), the needle-clawed bushbaby (*Euoticus elegantulus*), and the Senegal bushbaby (*Galago senegalensis*), which specialize in harvesting gums of species such as *Acacia*, produced by the activities of wood-boring insects, and using their tooth-comb as a scoop (Bearder and Martin 1980; Hladik 1979; Nash and Whitten 1989). Two genera of callitrichids, *Cebuella* and *Callithrix*, specialize in obtaining gums through tree-gouging, using their elongate lower incisors and incisorlike canines (Coimbra-Filho and Mittermeier 1977). This behavior guarantees the availability of gums year-round and enables some of the species (the common marmoset, *Callithrix jacchus*, for example) to occupy relatively harsh, seasonal habitats (Rylands 1984; Rylands and de Faria 1993). Nectar can be an important source of carbohydrate for some of the smaller and (at times when fruits are scarce) even for some of the larger primates, which may, as a result, play a role as pollinators (Ferrari and Strier 1992; Janson et al. 1981; Sussman 1978).

In primates, food is ingested using the anterior dentition (incisors and canines) and the lips and hands. Besides being important for such uses as grooming and fighting, the anterior dentition can be related to the diets of many primates. The use of the incisors and canines for tree-gouging by marmosets and the dental-comb for scooping gums mentioned above are two examples. The incisor teeth of frugivorous primates, typically the cercopithecids such as the mangabeys, macaques, and baboons, but also the spider monkeys (*Ateles*) and the chimpanzee (*Pan*), act as scoops and tend to be relatively broader and larger than those of the leaf-eating colobines, the howling monkeys (*Alouatta*), and the gorillas (*Gorilla*), for example. The anterior dentition is most important for the pithecines (the sakis and uakaris), which are specialist seed-eaters, using their large procumbent incisors and robust canines to break open tough, usually immature fruits to obtain them. The cheek teeth, used to process the food, are especially indicative of diet and are used in three main ways: puncturing and crushing with sharp high cusps (small animal prey); shearing between enamel blades formed by lophs (crests linking cusps) (insectivorous and folivorous diets); and crushing or grinding, associated with rounded cusps and flat basins, the size and robustness of the cheek teeth depending on the hardness of the foods they are eating (diets of seeds and hard materials or durophagy). In pithecines, the cheek teeth are relatively small and flat, used to process soft seeds once they have been extracted with the strong anterior dentition. The capuchin monkeys (*Cebus*), on the other hand, have large robust cheek teeth, with thick enamel, which they use to break open hard fruits. The folivorous gorillas have high crests on their cheek teeth for slicing up leaves, whereas the frugivorous chimpanzees have broader basins and lower rounded cusps for crushing fruit pulp. For reviews of dental adaptations see Kay and Hylander (1978) and Fleagle (1988).

Reproduction and Life History Characteristics

Considering their body size, and compared with other mammals, primates are slow breeders; they produce few offspring, and these develop relatively slowly. The majority of species produce single offspring, multiple offspring (2–3) being typical only of some of the prosimians, along with the New World marmosets and tamarins, which produce twins. Sexual maturity is delayed, and gestation periods are generally quite long, as are interbirth intervals. Considerable and prolonged infant care is an important characteristic of the primates.

Although adult primates tend to have relatively large brains, slightly larger than in other orders of mammals (Martin and Harvey 1985), there is considerable overlap, and this is not a distinguishing character. What does distinguish the primates is the fetal brain size: neonatal primates have larger brains relative to their body weight than other mammals (Martin 1995). This is a reflection of the fetal development of the brain, energetically costly and implying that primate mothers invest more energy in a fetus or neonate of any given body size. During gestation, which, with relatively slow fetal growth, tends to be long in primates, the brain grows relatively rapidly in comparison to the body, though soon after birth, brain growth slows down. The relatively slow fetal growth rate means that the proportion of the mother's daily food intake devoted to the fetus is relatively small when compared to rapid growing altricial species, and the fetus is therefore "buffered" against temporary food shortages which the mother may experience—an important feature of flexibility for variation in food supply during pregnancy (Martin 1995). Due to the long gestation, and relatively large brain size, primate neonates are considered to be precocial (associated with a slow reproductive turnover) despite the fact that they are born relatively helpless and develop relatively slowly, demanding extended periods of maternal care (a long lactation period and late weaning). Weaning may occur when the infant is 18–30 months old in muriquis, *Brachyteles* (see Strier 1992) and almost three years old in spider monkeys, *Ateles* (see van Roosmalen 1985). In chimpanzees, *Pan troglodytes*, weaning, though begun in the infant's second year, may be completed only when it is five years old, with interbirth intervals of five and a half years or more (Graham 1981). In the large majority of the primates, the newborn infant is carried by its mother, and for this has a well developed capacity to grasp her fur, using not only the hands but also prehensile feet and, in a number of neotropical species, a prehensile tail. The production of litters in prosimians is associated with the use of nests (Cheirogaleidae, Galaginae, and the ruffed lemur, *Varecia variegata*). In the marmosets and tamarins, which typically produce twins, carrying is shared by the parents as well as other members of their social groups. The occurrence of multiple offspring in prosimians indicates that this, along with the use of nests, is a primitive trait for the primates. Carrying infants is important for thermoregulation and also provides for intimate behavioral interactions between the mother and infant and other group members, which facilitates social learning (Martin 1995). As pointed out by Martin (1995), the fact that infants are carried means that suckling occurs according to the infant's demands rather than a schedule determined by the mother. This is associated with the composition of the relatively dilute milk: the fat and protein contents are typically low (high in scheduled sucklers), but there is a high proportion of carbohydrate (Shaul 1962).

Two important features separate the strepsirhine from the haplorhine primates: the form of the placenta and the relative size of the newborn infants (Martin 1992, 1995). In strepsirhine primates the placenta is endotheliochorial and diffuse, with the outer embryonic membrane (chorion) remaining in contact with the intact layer of uterine epithelium. In haplorhine primates, on the other hand, the placenta is hemochoriol and diskoid. The uterine epithelium and the maternal capillaries supplying the placenta break down, and the chorion is bathed directly by the maternal blood. Fetal growth rates are generally lower in strepsirhine than in haplorhine primates, with the latter producing neonates about three times larger than the strepsirhines, despite similar gestation lengths, when body size is taken into account (Martin and Maclarnon 1988).

The form of the uterus is also distinct between the prosimians (strepsirhine primates and the tarsiers) and the anthropoids. In prosimians the uterus is bicornuate (two separate uterine chambers), a primitive feature of placental mammals. In the anthropoid primates, however, the uterus is simplex (a single uterine chamber), a rare condition in mammals and probably associated with a reduction in litter size in this group to generally single offspring (twinning in marmosets and tamarins evolved secondarily).

Primate Social Structures

A striking feature of primates is their sociality. The social systems of many primates are unusual among mammals in a number of ways: they are very complex, they involve various kinds of long-term social relationships (Cheney et al. 1987; Cheney and Seyfarth 1990), and they have highly diverse forms of social communication, involving auditory/vocal, tactile, chemical, and visual systems (Snowdon et al. 1982). Epidermal scent glands are especially important in prosimians and many of the New World monkeys, associated with specific scent-

The male proboscis monkey (*Nasalis larvatus*) has a very long, pendulous nose unlike that of any other primate. Photo by Russell A. Mittermeier.

marking behaviors (Epple 1986). Urine and feces are also important in olfactory communication in a number of species. Some prosimians show a stereotyped behavior of self-anointing or urine washing (Zeller 1987), and Heymann (1995) has also observed this for the tamarin, *Saguinus mystax*. Prosimians and some of the New World monkeys (marmosets and tamarins, for example) have vomeronasal organs. Body postures, gestures, piloerection patterns, and facial expressions are important forms of visual communication, especially for the haplorhines. Muscles of facial expression are more highly developed and complex in the primates than any in any other mammalian order. In prosimians, however, the longer noses, tethered lips, and reduced facial enervation limit the variety and complexity of facial expressions. Tarsiers, on the other hand, have a muscular upper lip associated particularly with facial expression. Touch as a means of communication is notably significant and sophisticated in the social apes (Fossey 1983; Goodall 1996) but is manifested most importantly in the large majority of primates in the form of grooming, the most common affinitive behavior. Perhaps the most complex and sophisticated of communication mechanisms in primates is to be found in their vocalizations (Cheney and Seyarth 1990; Snowdon 1989; Snowdon et al. 1982). Calls concerned with territoriality are notable in such as the indri, titi monkeys, howling monkeys, the mangabeys, and gibbons. Besides close-range social interactions, long distance calls are especially important for contact in nocturnal and forest-dwelling species, and alarm calls have been found to identify different predators. While many researchers strive to understand the structure and information content of primate vocalizations (see, for example, Snowdon 1989), Seyfarth (1987) and Cheney and Seyfarth (1990) emphasize the relations between vocal communication, cognition, and the social interactions of primates.

Most primates live in stable groups, though the size and structure of the associations can vary significantly from one species to the next. A small number of primates (the majority of the nocturnal species) are considered to be solitary foragers. They include lorises and galagos, several of Madagascar's lemurs, and the tarsiers (Richard 1987). However, we are only now beginning to study the behavior of some of the more obscure nocturnal primates, and their being solitary foragers does not necessarily mean they do not have quite complex and varied systems of social organization (see, for example, Bearder 1987; Harcourt and Nash 1986; Nash and Harcourt 1986). Several of the tarsiers, as well as Coquerel's dwarf lemur (*Mirza coquereli*) and the fork-marked lemur (*Phaner furcifer*) of Madagascar, though regarded as being primarily solitary, also appear to congregate at times in larger social groupings (Klopfer and Boskoff 1979; Niemitz 1979; Pages 1980), and in galagos the range of each male typically overlaps that of several females in what is referred to as the *noyau* social system. Bearder (1987) reviews social systems in the nocturnal prosimians and identifies five main types of social organization based on a number of parameters, including range overlap, social contacts at night, sleeping associations during the day, and mating patterns.

The orangutan (*Pongo pygmaeus*) also exhibits the *noyau* system and appears to be the only diurnal primate with a largely solitary lifestyle (Rodman and Mitani 1987). Large adult male orangs defend individual territories from rivals and subadult males and are typically seen in association with females only for purposes of copulation (Rijksen 1978). Females and their young remain together as a unit for several years, until the infant is fully weaned and the mother again becomes pregnant, at which point her young usually establishes a territory somewhere near her own (Watts and Pusey 1993).

By comparison, strong pair bonds and small, monogamous family groups occur in a number of distantly related primates: the titi monkeys (*Callicebus*) and night monkeys (*Aotus*) in the Neotropics, several indriids and diurnal lemurs in Madagascar, De Brazza's monkey (*Cercopithecus neglectus*) among the African cercopithecines, and the gibbons (*Hylobates*) among the apes. Titi and night monkey groups characteristically develop around a monogamous breeding pair (Kinzey 1981; Wright 1981, 1986). These pairs are territorial, and intergroup interactions are highly vocal involving, as in the gibbons, "duetting" (Robinson et al. 1987). The male participates in infant care by carrying the young (Dixson and Fleming 1981; Wright 1984). Madagascar's indri (*Indri indri*) and avahi (*Avahi laniger*) both live in small family groups of two to five animals centered around a monogamous breeding pair; the former species is diurnal and the latter nocturnal (Kappeler 1991; Mittermeier et al. 1994; Pollock 1977). In addition, the mongoose lemur (*Eulemur mongoz*), red-bellied lemur (*E. rubriventer*), golden bamboo lemur (*Hapalemur aureus*), ruffed lemur (*Varecia variegata*), and Verreaux's sifaka (*Propithecus verreauxi verreauxi*) all have been reported to occur in single-male/single-female groups at some study sites, though multimale/multifemale associations are perhaps more commonly encountered (Jolly 1994; Klopfer and Boskoff 1979; Wright and Randrimanantena 1989). The monotypic De Brazza's monkey of central Africa is the only cercopithecine monkey known to be monogamous, although it also exhibits multimale/multifemale associations (Rowell 1988). The only colobine known to be monogamous is the Mentawai Island leaf monkey (*Presbytis potenziani*), living in what appear to be highly cohesive and territorial groups, although ranges of neighboring troops have been observed to overlap significantly (Fuentes 1994; Watanabe 1981). Gibbons live in small monogamous family groups that aggressively defend their territories. Early warnings of this defense are most evident in the melodious vocal duets of the adult breeding pair, a characteristic behavior of the Hylobatidae that has evolved to produce distinctive songs among the different species (Leighton 1987). Among the primates, these incredible songs are rivaled only by those of

Madagascar's indri, also renowned for its male-female chorusing. The indri's eerie lament typically occurs in early morning just after a family group rises from sleep and then again in late afternoon before they retire for the night, and it plays a role in territorial demarcation similar to that of the gibbon's call (Pollock 1977).

Monogamy was long thought to be the rule for the callitrichids (excepting possibly *Callimico*). While groups may contain only a single adult breeding pair, in many there are additional adult males and females, and group sizes can reach 15 or more in the wild, at least in the marmosets (*Callithrix*) (Ferrari and Lopes Ferrari 1989; Snowdon and Soini 1988; Stevenson and Rylands 1988). In both marmosets and tamarins, however, there is a unique system of reproductive suppression by the dominant female on other females in the group, and generally only one female breeds at any one time (Abbott and Hearn 1978). The behavioral and physiological mechanisms involved in this suppression vary between the genera (Abbott et al. 1993). Polyandry and polygyny have also been suggested as alternative mating systems (Ferrari and Lopes Ferrari 1989; Rylands 1996). Infant carrying by the males and other group members, as well as food sharing, are notable features in the social organization of these primates (Box 1977; Feistner and Price 1991; Goldizen 1987; Sussman and Kinzey 1984; Yamamoto and Box 1997).

As primate social units increase in size, the structures, hierarchies, and group dynamics become more complex. In some cases, the social system can characterized at the generic level, but in others it varies between species of the same genus and even between populations, which is to be expected considering the importance of diet and food availability for demographics and mating systems. In broad terms, however, it is possible to indicate two general categories with respect to group structure. The first is the one-male group, where mating is believed to be polygynous and generally restricted to a single male, even though in some cases other subadult or subordinate males may be resident. In this case, males not living with females may form all-male groups, as in some langurs, or live alone, as, for example, in howling monkeys (*Alouatta*). As discussed, however, by Cords (1987), periods of promiscuous mating have also been recorded for one-male groups of, for instance, patas monkeys (*Erythrocebus patas*), the red-tailed monkey (*Cercopithecus ascanius*), and the gentle monkey (*Cercopithecus mitis*), due to a rapid turnover of adult males and male influxes during mating seasons. One-male systems are characterized by strong competition by males for females with often violent male takeovers resulting in infanticide on the part of a successful new male (Crockett and Sekulic 1984; Hrdy 1980). The second major category is the large bisexual group. These groups are referred to as multimale, or multimale/multifemale, with at least two important variations in terms of group dynamics: fission-fusion societies, for example, in the spider monkeys (*Ateles*) and the chimpanzees (*Pan*), and harems (effectively one-male subgroups), typical, for example, of the hamadryas and gelada baboons and the drills, even though these species occur in large groups (Stammbach 1987). In hamadryas baboons (*Papio hamadryas*) two to three harems of related males tend to associate (forming clans). The large hamadryas groups, as such, are composed of clans of harems and single males (Kummer 1968).

The classification of "one-male" and "multimale" reflects the females' reproductive opportunities as much as it does that of the males. In some cases, the females are dominant to the males. In all the lemurs studied to date, the females are individually dominant to the males (Richard 1987). This is true, for example, for sifakas (*Propithecus verreauxi*), indri (*Indri indri*), the mouse lemur (*Microcebus murinus*), and the fork-marked lemur (*Phaner furcifer*). The alpha female in a ring-tailed lemur (*Lemur catta*) troop is unquestionably its leader, being of higher status than the resident males and sharing her power with other females rather than dominating them in a rigid hierarchy (Pereira 1993; Sauther and Sussman 1993). Female ring-tails also have the final word regarding which males will receive mating privileges, and they often choose the newest troop members. Female squirrel monkeys (*Saimiri*) are dominant to males (Baldwin and Baldwin 1981). Female howling monkeys (*Alouatta*), by comparison, establish hierarchies in which dominance increases with decreasing age, that is, the youngest female holds the highest rank (Neville et al. 1988), but as a group the females remain subordinate to males. This contrasts with a situation common to Asian colobine monkeys, first observed in the Hanuman langur (*Presbytis entellus*) and now known to occur in a number of species. Here the social structure revolves around territorial groups of related females (natal groups) who appear to safeguard availability to resources and allow preferential access to a single male, but usually for no more than two years. Each change of male membership is routinely followed by a bout of infanticide, with the new male methodically killing all unweaned infants in the troop, thus decreasing the genetic contribution of his predecessor and bringing the affected females into estrus so that they might bear his offspring instead (Hrdy 1980; Van Schaik et al. 1992). Curiously, infants of the langur genus *Trachypithecus* are born with a characteristic bright orange natal coat that contrasts sharply with that of the mother, but which eventually changes to the adult coloration at a few months of age. This developmental feature would seem to target logical victims for the incoming, infanticidal male.

Among the other Asian Cercopithecinae, one-male/multifemale social groups have been recorded in a couple of the macaques, *Macaca cyclopis* (see Wu and Lin 1994), and *Macaca silenus* (see Roonwal and Mohnot 1977), in most *Presbytis* (see Bennett and Davies 1994), in at least three of the four *Rhinopithecus* species, and in the proboscis monkey (*Nasalis larvatus*) and snub-nosed langur (*Simias concolor*). This social unit is also the type most commonly seen in the African guenons (*Cercopithecus*) (see Gautier-Hion 1988) and the patas monkey (*Erythrocebus*) (see Cords 1987), in the chacma baboon (*Papio ursinus*) (see Anderson 1990) and drill (*Mandrillus leucophaeus*) (see Gartlan 1970), and in a handful of the African colobines (*Colobus* spp.) (see Groves 1973; Struhsaker and Oates 1975) and *Procolobus* spp. (see Napier 1985). In most cases, the primary breeding male only has access to the breeding group of females for a period of a few years, and young males either emigrate to form bachelor troops or otherwise congregate in bachelor subgroups within the main group. As a result of male emigration, the female:male ratio observed in these primates can be quite large, reaching perhaps 20:1 or more in the drill, for example (Gartlan 1970). In the genus *Rhinopithecus*, on the other hand, a large troop forms through associations of several one-male/multifemale groups, and the males seem to tolerate one another within this larger unit. It also should be noted that many of these species also exhibit a more complex, multimale/multifemale social structure.

Of all the world's primates, the gorilla (*Gorilla gorilla*) serves as the icon for the one-male/multifemale group. A "silverback," the largest and most impressive male, adorned with a distinctive band of silver-gray hair along his back and flanks, serves as both leader and defender of the group and the sole breeding male (Fossey 1983; Schaller 1963). This image, however, has become slightly clouded by reports of troops in which a second male is tolerated by the silverback (Jenkins 1990).

There is a rather blurred distinction between single-male and multimale organizations in some cases. In the tufted capuchins (*Cebus apella*), there are often

The silvery gibbon (*Hylobates moloch*) is endemic to the island of Java in Indonesia and is one of the most endangered members of its family. Photo by Russell A. Mittermeier.

several males in a group, but mating is largely restricted to one that is dominant (Janson 1984). Similar hierarchical situations occur in, for example, woolly monkeys (*Lagothrix*) and howling monkeys (*Alouatta*). Eisenberg et al. (1972) refers to this form of grouping as age-graded. Multimale groups, in which a number of males have access to females, are typical of the untufted capuchins such as *C. albifrons* and *C. olivaceus*, and also the uakaris (*Cacajao*) and bearded sakis (*Chiropotes*). Among the lemurs, multimale/multifemale associations are seen in *Lemur catta* (described above) and in at least one species in each of the following genera: *Eulemur, Hapalemur, Varecia*, and *Propithecus* (Jolly 1994; Richard 1979; Wilson et al. 1989). Hierarchies are not typically observed, except in *Lemur catta*. Large primate groups with many males and females occur in the majority of the macaques (*Macaca*), where one also usually finds rigid matrilineal hierarchies and the emigration of young males (Roonwal and Mohnot 1977), in savanna baboons (*Papio cynocephalus*), savanna monkeys (*Cercopithecus* spp.), the gray-cheeked mangabey (*Lophocebus albigena*), and the talapoins (*Miopithecus*) (for a review, see Melnick and Pearl 1987).

Perhaps the most complex of primate societies, however, is the fission-fusion society. Whereas large groups in species such as the woolly monkeys (*Lagothrix*), the muriqui (*Brachyteles*), uakaris (*Cacajao*), and the bearded saki (*Chiropotes*) may split temporarily (for a few hours to the whole day) into two or more smaller groups at certain times of the year due to restricted food patch sizes, a fission-fusion society classification refers strictly to the New World spider monkeys (*Ateles*) and the chimpanzees (*Pan paniscus* and *P. troglodytes*) of central and west Africa (Goodall 1996; McFarland 1989; van Roosmalen 1985; van Roosmalen and Klein 1988). In these species the males are philopatric, and communities comprise a core group of related males which variously, and in different combinations, forage together, groom one another, and patrol a common territory covering the ranges of unrelated females (females disperse) and their young. Females (and their young) tend to forage independently though sometimes joining up with other females and their young at feeding sites. Males associate with females primarily for the purpose of mating.

Changes in the membership of primate groups arise from deaths, births, and dispersal (emigration and immigration). The latter has been increasingly documented in primates. Individuals leave a group either for survival (forced to leave) or as an adaptive strategy to find reproductive opportunities elsewhere (can be both). They may leave alone or with other individuals, and in some cases dispersal may be effected by new group formation rather than transfer to established groups. The rate of transfers or dispersal events determines group stability, which can vary between species and populations and between groups in the same population. In the majority of mammals, dispersal is typically by males rather than females, and this is also thought to be true of primates in general (Pusey and Packer 1987), but Strier (1994) pointed out that this could only be said of the prosimians (Kappeler and Ganzhorn 1993) and cercopithecines. Both sexes disperse in all the New World monkeys, the colobines, and the apes, and females sometimes disperse even in the cercopithecines. In the red colobus (*Piliocolobus badius*), chimpanzee (*Pan*), the gorilla (*Gorilla*), and all the atelines, the muriqui (*Brachyteles*), the woolly monkey (*Lagothrix*), the spider monkeys (*Ateles*), and the howling monkeys (*Alouatta*), dispersal is female biased (Rosenberger and Strier 1989; Strier 1994). Dispersal patterns are fundamental for an understanding of primate social dynamics as they determine the kin relations between group members, influencing such factors as grooming relationships, coalitions, conflicts, hierarchical rank, infant development, and mate choice (Gouzoules and Gouzoules 1987; Walters and Seyfarth 1987). The relationships between kinship and social dynamics are more easily understood in species in which sex-biased dispersal is strong, but less in those, such as howling monkeys, in which both sexes disperse. Problems regarding generalizations concerning the effects of dispersal and kinship on the social structures of primate groups are reviewed in Moore (1984, 1992) and Strier (1994).

Primate Conservation

During the last few decades, interest in nonhuman primates has grown significantly on a number of fronts and has helped build support for their conservation. Pioneering and long-term field studies of the great apes (e.g., Fossey 1983; Galdikas 1995; Goodall 1968; Schaller 1963; Van Lawick-Goodall 1971) have dispelled age-old myths about mankind's closest living relatives—greatly narrowing the gaps between these species and our own—and provided new insights into human origins and behavior. The continuing search for drugs to treat global maladies such as cancer, AIDs, and malaria has required large numbers of nonhuman primates as experimental subjects (Mack and Mittermeier 1984). For some species, this use has contributed to serious declines of wild populations and ultimately forced issues of conservation and captive breeding as part of the long-term strategy for biomedical research. In other cases, little-known and formerly obscure primate species have emerged as prominent "flagships" for conserving their tropical forest habitats, the richest natural ecosystems remaining on our planet.

Research on the behavior and ecology of nonhuman primates has grown dramatically since the first field studies in the 1930s (Carpenter 1934), as has our awareness of the widespread and severe threats to their survival. Varying combinations of habitat destruction, hunting, and live capture have driven dozens of primate species to the brink of extinction, to the point that many taxa now number only in the low thousands, and a few probably no more than a few hundred individuals. Such populations are doomed without long-term protection, monitoring, and a heightened understanding of their plight by local human populations. With the increasing fragmentation and isolation of primate populations, an understanding of their viability over the long-term is fundamental. While primate field studies in the past have generated concern for, and a knowledge of, the threatened status of numerous species, the emphasis through the 1980s and 1990s has been on the study of groups, of individual social and reproductive mechanisms and strategies, and of feeding behavior and ecology. This is, of course, important information but, as pointed out by Strier (1997; see also Caughley 1994; Harcourt 1995), even the few most comprehensive long-term field studies rarely supply the demographic data required for the increasingly sophisticated methods available for analyzing population viability—critical tools in conservation biology (Lacy 1993/1994; Rylands et al. 1998). As cogently argued by Strier (1997), the challenge for the future will be a unified agenda for primate behavioral ecology and the rapidly developing science of conservation biology, combining scientifically based, informed management procedures in the wild as well as in captive populations (IUDZG/CBSG 1993; Mallinson 1995; Olney et al. 1994).

In the face of continuing threat, however, primate conservationists can also look back upon the last century and see with some degree of pride that, to the best of anyone's knowledge, not a single primate taxon has gone extinct during that period. The year 2000 could conceivably witness the loss of such critically endangered primates as the Tonkin snub-nosed monkey (*Rhinopithecus avunculus*) of Viet Nam, reduced to between 130 and 350 animals (Ren et al. 1996/1997), or Miss Waldron's red colobus (*Procolobus badius waldroni*) of West Africa, which has not been seen in some 20 years and which researchers failed to locate in several recent surveys in its minute range in Ghana (Oates et al. 1996/1997). Reaching the new millennium with the survival of these and other threatened taxa is an achievable goal, providing that existing primate conservation programs are sustained and new ones created according to a global strategy. Fortunately, support for global primate conservation appears to be increasing in the latter half of the 1990s after having suffered something of a dry spell, and the expertise is at hand to direct available resources to the highest priority species, habitats, and projects.

Threats to the survival of nonhuman primates can be divided into three major categories: habitat destruction, hunting for food and a variety of other purposes, and live capture for export or local trade (Mittermeier et al. 1986). The effects of these threats vary significantly from species to species and from region to region, and are influenced by the extent of remaining habitat, the nature and degree of human activity within the range of a particular species, local hunting traditions, the size and desirability of different species as food items or as sources of other products useful to man, the demand for a given species in research or the pet trade, enforcement of existing wildlife laws, and regulation of commercial animal dealers (Eudey 1994). However, one or more of the three major threats affect almost all primate populations.

On a global scale, habitat destruction is without a doubt the principal factor contributing to the disappearance of wild primate populations. The continuing growth of the human population and its ever-expanding need for natural resources have contributed greatly to the destruction or alteration of natural habitats on an almost unimaginable scale, and nowhere has this problem been more acute than in the tropical regions of the world. More than 90 percent of all non-human primates inhabit the tropical forests of Africa, Asia, and South and Central America, and these forests are being cut at a rate of more than 10 million hectares per year (Bryant et al. 1997). The immediate effects of habitat destruction on nonhuman primates vary significantly from one region to the next. For example, in Madagascar and the Atlantic forest region of eastern Brazil, so little suitable forest habitat remains that any further loss constitutes a grave threat to primates and other wildlife. In contrast to this situation, in the vast forest regions of Amazonia and the Congo basin, which with the island of New Guinea represent the three remaining major tropical wilderness areas of our planet, the effects of habitat destruction are only starting to be felt.

Primates are hunted for a variety of reasons, but by far the most important is to acquire food. Although such hunting is prohibited in many countries, enforcement of such protective legislation is typically rare and sometimes nonexistent in the remote areas where this activity almost always takes place. Hunting of primates as a source of food is a significant threat in the Amazon region of South America, West Africa, Central Africa, and many parts of Asia. In each region, primates are among the animals most frequently hunted, and they are regularly sold in markets, except where this is prohibited by law. However, even in areas where primate hunting is common, it by no means affects all species equally. In Amazonia, for example, the larger monkeys such as *Lagothrix, Ateles, Alouatta,* and *Cebus* are among the more desirable food species, while smaller monkeys such as *Saguinus* and *Saimiri* are rarely shot for food because they barely provide enough meat to recompense the hunter for the cost of his shotgun shell. The same situation holds true in West Africa, where hunters prefer to shoot the larger-bodied *Colobus* than the smaller *Cercopithecus* species.

In areas where the hunting of primates for food is common, it can sometimes represent a threat even more severe than forest destruction. For example, in some parts of Amazonia there are large tracts of primary forest remaining where populations of *Lagothrix, Ateles, Alouatta,* and *Cebus* have effectively been exterminated by excessive hunting (Mittermeier and Coimbra-Filho 1977; Peres 1990; Soini 1982). In areas where food hunting and deforestation both are prevalent, populations of the larger forest primates and other game species can disappear very quickly.

This is becoming increasingly apparent in the rain forests of Central and West Africa where a number of primate species are now threatened by new logging concessions and the commercialized bushmeat trade that they spawn. Logging roads open up formerly remote areas to hunting well above subsistence levels of the indigenous populations. Hunting provides the primary source of nutrition for logging crews, as well as delicacies to distant markets. Primates are highly sought after, especially the larger-bodied monkeys and apes, even though many species are protected by law. Commercial bushmeat hunting has reached crisis proportions in several African countries and is now considered one of the most serious threats to already endangered monkeys and apes.

It is important to note that, in some parts of the world, religious restrictions or other cultural factors prohibit (or inhibit) the killing and eating of primates. In India, for example, primates are rarely hunted for food because they are linked to the monkey god Hanuman, an important figure in the Hindu religion, while in strictly Muslim countries primates are not eaten because their flesh is considered unclean and unfit for human consumption. Indeed, in India, Hindu people refuse to kill rhesus monkeys or even resist translocating them even when populations have become so high that they constitute a menace to humans (Richard et al. 1989). In other countries, such as Madagascar, local taboos may exist against eating certain primates (*Indri,* for example), while other species (*Eulemur* and *Varecia*) may be the most popular food items for a given tribe or village.

Primates are also hunted to supply a number of other products in addition to food: traditional medicines, bait, body parts for ornamentation, and trophies. Primate hunting to supply medicinal products may be nothing more than a by-product of food hunting in most cases and usually involves the use of specific body parts for their supposed medicinal value. In south India, for example, the meat of the Nilgiri langur (*Trachypithecus johnii*) and the endangered lion-tailed macaque (*Macaca silenus*) is regarded as an aphrodisiac and thought to contain other medicinal properties. The blood of leaf monkeys, such as Phayre's langur (*Trachypithecus phayrei*) in Thailand, is believed to impart vigor to the drinker, especially when mixed with local whisky. In various South American countries, drinking from the cup-shaped hyoid apparatus of an adult male howling monkey (*Alouatta*) is reported to cure goiters, coughs, and stuttering, as well as to ease a mother's labor pains during birth. Although the hunting of primates for medicinal purposes is considered a relatively minor factor overall in the global decline of wild primate populations, when it involves endangered species, such as the lion-tailed macaque, it can be a serious threat indeed.

Primates are also shot to provide bait for capturing and killing other animals, mainly in remote corners of the Amazon region. There, spotted-cat hunters preferentially shoot larger monkeys such as *Lagothrix* and *Ateles* to bait traps set for jaguars and ocelots (Mittermeier and Coimbra-Filho 1977). Any number of Amazonian primates may also be shot for fish or turtle bait, and in Sri Lanka, monkeys often serve as bait for crocodiles (R. Rudran, pers. comm.). While the use of primates as bait is a relatively minor threat, comparable to hunting for medicinal products, it can and does add to the pressures faced by over-exploited, large-bodied species such as *Lagothrix* and *Ateles.*

In some countries, primates may be killed for their skins or to provide other body parts used in ornamentation. Perhaps the most striking case of this is in Africa, where the skins of black-and-white colobus (*Colobus guereza*) and related species have been used to fashion cloaks and headdresses for native African peoples, but have also figured significantly in the international fur trade. For example, in 1899, a reported 223,599 monkey skins were auctioned in London alone, and at least 2.5 million probably were exported to Europe between 1880 and 1900 (Brass 1925; Oates 1977), especially in Germany, where they were used to make capes, muffs, and rugs. As recently as the early 1970s, colobus monkey rugs were still common in East African tourist shops, and colobus coats were still being sold in Europe and Japan (Mittermeier 1973; Oates 1977). Throughout much of Amazonia, tourist shops still offer stuffed monkeys, monkey skulls, monkey-skin hats, monkey-tail dusters, and necklaces fashioned from monkey teeth, bones, hands, feet, or tails. However, these activities are typically carried out on a small scale and almost always as a by-product of hunting for food. Nonetheless, the demand for primate body parts for sale to tourists can be a very serious matter indeed if it involves endangered species. The most striking example of this would be the slaughter of mountain gorillas in Rwanda and the Democratic Republic of the Congo, which produces hands and skulls for sale to European tourists (Fossey 1983). Although relatively rare, this practice still occurs despite effective, long-term conservation programs in the region.

Hunting primates for sport is fortunately rare and a minor threat to wild populations. It appears to be most prevalent around logging camps and within military zones in remote areas of developing countries, where arms are plentiful and law enforcement basically nonexistent. Children armed with slingshots and air rifles are often among the worst offenders. More prestigious trophy hunting has also played a role (albeit a minor one) in primate decline. Species such as the gorilla were especially desirable quarry for nineteenth-century and early–twentieth-century trophy hunters, and the tales of their exploits are recounted in a number of books (e.g., Burbridge 1928; Du Chaillu 1930; Gatti 1932; Hastings 1922; Merfield and Miller 1956). On the whole, however, such sport hunting must be considered a very minor factor, unless an endangered species happens to be involved, in which case the activity is almost always illegal as well.

A final reason for hunting primates considered here is as agricultural pests which, for some African and Asian species, can represent a significant drain on wild populations. The most striking example is that of government-sponsored "monkey drives" that were common in Sierra Leone several decades ago. Eleven of the country's 14 primate species were routinely shot or driven into nets and clubbed to death during such drives; only 3 species were considered harmless to farm crops. According to government records, close to a quarter of a million monkeys were destroyed in such drives between 1949 and 1962, and these were only the ones actually counted. Bounties were paid for primate heads or tails, and there was no control over the species killed (Jones 1950; Tappen 1964). The major primate crop raiders are usually the more adaptable and widespread species such as the savanna baboons (*Papio* spp.) in Africa and the macaques (*Macaca* spp.) in Asia, but there are also instances on record of orangutans being killed for raiding fruit trees and gorillas for destroying crops. The only neotropical species regarded as agricultural pests are the capuchins (*Cebus* spp.) and sometimes the squirrel monkeys (*Saimiri* spp.), and their common names sometime reflect their crop-raiding habits. For example, the tufted capuchin (*Cebus apella*) is called *maicero* in Colombia and in Suriname the weeper capuchin (*Cebus olivaceus*) is called *nyan-karu mongi*, both of which translate as "corn-eater" (Mittermeier 1977).

It is difficult to assess how much damage primates actually do to crops in different parts of the world. It is equally difficult to determine how effective pest control efforts have been or to what degree they have contributed to the decline of wild primate populations. However, as primate habitats continue to be encroached upon, resulting in shortages of other food sources, it is likely that the more adaptable primate species will continue to raid crops and perhaps become more dependent upon them as a regular food source. This, unfortunately, will likely result in increased conflict between man and nonhuman primates.

Primates routinely have been captured alive for export (the international trade to supply zoos, biomedical research, and pharmaceutical testing) or to serve local pet trades. The peak of the international primate trade began at the end of the 1950s and continued through the early 1960s, during which time hundreds of thousands of monkeys were taken from the wild each year (Mack and Mittermeier 1984). The trade consisted largely of rhesus macaques (*Macaca mulatta*) exported from India and used in laboratory tests as part of the effort to develop a vaccine for polio, and squirrel monkeys (*Saimiri sciureus*) imported by the United States from several Amazonian countries. Subsequently, the imposition of export bans by habitat countries, import restrictions by user countries, and a decreased demand from biomedical research and zoological parks contributed to a significant decline in the international traffic of primates. Recent evidence that the chimpanzees, *Pan troglodytes*, can carry, and be a reservoir for, AIDS viruses is causing concern regarding future demand for biomedical research (Gao et al. 1999).

In 1982, in recognition of the serious effect that live capture for export can have on wild primate populations, the IUCN/SSC Primate Specialist Group prepared a *Policy Statement on Use of Primates for Biomedical Purposes*, which included the recommendation that endangered, vulnerable, and rare species be considered for use in biomedical research projects *only* if they are obtained from existing, self-sustaining captive breeding colonies. This policy statement was subsequently adopted by the World Health Organization (WHO) and the Ecosystem Conservation Group (ECG) of the United Nations, which includes the United Nations Educational, Scientific and Cultural Organization (UNESCO), the Food and Agricultural Organization (FAO), the United Nations Environment Program (UNEP), and the World Conservation Union or International Union for Conservation of Nature and Natural Resources (IUCN). It is still valid to this day.

The most recent, comprehensive conservation status assessment of the world's primates is included in the *1996 IUCN Red List of Threatened Animals* (Baillie and Groombridge 1996), a collaborative effort of the World Conservation Union Species Survival Commission, the World Conservation Monitoring Centre, and BirdLife International. This document differs significantly from past Red Lists in its use of new categories and criteria for threat. All primate taxa were included in this assessment and have been identified either as *Threatened* (a designation that includes the categories *Critically Endangered, Endangered,* and *Vulnerable*), *Lower Risk: Conservation Dependent, Lower Risk: Near Threatened, Lower Risk: Least Concern, Extinct* and *Extinct in the Wild,* and *Data Deficient.* In general terms, a threatened taxon is defined as:

1. *Critically Endangered* if the extent of its occurrence is estimated to be less than 100 km^2, if its population is estimated to be less than 250 mature individuals, and/or if quantitative analysis indicates the probability of extinction in the wild is at least 50 percent within 10 years or three generations;
2. *Endangered* if the extent of its occurrence is estimated to be less than 5,000 km^2, if its population is estimated to number less than 2,500 individuals, and/or if quantitative analysis shows the probability of extinction in the wild is at least 20 percent within 20 years or five generations; and
3. *Vulnerable* if the extent of its occurrence is estimated to be less than 20,000 km^2, if its population is estimated to number less than 10,000 individuals, and/or if quantitative analysis shows the probability of extinction in the wild is at least 10 percent within 100 years.

As a result of this assessment, 204 (roughly one-third) of the world's 650 or so primate taxa are currently considered *Critically Endangered, Endangered,* or *Vulnerable* (Table 4). Of these, 103 taxa (16.6 percent) are listed as *Critically Endangered* or *Endangered*—34 in the Neotropics, 28 in Africa, 17 in Madagascar, and 24 in Asia (Table 5). All members of eight genera—*Allocebus, Varecia, Indri, Leontopithecus, Brachyteles, Simias, Pan,* and *Gorilla*—are considered *Critically Endangered* or *Endangered,* and the monotypic family Daubentoniidae is *Endangered.*

Although the *1996 IUCN Red List of Threatened Animals* is comprehensive with regard to primates and identifies a significant number of taxa as threatened, we feel that it is sometimes misleading by assigning the lowest level of threat to a given species when one or more its subspecies are, in fact, more endangered. For example, Verreaux's sifaka (*Propithecus verreauxi*) is listed as *Vulnerable* even though two of its subspecies, Coquerel's sifaka (*Propithecus verreauxi coquere-*

li) and the crowned sifaka (*Propithecus verreauxi coronatus*), are considered *Endangered* and *Critically Endangered*, respectively. Similar situations exist for the gentle lemur (*Hapalemur griseus*), black lemur (*Eulemur macaco*), ruffed lemur (*Varecia variegata*), diademed sifaka (*Propithecus diadema*), brown howling monkey (*Alouatta fusca*), night monkey (*Aotus lemurinus*), long-haired spider monkey (*Ateles belzebuth*), brown-headed spider monkey (*Ateles fusciceps*), bald uakari (*Cacajao calvus*), masked titi (*Callicebus personatus*), Central American squirrel monkey (*Saimiri oerstedii*), Diana monkey (*Cercopithecus diana*), red-eared monkey (*Cercopithecus erythrotis*), and gorilla (*Gorilla gorilla*).

The IUCN was established in 1948 to promote and carry out scientifically based action for the conservation and sustainable use of living natural resources. The IUCN enrolls sovereign states, governmental agencies, research institutions, and nongovernmental organizations to conserve the world's natural heritage. The Species Survival Commission (SSC), founded in 1949, is the largest of IUCN's six commissions with more than 7,000 volunteer member scientists, field researchers, government officials, and conservation leaders from 188 countries. The SSC works principally through its more than 100 Specialist Groups, of which the Primate Specialist Group is one of the largest (Butynski 1996/1997; Eudey 1996/1997; Ganzhorn et al. 1996/1997; Rylands et al. 1996/1997).

The mission of the Primate Specialist Group is to maintain the current diversity of the order Primates, with dual emphasis on: (1) ensuring the survival of endangered and vulnerable species wherever they occur and (2) providing effective protection for large numbers of primates in areas of high primate diversity and/or abundance.

Although activities under way in many parts of the world make it inevitable that a proportion of the world's forests and the primates living in them will disappear, the role of the Primate Specialist Group is to minimize this loss wherever possible by:

- setting aside special protected areas for critically endangered, endangered, and vulnerable species;
- creating national parks and reserves in areas of high primate diversity and/or abundance;
- maintaining parks and reserves that already exist and enforcing protective legislation in them;
- determining ways in which human and nonhuman primates can coexist in multiple-use areas;
- establishing conservation-oriented captive breeding programs for threatened taxa;
- ending all illegal and otherwise destructive traffic in primates;
- ensuring that research institutions using primates are aware of conservation issues and the status of species they use, that they use primates as prudently as possible, and that they make every attempt to breed in captivity most or all of the primates they require; and
- creating public awareness of the need for primate conservation and the importance of primates as a natural heritage in the countries in which they occur.

In late 1977, the chairman of the Primate Specialist Group, in collaboration with group members, put together a 325-page *Global Strategy for Primate Conservation* (Mittermeier 1978*a*). This document was an attempt to organize primate conservation activities based on the highest international priorities and to ensure that the limited funds available for primate conservation were put to the

TABLE 4. CRITICALLY ENDANGERED AND ENDANGERED PRIMATES

Critically endangered	Endangered
Hairy-eared dwarf lemur (*Allocebus trichotis*)	White-collared lemur (*Eulemur fulvus albocollaris*)
Sclater's lemur (*Eulemur macaco flavifrons*)	Black-and-white ruffed lemur (*Varecia variegata variegata*)
Alaotran gentle lemur (*Hapalemur griseus alaotrensis*)	Indri (*Indri indri*)
Golden bamboo lemur (*Hapalemur aureus*)	Diademed sifaka (*Propithecus diadema diadema*)
Broad-nosed gentle lemur (*Hapalemur simus*)	Milne-Edwards' sifaka (*Propithecus diadema edwardsi*)
Red ruffed lemur (*Varecia variegata rubra*)	Coquerel's sifaka (*Propithecus verreauxi coquereli*)
Silky sifaka (*Propithecus diadema candidus*)	Aye-aye (*Daubentonia madagascariensis*)
Perrier's sifaka (*Propithecus diadema perrieri*)	Buffy-tufted-ear marmoset (*Callithrix aurita*)
Tattersall's sifaka (*Propithecus tattersalli*)	Buffy-headed marmoset (*Callithrix flaviceps*)
Crowned sifaka (*Propithecus verreauxi coronatus*)	Golden-headed lion tamarin (*Leontopithecus chrysomelas*)
Black-faced lion tamarin (*Leontopithecus caissara*)	Pied tamarin (*Saguinus bicolor bicolor*)
Black lion tamarin (*Leontopithecus chrysopygus*)	Cotton-top tamarin (*Saguinus oedipus*)
Golden lion tamarin (*Leontopithecus rosalia*)	Coiba Island howling monkey (*Alouatta coibensis coibensis*)
Red-handed howling monkey (*Alouatta belzebul ululata*)	Night monkey (*Aotus lemurinus griseimembra*)
Coiba Island howling monkey (*Alouatta coibensis trabeata*)	White-bellied spider monkey (*Ateles belzebuth brunneus*)
Northern brown howling monkey (*Alouatta fusca fusca*)	Hybrid spider monkey (*Ateles belzebuth hybridus*)
Brown-headed spider monkey (*Ateles fusciceps fusciceps*)	Grizzled spider monkey (*Ateles geoffroyi grisescens*)
Azuero spider monkey (*Ateles geoffroyi azuerensis*)	Panamanian spider monkey (*Ateles geoffroyi panamensis*)
Southern muriqui (*Brachyteles arachnoides*)	White-whiskered spider monkey (*Ateles marginatus*)
Northern muriqui (*Brachyteles hypoxathus*)	Bald uakari (*Cacajao calvus calvus*)
Northern Bahian brown titi (*Callicebus personatus barbarabrownae*)	Bald uakari (*Cacajao calvus novaesi*)
White-fronted capuchin (*Cebus albifrons trinitatis*)	Red uakari (*Cacajao calvus rubicundus*)
Margarita Island tufted capuchin (*Cebus apella margaritae*)	Black bearded saki (*Chiropotes satanas satanas*)
Buffy-headed tufted capuchin (*Cebus xanthosternos*)	Central American squirrel monkey (*Saimiri oerstedii oerstedii*)
Yellow-tailed woolly monkey (*Lagothrix flavicauda*)	White-collared mangabey (*Cercocebus atys lunulatus*)
Colombian woolly monkey (*Lagothrix lagotricha lugens*)	Sanje mangabey (*Cercocebus galeritus sanjei*)
Cental American squirrel monkey (*Saimiri oersteadi citrinellus*)	Tana River mangabey (*Cercocebus galeritus galeritus*)
Miss Waldron's red colobus (*Procolobus badius waldroni*)	Roloway monkey (*Cercopithecus diana roloway*)
Mentawai macaque (*Macaca pagensis*)	Red-eared monkey (*Cercopithecus erythrotis erythrotis*)
Tonkin snub-nosed monkey (*Rhinopithecus avunculus*)	Golden monkey (*Cercopithecus mitis kandti*)
Delacour's langur (*Trachypithecus delacouri*)	Preuss's monkey (*Cercopithecus preussi insularis*)
Silvery gibbon (*Hylobates moloch*)	Preuss's monkey (*Cercopithecus preussi preussi*)
Mountain gorilla (*Gorilla gorilla beringei*)	Scalter's guenon (*Cercopithecus sclateri*)
	Drill (*Mandrillus leucophaeus leucophaeus*)
	Drill (*Mandrillus leucophaeus mundamensis*)
	Bouvier's red colobus (*Procolobus badius bouvieri*)
	Niger Delta red colobus (*Procolobus badius epieni*)
	Uhehe red colobus (*Procolobus badius gordonorum*)
	Zanzibar red colobus (*Procolobus badius kirkii*)
	Pennant's red colobus (*Procolobus badius pennanti*)
	Preuss's red colobus (*Procolobus badius preussi*)
	Tana River red colobus (*Procolobus badius rufomitratus*)
	Temminck's red colobus (*Procolobus badius temmincki*)
	Japanese macaque (*Macaca fuscata*)

continued

TABLE 4. CONTINUED

Critically endangered	Endangered
	Moor macaque (*Macaca maura*)
	Sulawesi black macaque (*Macaca nigra*)
	Lion-tailed macaque (*Macaca silenus*)
	Grizzled leaf monkey (*Presbytis comata*)
	Douc langur (*Pygathrix nemaeus*)
	Yunnan snub-nosed monkey (*Rhinopithecus bieti*)
	Guizhou snub-nosed monkey (*Rhinopithecus brelichi*)
	Pig-tailed snub-nosed monkey (*Simias concolor*)
	Francois' langur (*Trachypithecus poliocephalus*)
	Black gibbon (*Hylobates concolor*)
	Western lowland gorilla (*Gorilla gorilla gorilla*)
	Grauer's gorilla (*Gorilla gorilla graueri*)
	Eastern chimpanzee (*Pan troglodytes schweinfurthi*)
	Central chimpanzee (*Pan troglodytes troglodytes*)
	Western chimpanzee (*Pan troglodytes verus*)

best possible use. The first draft of the *Global Strategy* included 65 projects for Africa, Asia, and South and Central America. Each project was categorized as *highest priority, high priority, priority,* and *desirable,* based mainly on the status of focal species and how likely the project would be to bring about the desired conservation action. The *Global Strategy* quickly led to a substantial increase in funding for primate conservation activities and, in 1979, to the establishment of a special Primate Program and Primate Action Fund by the World Wildlife Fund–U.S. In addition to major projects supported as a result of this program, the Primate Action Fund provided rapid support for small primate conservation projects (ranging from $500 to $3,000). The Primate Action Fund functioned for more than a decade, contributing several hundred thousand dollars to more than 100 projects. Other key institutions that contributed significantly to primate conservation during this period include the New York Zoological Society (now the Wildlife Conservation Society), the Fauna and Flora Preservation Society (now Fauna and Flora International), the Rare Animal Relief Effort, Jersey Wildlife Preservation Trust, Wildlife Preservation Trust International, and the National Geographic Society.

Almost a decade after the *Global Strategy* was launched, the first regional primate conservation action plans were prepared by the IUCN/SSC Primate Specialist Group. First to be published was the *Action Plan for African Primate Conservation: 1986–90* (Oates 1986), which was quickly followed by the *Action Plan for Asian Primate Conservation: 1987–91* (Eudey 1987), and several years

TABLE 5. THREATENED PRIMATES BY REGION

Region	Taxa	Threatened	Percent	CR + E	Percent
Neotropics	202	69	34.2	34	16.8
Africa	190	40	21.1	29	14.7
Madagascar	51	35	68.6	17	33.3
Asia	176	60	34.1	24	13.6

Note: CR = critically endangered, E = endangered.

later by *Lemurs of Madagascar, an Action Plan for Their Conservation: 1993–1999* (Mittermeier et al. 1992). The last plan to appear was *African Primates: Status Survey and Conservation Action Plan* (Oates 1996), an update of the 1986 document. A draft action plan has also been drawn up for Mesoamerican primates (Rodríguez-Luna et al. 1996). These action plans have effectively focused conservation activities in three of the four major regions in which primates occur and are useful measures with regard to the success of proposed strategies.

The first vehicle for regular and effective communication among the world's primate conservationists was the *IUCN/SSC Primate Specialist Group Newsletter*, which was launched in 1981. Changed to *Primate Conservation* in 1985, it has appeared on more or less an annual basis ever since. In addition, the four regional sections of the Primate Specialist Group subsequently began publishing their own periodic newsletters to meet the growing need for more timely information. *Asian Primates* appeared in 1991, *Neotropical Primates* and *Lemur News* in 1993, and *African Primates* in 1995, with *Neotropical Primates* and *Asian Primates* appearing with the greatest frequency and regularity. In combination they have significantly increased the amount, quality, and timeliness of information available to primate conservationists throughout the world.

For the years ahead, there is a need to sustain conservation activities based on recommendations of the original *Global Strategy for Primate Conservation* and subsequent regional action plans, as well as to increase the focus on those primate taxa most seriously threatened with extinction. With luck, we will come through the twentieth century without having lost a single known, modern primate taxon—an enviable record indeed considering the number of reptiles, birds, and other mammals known to have disappeared already during this period—but we will do so only by the "skin of our teeth" and with several species and subspecies still in jeopardy.

A more focused conservation strategy is at present being prepared. Much of the groundwork has been done with the publication of the *1996 IUCN Red List of Threatened Animals* (Baillie and Groombridge 1996), which provides the starting point for identifying the highest priority taxa. Several more species and subspecies of conservation concern need to be added, and other considerations will include taxonomic uniqueness, population size, and apparent rate of decline in order to establish priority rankings for conservation action. The results of this analysis will be presented in 2000 in the form of a global action plan for the world's most endangered primates.

With such a plan in hand, serious work can begin to amass both the human and financial resources needed for implementation. Fortunately, several new sources of support for primate conservation have materialized over the last decade. While the World Wildlife Fund–U.S. Primate Program no longer exists, many other traditional nongovernmental sources still offer grants for field, captive, and laboratory programs, and academic institutions continue to provide funds for primate field studies that have significant conservation impact. Government-supported efforts such as the Indo-U.S. Primate Project provide excellent models for international cooperation, and a growing number of zoos have joined forces to focus on regional primate faunas, generating funds not only for captive breeding programs, but for support for *in situ* projects as well. In addition, at least two new significant sources of philanthropic support dedicated to primates were established in the 1990s: Primate Conservation, Inc., and the Margot Marsh Biodiversity Foundation. Together, these organizations and agencies represent the core of funding necessary to move ahead with a global action plan for the world's most endangered primates, and it is hoped that such a plan will help uncover new sources of support as well.

Literature Cited

Abbott, D. H., J. Barrett, and L. M. George. 1993. Comparative aspects of the social suppression of reproduction in female marmosets and tamarins. *In* A. B. Rylands, ed., Marmosets and Tamarins: Systematics, Behaviour, and Ecology, Oxford Univ. Press, Oxford, pp. 152–63.

Abbott, D. H., and J. P. Hearn. 1978. Physical and hormonal behavioural aspects of sexual development in the marmoset monkey, *Callithrix jacchus*. J. Reprod. Fert. 53:155–66.

Anderson, C. M. 1990. Desert, mountain and savanna baboons: a comparison with special references to the Suikerbosrand population. *In* M. T. de Mello, A. Whitten, and R. W. Byrne, eds., Baboons: Behaviour and Ecology, Use and Care, Brasília, Brazil, pp. 89–104.

Ayres, J. M. 1989. Comparative feeding ecology of the uakari and bearded saki, *Cacajao* and *Chiropotes*. J. Hum. Evol. 18:697–716.

Baillie, J., and B. Groombridge (compilers). 1996. IUCN Red List of Threatened Animals, The World Conservation Union (IUCN), Species Survival Commission (SSC), Gland, Switzerland.

Baldwin, J. D., and J. I. Baldwin. 1981. The squirrel monkeys, genus *Saimiri*. *In* A. F. Coimbra-Filho and R. A. Mittermeier, eds., The Ecology and Behavior of Neotropical Primates, Academia Brasileira de Ciências, Rio de Janeiro, vol. 1, pp. 277–330.

Bearder, S. K. 1987. Lorises, bushbabies, and tarsiers: diverse societies in solitary foragers. *In* B. B. Smuts, D. L. Cheney, R. M. Seyfarth, R. W. Wrangham, and T. T. Struhsaker, eds., Primate Societies, Univ. Chicago Press, pp. 11–24.

Bearder, S. K., and R. D. Martin. 1980. *Acacia* gum and its use by bushbabies *Galago senegalensis* (Primates, Lorisidae). Int. J. Primatol. 1:103–28.

Bennett, E. L., and A. G. Davies. 1994. The ecology of Asian colobines. *In* A. G. Davies and J. F. Oates, eds., Colobine Monkeys: Their Ecology, Behavior and Evolution, Cambridge Univ. Press, Cambridge, pp. 129–72.

Bourliére, F. 1979. Significant parameters of environmental quality for nonhuman primates. *In* I. S. Bernstein and E. O. Smith, eds., Primate Ecology and Human Origins, Garland Press, New York, pp. 23–46.

———. 1985. Primate communities: their structure and role in tropical ecosystems. Int. J. Primatol. 6(1):1–26.

Box, H. O. 1977. Quantitative data on the carrying of young captive monkeys (*Callithrix jacchus*) by other members of their family groups. Primates 18:475–84.

Brandon-Jones, D. 1996. The Asian Colobinae (Mammalia: Cercopithecidae) as indicators of Quaternary climatic change. Biol. J. Linn. Soc. 59:327–50.

Brass, E. 1925. Aus dem Reiche der Pelse, Neuen Pelwarzen-Zeitung und Kürschner-Zeitung, Berlin.

Bryant, D., D. Nielsen, and L. Tangley, eds. 1997. The Last Frontier Forests: Ecosystems and Economies on the Edge, World Resources Institute, Washington, D.C.

Buchanan-Smith, H. M. 1990. Polyspecific association of two tamarin species, *Saguinus labiatus* and *Saguinus fuscicollis*, in Bolivia. Amer. J. Primatol. 22:205–14.

Burbridge, B. 1928. Gorilla. Tracking and Capturing the Ape-Man of Africa, Century, New York.

Butynski, T. M. 1996/1997. African primate conservation—the species and IUCN/SSC Primate Specialist Group Network. Primate Conservation (17):87–100.

Caldecott, J. O. 1986. An ecological and behavioural study of the pig-tailed macaque. Contrib. Primatol. 21:1–259. S. Karger, Basel.

Carpenter, C. R. 1934. A field study of the behavior and social relations of howling monkeys. Comp. Psychol. Monogr. 48:1–168.

Cartelle, C., and W. C. Hartwig, 1996. A new extinct primate among the Pleistocene megafauna of Bahia, Brazil. Proc. Natl. Acad. Sci. USA 93:6405–409.

Caughley, G. 1994. Directions in conservation biology. J. Anim. Ecol. 63:215–44.

Charles-Dominique, P. 1977. Ecology and Behaviour of Nocturnal Primates, Duckworth, London.

Cheney, D. L., and R. M. Seyfarth. 1990. How Monkeys See the World: Inside the Mind of Another Species, Univ. Chicago Press.

Cheney, D. L., R. M. Seyfarth, B. B. Smuts, and R. W. Wrangham. 1987. The study of primate societies. In B. B. Smuts, D. L. Cheney, R. M. Seyfarth, R. W. Wrangham, and T. T. Struhsaker, eds., Primate Societies, Univ. Chicago Press, pp. 1–8.

Chivers, D. J., ed. 1980. Malayan Forest Primates: Ten Years' Study in Tropical Rain Forest, Plenum Press, New York.

Chivers, D. J. 1992. Diets and guts. In S. Jones, R. D. Martin, and D. Pilbeam, eds., The Cambridge Encyclopaedia of Human Evolution, Cambridge Univ. Press, Cambridge, pp. 60–64.

Chivers, D. J., and C. M. Hladik. 1980. Morphology of the gastrointestinal tract in primates: comparisons with other mammals in relation to diet. J. Morphol. 166:337–86.

Coimbra-Filho, A. F., and R. A. Mittermeier. 1977. Tree-gouging, exudate-eating and the "short-tusked" condition in Callithrix and Cebuella. In D. G. Kleiman, ed., The Biology and Conservation of the Callitrichidae, Smiths. Inst. Press, Washington, D.C., pp. 105–15.

Coimbra-Filho, A. F., N. Rocha, and A. Pissinatti. 1980. Morfofisiologia do ceco e sua correlação com o tipo odontológico em Callitrichidae (Platyrrhini, Primates). Rev. Brasil. Biol. 40:177–85.

Colquhoun, I. C. 1998. Cathemeral behavior of Eulemur macaco macaco at Ambato Massif, Madagascar. Folia Primatol. 69(suppl. 1):22–34.

Cords, M. 1987. Forest guenons and patas monkeys: male-male competition in one-male groups. In B. B. Smuts, D. L. Cheney, R. M. Seyfarth, R. W. Wrangham, and T. T. Struhsaker, eds., Primate Societies, Univ. Chicago Press, pp. 98–111.

Crockett, C. M., and R. Sekulic. 1984. Infanticide in red howler monkeys (Alouatta seniculus). In G. Hausfater and S. B. Hrdy, eds., Infanticide: Comparative and Evolutionary Perspectives, Aldine, New York, pp. 173–94.

Cronin, J. E., and V. M. Sarich. 1978. Marmoset evolution: the molecular evidence. Prim. Med. 10:12–19.

Curtis, D. 1997. A third activity pattern in primates: cathemerality in lemurs. Primate Eye (63):19.

Dixson, A. F., and Fleming, D. 1981. Parental behaviour and infant development in owl monkeys (Aotus trivirgatus griseimembra). J. Zool., Lond. 194:25–39.

Du Chaillu, P. 1930. Paul Du Chaillu: Gorilla Hunter, Harper, New York.

Eisenberg, J. F. 1979. Habitat, economy, and society: some correlations and hypotheses for the Neotropical primates. In I. S. Bernstein and E. O. Smith, eds., Primate Ecology and Human Origins, Garland Press, New York, pp. 215–62.

Eisenberg, J. F., N. A. Muckenhirn, and R. Rudran. 1972. The relation between ecology and social structure in primates. Science 176:863–74.

Epple, G. 1986. Communication by chemical signals. In G. Mitchell and J. Erwin, eds., Comparative Primate Biology, vol. 2A, Behaviour, Conservation and Ecology, Alan R. Liss, New York, pp. 531–80.

Erickson, C. J. 1995. Feeding sites for extractive foraging by the aye-aye, Daubentonia madagascariensis. Amer. J. Primatol. 35:235–40.

Eudey, A. A. 1987. Action Plan for Asian Primate Conservation: 1987–91, IUCN/SSC Primate Specialist Group, Gland, Switzerland.

———. 1994. Temple and pet primates in Thailand. Rev. Ecol. (Terre Vie) 49:273–80.

———. 1996/1997. Asian primate conservation—the species and the IUCN/SSC Primate Specialist Group network. Primate Conservation (17):101–10.

Feistner, A. T. C., and E. C. Price. 1991. Food offering in New World primates: two species added. Folia Primatol. 57:165–68.

Ferrari, S. F. 1993. Ecological differentiation in the Callitrichidae. In A. B. Rylands, ed.,

Marmosets and Tamarins: Systematics, Behaviour and Ecology, Oxford Univ. Press, Oxford, pp. 314–28.

Ferrari, S. F., and M. A. Lopes Ferrari. 1989. A re-evaluation of the social organisation of the Callitrichidae, with special reference to the ecological differences between genera. Folia Primatol. 52:132–47.

Ferrari, S. F., and M. A. Lopes. 1996. Primate populations in eastern Amazonia. In M. A.. Norconk, A. L. Rosenberger, and P. A. Garber, eds., Adaptive Radiations of the Neotropical Primates, Plenum Press, New York, pp. 53–67.

Ferrari, S. F., and E. S. Martins. 1992, Gummivory and gut morphology in two sympatric callitrichids (*Callithrix emiliae* and *Saguinus fuscicollis weddelli*) from western Brazilian Amazonia. Amer. J. Phys. Anthropol. 88:97–103.

Ferrari, S. F., and Strier, K. B. 1992. Exploitation of *Mabea fistulifera* nectar by marmosets (*Callithrix flaviceps*) and muriquis (*Brachyteles arachnoides*) in south-east Brazil. J. Trop. Ecol. 8:225–39.

Fleagle, J. G. 1980. Locomotion and posture. In D. J. Chivers, ed., Malayan Forest Primates: Ten Years' Study in Tropical Rain Forest, Plenum Press, New York, pp. 191–207.

———. 1984. Primate locomotion and diet. In D. J. Chivers, B. A. Wood, and A. L. Bilsborough, eds., Food Acquisition and Processing in Primates, Plenum Press, New York, pp. 105–17.

———. 1988. Primate Adaptation and Evolution, Academic Press, San Diego.

———. 1992. Primate locomotion and posture. In S. Jones, R. D. Martin, and D. Pilbeam, eds., The Cambridge Encyclopaedia of Human Evolution, Cambridge Univ. Press, Cambridge, pp. 86–90.

Fleagle, J. G., and R. A. Mittermeier. 1980. Locomotor behaviour, body size, and comparative ecology of seven Surinam monkeys. Amer. J. Phys. Anthropol. 52:301–22.

Fleagle, J. G., Mittermeier, R. A., and Skopec, A. L. 1981. Differential habitat use by *Cebus apella* and *Saimiri sciureus* in central Surinam. Primates 22(3):361–67.

Fleagle, J. G., and K. E. Reed. 1996. Comparing primate communities: a multivariate approach. J. Hum. Evol. 30:489–510.

Ford, S. M., and L. C. Davis. 1992. Systematics and body size: implications for feeding in the New World monkeys. Amer. J. Phys. Anthropol. 88:469–82.

Fossey, D. 1983. Gorillas in the Mist, Houghton Mifflin, Boston.

Fuentes, A. 1994. Social organization of the Mentawai langur (*Presbytis potenziani*). Amer. J. Phys. Anthropol. 18(suppl.):90.

Galdikas, B. M. F. 1995. Reflections of Eden: My Years with the Orangutans of Borneo, Little, Brown, Boston.

Ganzhorn, J. U. 1998. Nested patterns of species composition and its implications for lemur biogeography in Madagascar. Folia Primatol. 69(suppl. 1):332–41.

Ganzhorn, J. U., O. Langrand, P. C. Wright, S. O'Connor, B. Rakotosamimanana, A. T. C. Feistner, and Y. Rumpler. 1996/1997. The state of lemur conservation in Madagascar. Primate Conservation (17):70–86.

Ganzhorn, J. U., S. Malcomber, O. Andrianantoanina, and S. M. Goodman. 1997. Habitat characteristics and lemur species richness in Madagascar. Biotropica 29(3):331–43.

Gao, F., E. Bailes, D. L. Robertson, Y. Chen, C. M. Rodenburg, S. F. Michael, L. B. Cummins, L. O. Arthur, M. Peeters, G. M. Shaw, P. M. Sharp, and B. H. Hahn. 1999. Origin of HIV-1 in the chimpanzee *Pan troglodytes troglodytes*. Nature, Lond. 397:436–41.

Gartlan, J. S. 1970. Preliminary notes on the ecology and behavior of the drill *Mandrillus leucophaeus* Ritgen, 1824. In J. R. Napier and P. H. Napier, eds., Old World Monkeys: Evolution, Systematics and Behavior, Academic Press, New York, pp. 445–80.

Gatti, A. 1932. The King of the Gorillas, Doubleday, Garden City, New York.

Gautier-Hion, A. 1978. Food niches and coexistence in sympatric primates in Gabon. In D. J. Chivers and J. Herbert, eds., Recent Advances in Primatology, vol. 1, Behaviour, Academic Press, London, pp. 269–86.

———. 1988. Polyspecific associations among forest guenons: ecological, behavioural and evolutionary aspects. In A. Gautier-Hion, F. Bourlière, J.-P. Gautier, and J. Kingdon, eds.,

A Primate Radiation: Evolutionary Biology of the African Guenons, Cambridge Univ. Press, Cambridge, pp. 452–76.

Gautier-Hion, A., F. Bourlière, J.-P. Gautier, and J. Kingdon, eds. 1988. A Primate Radiation: Evolutionary Biology of the African Guenons, Cambridge Univ. Press, Cambridge.

Gautier-Hion, A., and J.-P. Gautier. 1974. Les associations polyspécifiques de cercopithèques du plateau de M'passa (Gabon). Folia Primatol. 22:134–77.

Gautier-Hion, A., R. Quris, and J.-P. Gautier. 1983. Monospecific vs polyspecific life: a comparative study of foraging and antipredatory tactics in a community of *Cercopithecus* monkeys. Behav. Ecol. Sociobiol. 12:325–35.

Gebo, D. L., and Chapman, C. A. 1995. Positional behavior in five sympatric Old World monkeys. Amer. J. Phys. Anthropol. 97:49–76.

Glander, K. E. 1982. The impact of plant secondary compounds on primate feeding behavior. Yearb. Phys. Anthropol. 25:1–18.

Goldizen, A. W. 1987. Tamarins and marmosets: communal care of offspring. *In* B. B. Smuts, D. L. Cheney, R. M. Seyfarth, R. W. Wrangham, and T. T. Struhsaker, eds., Primate Societies, Univ. Chicago Press, pp. 34–43.

Goodall, J. 1968. The behaviour of free-living chimpanzees in the Gombe Stream area. Anim. Behav. Monogr. 1(3):161–311.

———. 1996. The Chimpanzees of Gombe: Patterns of Behavior, Harvard Univ. Press, Belknap Press, Cambridge.

Gouzoules, S., and H. Gouzoules. 1987. Kinship. *In* B. B. Smuts, D. L. Cheney, R. M. Seyfarth, R. W. Wrangham, and T. T. Struhsaker, eds., Primate Societies, Univ. Chicago Press, pp. 299–305.

Graham, C. E. 1981. Reproductive Biology of the Great Apes: Comparative and Biomedical Perspectives, Academic Press, New York.

Groves, C. P. 1973. Notes on the ecology and behaviour of the Angola colobus (*Colobus angolensis*, P. L. Sclater 1860) in N. E. Tanzania. Folia Primatol. 20:12–26.

———. 1993. Order primates. *In* D. E. Wilson and D. M. Reeder, eds., Mammal Species of the World: A Taxonomic and Geographic Reference, 2d ed., Smiths. Inst. Press, Washington, D.C., pp. 243–77.

———. 1998. Systematics of tarsiers and lorises. Primates 39:13–27.

Harcourt, A. H. 1995. Population viability estimates: theory and practice for a wild gorilla. Conserv. Biol. 9:134–42.

Harcourt, C. H., and L. T. Nash. 1986. Social organization of galagos in Kenyan coastal forests. I. *Galago zanzibaricus*. Amer. J. Primatol. 10:339–55.

Hartwig, W. C., and C. Cartelle. 1996. A complete skeleton of the giant South American primate *Protopithecus*. Nature, Lond. 381:307–11.

Hastings, M. H. 1922. On the Gorilla Trail, D. Appleton, New York.

Hershkovitz, P. 1977. Living New World Monkeys (Platyrrhini), with an Introduction to the Primates, Univ. Chicago Press, vol. 1.

Heymann, E. W. 1990. Interspecific relations in a mixed species troop of moustached tamarins, *Saguinus mystax*, and saddle-back tamarins, *Saguinus fuscicollis* (Platyrrhini: Callitrichidae), at the Rio Blanco, Peruvian Amazonia. Amer. J. Primatol. 21:115–27.

———. 1995. Urine washing and related behaviour in wild moustached tamarins, *Saguinus mystax* (Callitrichidae). Primates 36:259–64.

———. 1998. Giant fossil New World primates: arboreal or terrestrial? J. Hum. Evol. 34:99–101.

Hladik, C. M. 1979. Diet and ecology of prosimians. In G. A. Doyle and R. D. Martin, eds., The Study of Prosimian Behavior, Academic Press, New York, pp. 307–58.

———. 1988. Seasonal variations in food supply for wild primates. *In* I. de Garine and G. A. Harrison, eds., Coping with Uncertainty in Food Supply, Clarendon Press, Oxford, pp. 1–25.

Hladik, C. M., P. Charles-Dominique, and J. J. Petter. 1980. Feeding strategies of nocturnal prosimians in the dry forest of the west coast of Madagascar. *In* P. Charles Dominique,

H. M. Cooper, A. Hladik, C. M. Hladik, E, Pages, G. F. Pariente, A. Petter-Rousseaux, and A. Schilling, eds., Nocturnal Malagasy Primates, Academic Press, New York, pp. 41–74.

Hrdy, S. B. 1980. The Langurs of Abu: Female and Male Strategies of Reproduction, Harvard Univ. Press, Cambridge.

IUDZG/CBSG. 1993. The World Zoo Conservation Strategy: The Role of the Zoos and Aquaria of the World in Global Conservation, The World Zoo Organization (IUDZG) and the IUCN/SSC Captive Breeding Specialist Group, CBSG, Chicago Zoological Society, Chicago.

Janson, C. H. 1984. Female choice and mating system of the brown capuchin monkey Cebus apella (Primates: Cebidae). Z. Tierpsychol. 65:177–200.

Janson, C. H., J. Terborgh, and L. H. Emmons. 1981. Non-flying mammals as pollinating agents in the Amazonian forest. Biotropica 13(suppl.):1–6.

Jenkins, P. A. 1990. Catalogue of Primates in the British Museum (Natural History) and Elsewhere in the British Isles, Part 5, British Museum (Natural History) Publications, London.

Jolly, A. 1994. Female dominance and social structure in lemurs. In Handbook and Abstracts: 15th Congress of the International Primatological Society, Kuta, Bali, Indonesia, 1994, p. 285.

Jones, T. S. 1950. Notes on the monkeys of Sierra Leone. Sierra Leone Agriculture Notes 22:1–8.

Kappeler, P. M. 1991. Patterns of sexual dimorphism in body weight among prosimian primates. Folia Primatol. 57:132–46.

Kappeler, P. M., and J. U. Ganzhorn. 1993. The evolution of primate communities and societies in Madagascar. Evol. Anthropol. 2:159–71.

Kappeler, P. M., and E. W. Heymann. 1996. Nonconvergence in the evolution of primate life history and socio-ecology. Biol. J. Linn. Soc. 59:297–326.

Kay, R. F. 1984. On the use of anatomical features to infer foraging behavior in extinct primates. In P. S. Rodman and J. G. H. Cant, eds., Adaptations for Foraging in Nonhuman Primates, Columbia Univ. Press, New York, pp. 21–53.

Kay, R. F., and W. L. Hylander. 1978. The dental structure of mammalian folivores, with special reference to primates and Phalangeroidea. In G. G. Montgomery, ed., The Ecology of Arboreal Folivores, Smiths. Inst. Press, Washington, D.C., pp. 173–91.

Kingdon, J. 1997. The Kingdon Field Guide to African Mammals, Academic Press, New York.

Kinzey, W. G. 1981. The titi monkeys, genus Callicebus. In A. F. Coimbra-Filho and R. A. Mittermeier, eds., Ecology and Behavior of Neotropical Primates, Academia Brasileira de Ciências, Rio de Janeiro, vol. 1, pp. 241–76.

Kinzey, W. G., A. L. Rosenberger, and M. Ramirez. 1975. Vertical clinging and leaping in a Neotropical anthropoid. Nature, Lond. 255:327–28.

Klopfer, P. H., and K. J. Boskoff. 1979. Maternal behavior in prosimians. In G. A. Doyle and R. D. Martin, eds., The Study of Prosimian Behavior, Academic Press, New York, pp. 123–57.

Krishtalka, L., and J. H. Schwartz. 1978. Phylogenetic relationships of plesiadapiform-tarsiiform primates. Ann. Carnegie Mus. Nat. Hist. 47:515–40.

Kummer, H. 1968. Social Organization of Hamadryas Baboons: A Field Study, Univ. Chicago Press.

Lacy, R. C. 1993/1994. What is population (and habitat) viability analysis? Primate Conservation (14–15):27–33.

Leigh, E. G., Jr., A. S. Rand, and D. M. Windsor. 1982. The Ecology of a Tropical Forest: Seasonal Rhythms and Long-term Changes, Smiths. Inst. Press, Washington, D.C.

Leighton, D. R. 1987. Gibbons: territoriality and monogamy. In B. B. Smuts, D. L. Cheney, R. M. Seyfarth, R. W. Wrangham, and T. T. Struhsaker, eds., Primate Societies, Univ. Chicago Press, pp. 135–45.

Lindburg, D. G., ed. 1980. The Macaques: Studies in Ecology, Berhavior and Evolution, Van Nostrand Reinhold, New York.

Mack, D., and R. A. Mittermeier. 1984. The International Primate Trade, TRAFFIC (USA), Washington, D.C., vol. 1.

MacKinnon, J. R., and K. S. MacKinnon. 1980. Niche differentiation in a primate community. *In* D. J. Chivers, ed., Malayan Forest Primates: Ten Years' Study in Tropical Rain Forest, Plenum Press, New York, pp. 167–90.

Mallinson, J. J. C. 1995. Conservation breeding programmes: an important ingredient for species survival. Biodiv. Conserv. 4:617–35.

Martin, R. D. 1978. Major features of prosimian evolution: a discussion in the light of chromosomal evidence. *In* D. J. Chivers and K. A. Joysey, eds., Recent Advances in Primatology, vol. 3, Evolution, Academic Press, London, pp. 3–26.

———. 1979. Phylogenetic aspects of prosimian behavior. In G. A. Doyle and R. D. Martin, eds., The Study of Prosimian Behavior, Academic Press, New York, pp. 45–77.

———. 1992. Primate reproduction. *In* S. Jones, R. D. Martin, and D. Pilbeam, eds., The Cambridge Encyclopaedia of Human Evolution, Cambridge Univ. Press, Cambridge, pp. 86–90.

———. 1995. Phylogenetic aspects of primate reproduction: the context of advanced maternal care. *In* C. R. Pryce, R. D. Martin, and S. D. Skuse, eds., Motherhood in Human and Nonhuman Primates, S. Karger, Basel, pp. 16–26.

Martin, R. D., and P. H. Harvey. 1985. Brain size allometry: ontogeny and phylogeny. *In* W. L. Jungers, ed., Size and Scaling in Primate Biology, Plenum Press, New York, pp. 147–73.

Martin, R. D., and A. M. Maclarnon. 1988. Comparative quantitative studies of growth and reproduction. Symp. Zool. Soc. Lond. 60:39–80.

McFarland, M. J. 1989. Ecological determinants of fission-fusion sociality in *Ateles* and *Pan*. *In* J. G. Else and P. C. Lee, Primate Ecology and Conservation, Cambridge Univ. Press, Cambridge, pp. 181–90.

Meier, B., and R. Albignac. 1991. Rediscovery of *Allocebus trichotis* Günther 1875 (Primates) in North East Madagascar. Folia Primatol. 56(1):57–63.

Meier, B., R. Albignac, A. Peyriéras, Y. Rumpler, and P. Wright. 1987. A new species of *Hapalemur* (Primates) from south-east Madagascar. Folia Primatol. 48:211–15.

Melnick, D. J., and M. C. Pearl. 1987. Cercopithecines in multimale groups: genetic diversity and population structure. *In* B. B. Smuts, D. L. Cheney, R. M. Seyfarth, R. W. Wrangham, and T. T. Struhsaker, eds., Primate Societies, Univ. Chicago Press, pp. 121–34.

Merfield, F. G., and H. Miller. 1956. Gorilla Hunter, Farrar Straus, New York.

Mittermeier, R. A. 1973. Colobus monkeys and the tourist trade. Oryx 12:113–17.

———. 1977. Distribution, Synecology and Conservation of Surinam primates. Ph.D. dissertation, Harvard Univ. Press, Cambridge.

———. 1978*a*. A Global Strategy for Primate Conservation, IUCN/SSC Primate Specialist Group, Washington, D.C.

———. 1978*b*. Locomotion and posture in *Ateles geoffroyi* and *Ateles paniscus*. Folia Primatol. 30:161–93.

Mittermeier, R. A., and A. F. Coimbra-Filho. 1977. Primate conservation in Brazilian Amazonia. *In* H. S. H. Prince Rainier III of Monaco and G. H. Bourne, eds., Primate Conservation, Academic Press, New York, pp. 117–66.

Mittermeier, R. A., and W. R. Konstant. 1996/1997. Primate conservation: a retrospective and a look into the 21st century. Primate Conservation (17):7–17.

Mittermeier, R. A., W. R. Konstant, M. E. Nicoll, and O. Langrand. 1992. Lemurs of Madagascar: An Action Plan for Their Conservation (1993–1999), IUCN/SSC Primate Specialist Group, Gland, Switzerland.

Mittermeier, R. A., H. de Macedo-Ruiz, A. Luscombe, and J. Cassidy. 1977. Rediscovery and conservation of the Peruvian yellow-tailed woolly monkey (*Lagothrix flavicauda*). *In* H. S. H. Prince Rainier III of Monaco and G. H. Bourne, eds., Primate Conservation, Academic Press, New York, pp. 95–115.

Mittermeier, R. A., J. F. Oates, A. E. Eudey, and J. Thornback. 1986. Primate conservation.

In G. Mitchell and J. Erwin, eds., Comparative Primate Biology, vol. 2A, Behavior, Conservation and Ecology, Alan R. Liss, New York, pp. 3–72.

Mittermeier, R. A., I. Tattersall, W. R. Konstant, D. M. Meyers, and R. B. Mast. 1994. Lemurs of Madagascar, Conservation International Tropical Field Guide Series, Conservation International, Washington, D.C.

Mittermeier, R. A., and M. G. M. van Roosmalen. 1981. Preliminary observations on habitat utilization and diet in eight Surinam monkeys. Folia Primatol. 36:1–39.

Moore, J. 1984. Female transfer in primates. Int. J. Primatol. 5:537–89.

———. 1992. Dispersal, nepotism and primate social behavior. Int. J. Primatol. 12: 361–78.

Nadler, T. 1997. A new subspecies of douc langur, *Pygathrix nemeaus cinereus* ssp. nov. Zool. Garten N. F. 67:165–76.

Napier, P. H. 1985. Catalogue of Primates in the British Museum (Natural History) and Elsewhere in the British Isles, Part 3: Family Cercopithecidae, Subfamily Colobinae, British Museum (Natural History), London.

Nash, L. T. 1986. Dietary, behavioral, and morphological aspects of gummivory in primates. Yearb. Phys. Anthropol. 29:113–37.

Nash, L. T., and C. H. Harcourt. 1986. Social organization of galagos in Kenyan coastal forests. II. *Galago garnetti.* Amer. J. Primatol. 10:357–69.

Nash, L. T., and P. L. Whitten. 1989. Preliminary observations on the role of *Acacia* gum chemistry in *Acacia* utilization by *Galago senegalensis* in Kenya. Amer. J. Primatol. 17:27–39.

Neville, M. K., K. E. Glander, F. Braza, and A. B. Rylands. 1988. The howling monkeys, genus *Alouatta. In* R. A. Mittermeier, A. B. Rylands, A. F. Coimbra-Filho, and G. A. B. da Fonseca, eds., Ecology and Behavior of Neotropical Primates, World Wildlife Fund, Washington, D.C., vol. 2, pp. 349–453.

Niemitz, C. 1979. Outline of the behavior of *Tarsius bancanus. In* G. A. Doyle and R. D. Martin, eds., The Study of Prosimian Behavior, Academic Press, New York, pp. 631–60.

Oates, J. F. 1977. The guereza and man. *In* H. S. H. Prince Rainier III of Monaco and G. H. Bourne, eds., Primate Conservation, Academic Press, New York, pp. 419–67.

———. 1986. Action Plan for African Primate Conservation:1986–1990, IUCN/SSC Primate Specialist Group, Gland, Switzerland.

———. 1996. African Primates: Status Survey and Conservation Action Plan, IUCN/SSC Primate Specialist Group, Gland, Switzerland.

Oates, J. F., and G. Davies. 1994. Conclusions: the past, present and future of the colobines. *In* A. G. Davies and J. F. Oates, eds., Colobine monkeys: Their Ecology, Behaviour and Evolution, Cambridge Univ. Press, Cambridge, pp. 347–58.

Oates, J. F., T. T. Struhsaker, and G. H. Whitesides. 1996/1997. Extinction faces Ghana's red colobus monkey and other locally endemic subspecies. Primate Conservation (17):138–44.

Oates, J. F., G. H. Whitesides, A. G. Davies, P. G. Waterman, S. M. Green, G. L. Dasilva, and S. Mole. 1990. Determinants of variation in tropical forest primate biomass: new evidence from West Africa. Ecology 71:328–43.

Olney, P. J. S., G. M. Mace, and A. T. C. Feistner, eds. 1994. Creative Conservation: Interactive Management of Wild and Captive Animals, Chapman and Hall, London.

Oppenheimer, J. R. 1977. *Presbytis entellus,* the Hanuman langur. *In* H. S. H. Prince Rainier III of Monaco and G. H. Bourne, eds., Primate Conservation, Academic Press, New York, pp. 469–512.

Pages, E. 1980. Ethoecology of *Microcebus coquereli. In* P. Charles-Dominique, H. M. Cooper, and A. Hladik, eds. Nocturnal Malagasy Primates: Ecology, Physiology and Behavior, Academic Press, New York, pp. 97–115.

Pastorini, J., M. R. J. Forstner, R. D. Martin, and D. J. Melnick. 1998. A reexamination of the phylogenetic position of *Callimico* (Primates) incorporating new mitochondrial DNA sequence data. J. Mol. Evol. 47:32–41.

Pastor-Nieto, R., and D. K. Wiiliamson. 1998. The effect of rainfall seasonality on the geographic distribution of Neotropical primates, Neotropical Primates 6:7–14.

Pereira, M. E. 1993. Agonistic interaction, dominance, relation, and ontogenetic trajectories in ring-tailed lemurs. *In* M. E. Pereira and L. A. Fairbanks, eds., Juvenile Primates: Life History, Development and Behavior, Oxford Univ. Press, New York, pp. 285–308.

Peres, C. A. 1990. Effects of hunting on western Amazonian primate communities. Biol. Conserv. 54:47–59.

———. 1992*a*. Consequences of joint territoriality in a mixed species group of tamarin monkeys. Behaviour 123(3–4):220–46.

———. 1992*b*. Prey capture benefits in a mixed species group of Amazonian tamarins, *Saguinus fuscicollis* and *S. mystax*. Behav. Ecol. Sociobiol. 31:339–47.

———. 1993. Anti-predation benefits in a mixed species group of Amazonian tamarins. Folia Primatol. 61:61–76.

———. 1997*a*. Primate community structure at twenty western Amazonian flooded and unflooded forests. J. Trop. Ecol. 13:381–405.

———. 1997*b*. Effects of habitat quality and hunting pressure on arboreal folivore densities in Neotropical forests: a case study of howler monkeys (*Alouatta* spp.). Folia Primatol. 68(3–5):199–222.

Podolsky, R. D. 1990. Effects of mixed-species association on resource use by *Saimiri sciureus* and *Cebus apella*. Amer. J. Primatol. 21:147–58.

Pollock, J. I. 1977. The ecology and sociology of feeding in *Indri indri*. *In* T. H. Clutton-Brock, ed., Primate Ecology: Studies of Feeding and Ranging Behaviour in Lemurs, Monkeys and Apes, Academic Press, London, pp. 37–69.

Pook, A. G., and G. Pook. 1982. Polyspecific association between *Saguinus fuscicollis, Saguinus labiatus, Callimico goeldii* and other primates in north-western Bolivia. Folia Primatol. 38:196–216.

Pusey, A. E., and C. Packer. 1987. Dispersal and philopatry. *In* B. B. Smuts, D. L. Cheney, R. M. Seyfarth, R. W. Wrangham, and T. T. Struhsaker, eds., Primate Societies, Univ. Chicago Press, pp. 250–66.

Quris, R. 1976. Données comparatives sur la socio-écologie de huit espèces de Cercopithecidae vivant dans une même zone de forêt primitive périodiquement inondée (Nord-est du Gabon). Terre Vie 30:193–209.

Ramirez, M. 1988. The woolly monkeys, genus *Lagothrix*. *In* R. A. Mittermeier, A. B. Rylands, A. F. Coimbra-Filho, and G. A. B. da Fonseca, eds., Ecology and Behavior of Neotropical Primates, World Wildlife Fund, Washington, D.C., vol. 2, pp. 539–75.

Reed, K. E., and J. G. Fleagle. 1995. Geographic and climatic control of primate diversity. Proc. Natl. Acad. Sci. USA. 92:7874–76.

Ren, R. M., R. C. Kirkpatrick, N. G. Jablonski, W. V. Bleitsch, and X. C. Le. 1996/1997. Conservation status and prospects for the snub-nosed langurs (Colobinae: *Rhinopithecus*). Primate Conservation (17):152–59.

Richard, A. F. 1979. Patterns of mating in *Propithecus verreauxi verreauxi*. *In* R. D. Martin, G. A. Doyle, and A. C. Walker, eds., Prosimian Biology, Univ. Pittsburgh Press, pp. 49–74.

———. 1987. Malagasy prosimians: female dominance. *In* B. B. Smuts, D. L. Cheney, R. M. Seyfarth, R. W. Wrangham, and T. T. Struhsaker, eds., Primate Societies, Univ. Chicago Press, pp. 25–33.

Richard, A. F., S. J. Goldstein, and R. E. Dewar. 1989. Weed macaques: the evolutionary implications of macaque feeding ecology. Int. J. Primatol. 10:569–94.

Rijksen, H. D. 1978. A Field Study on Sumatran Orangutans: Ecology, Behaviour and Conservation, H. Veenman and Zonig, Wageningen.

Robinson, J. G., P. C. Wright, and W. G. Kinzey. 1987. Monogamous Cebids and Their Relatives: Intergroup Calls and Spacing. *In* B. B. Smuts, D. L. Cheney, R. M. Seyfarth, R. W. Wrangham, and T. T. Struhsaker, eds., Primate Societies, Univ. Chicago Press, pp. 4–53.

Rodman, P. S. 1978. Diets, densities, and distributions of Bornean primates. *In* G. G. Montgomery, ed., The Ecology of Arboreal Folivores, Smiths. Inst. Press, Washington, D.C., pp. 465–78.

Rodman, P. S., and C. J. Mitani. 1987. Orangutans: sexual dimorphism in a solitary species. *In* B. B. Smuts, D. L. Cheney, R. M. Seyfarth, R. W. Wrangham, and T. T. Struhsaker, eds., Primate Societies, Univ. Chicago Press, pp. 146–54.

Rodríguez-Luna, E., L. Cortés-Ortiz, R. A. Mittermeier, and A. B. Rylands. 1996. Plan de Acción para Los Primates Mesoamericanos. Borrador de Trabajo, IUCN/SSC Primate Specialist Group, Neotropical Section, Xalapa, Veracruz.

Roonwal, M. L., and S. M. Mohnot. 1977. Primates of South Asia: Ecology, Sociology and Behavior, Harvard Univ. Press, Cambridge.

Rosenberger, A. L. 1981. Systematics: the higher taxa. *In* A. F. Coimbra-Filho and R. A. Mittermeier, eds., Ecology and Behavior of Neotropical Primates, Academia Brasileira de Ciências, Rio de Janeiro, vol. 1, pp. 9–27.

Rosenberger, A. L., and K. B. Strier. 1989. Adaptive radiation of the ateline primates. J. Hum. Evol. 18:717–50.

Rowe, N. 1996. The Pictorial Guide to Living Primates, Pogonias Press, East Hampton, New York.

Rowell, T. E. 1988. The social system of guenons, compared with baboons, macaques and mangabeys. *In* A. Gautier-Hion, A., F. Bourlière, J. P. Gautier, and J. Kingdon, eds., A Primate Radiation: Evolutionary Biology of the African Guenons, Cambridge Univ. Press, Cambridge, pp. 439–51.

Rylands, A. B. 1984. Exudate-eating and tree-gouging by marmosets (Callitrichidae, Primates). *In* A. C. Chadwick and S. L. Sutton, eds., Tropical Rain Forest: The Leeds Symposium, Leeds Philosophical and Literary Society, Leeds, pp. 155–68.

———. 1987. Primate communities in Amazonian forests: their habitats and food resources. Experientia 43(3):265–79.

———. 1996. Habitat and the evolution of social and reproductive behavior in Callitrichidae. Amer. J. Primatol. 38:5–18.

Rylands, A. B., and D. S. de Faria. 1993. Habitats, feeding ecology, and home range size in the genus *Callithrix. In* A. B. Rylands, ed., Marmosets and Tamarins. Systematics, Behaviour and Ecology, Oxford Univ. Press, Oxford, pp. 262–72.

Rylands, A. B., R. A. Mittermeier, and E. Rodríguez-Luna. 1995. A species list for the New World primates (Platyrrhini): Distribution by country, endemism, and conservation status according to the Mace-Lande system. Neotropical Primates 3(suppl.): 113–60.

Rylands, A. B., G. A. B. da Fonseca, Y. L. R. Leite, and R. A. Mittermeier. 1996. Primates of the Atlantic forest: origin, endemism, distributions and communities. *In* M. A. Norconk, A. L. Rosenberger, and P. A. Garber, eds., Adaptive Radiations of the Neotropical Primates, Plenum Press, New York, pp. 21–51.

Rylands, A. B., E. Rodríguez-Luna, and L. Cortés-Ortiz. 1996/1997. Neotropical primate conservation—the species and the IUCN/SSC Primate Specialist Group network. Primate Conservation (17):46–69.

Rylands, A. B., K. B. Strier, R. A. Mittermeier, J. Borovansky, and U. S. Seal, eds. 1998. Conserving Brazil's Muriqui: Population and Habitat Viability Assessment (PHVA) for *Brachyteles arachnoides*, IUCN/SSC Conservation Breeding Specialist Group Apple Valley, Minn.

Sauther, M. L., and R. W. Sussman. 1993. A new interpretation of the social organization and mating system of the ring-tailed lemur, *Lemur catta. In* P. M. Kappeler and J. U. Ganzhorn, eds., Lemur Social Systems and Their Ecological Basis, Plenum Press, New York, pp. 111–22.

Schaller, G. B. 1963. The Mountain Gorilla: Ecology and Behavior, Univ. Chicago Press.

Schneider, H., and A. L. Rosenberger. 1996. Molecules, morphology, and platyrrhine systematics. *In* M. A. Norconk, A. L. Rosenberger, and P. A. Garber, eds., Adaptive Radiations of Neotropical Primates, Plenum Press, New York, pp. 1–19.

Schwartz, J. H. 1978. If *Tarsius* is not a prosimian, is it a haplorhine? *In* Recent Advances in Primatology, vol. 3, Evolution, D. J. Chivers and K. A. Joysey, eds. Academic Press, London , pp. 195–202.

Seyfarth, R, M. 1987. Vocal communication and its relation to language. *In* B. B. Smuts,

D. L. Cheney, R. M. Seyfarth, R. W. Wrangham, and T. T. Struhsaker, eds., Primate Societies, Univ. Chicago Press, pp. 440–51.

Shaul, D. M. B. 1962. The composition of milk of wild animals. Int. Zoo Yearb. 4:333–45.

Smith, A. P., and Ganzhorn, J. U. 1996. Convergence and divergence in community structure and dietary adaptation in Australian possums and gliders and Malagasy lemurs. Aust. J. Ecol. 21:31–46.

Snowdon, C. T. 1989. Vocal communication in New World primates. J. Hum. Evol. 18:611–33.

Snowdon, C. T., C. H. Brown, and M. R. Peterson, eds.. 1982. Primate Communication, Cambridge Univ. Press, Cambridge.

Snowdon, C. T., and P. Soini. 1988. The tamarins, genus Saguinus. In R. A. Mittermeier, A. B. Rylands, A. F. Coimbra-Filho, and G. A. B. da Fonseca, eds., Ecology and Behavior of Neotropical Primates, World Wildlife Fund, Washington, D.C., vol. 2, pp. 223–98.

Soini, P. 1982. Primate conservation in Peruvian Amazonia. Int. Zoo Yearb. 22:37–47.

Stammbach, E. 1987. Desert, forest and montane baboons: multilevel societies. In B. B. Smuts, D. L. Cheney, R. M. Seyfarth, R. W. Wrangham, and T. T. Struhsaker, eds., Primate Societies, Univ. Chicago Press, pp. 112–20.

Sterling, E. 1994. Taxonomy and distribution of Daubentonia: a historical perspective. Folia Primatol. 62:8–13.

Stevenson, M. F., and A. B. Rylands. 1988. The marmosets, genus Callithrix. In R. A. Mittermeier, A. B. Rylands, A. F. Coimbra-Filho, and G. A. B. da Fonseca, eds.. Ecology and Behavior of Neotropical Primates, World Wildlife Fund, Washington, D.C., vol. 2, pp. 131–222.

Strier, K. B. 1992. Faces in the Forest: The Endangered Muriqui Monkeys of Brazil, Oxford Univ. Press, New York.

———. 1994. The myth of the typical primate. Yearb. Phys. Anthropol. 37:233–71.

———. 1997. Behavioral ecology and conservation biology of primates and other animals. Advances in the Study of Behavior, Academic Press, New York, vol. 26, pp. 101–58.

Struhsaker, T. T. 1975. The Red Colobus Monkey, Univ. Chicago Press.

———. 1978. Food habits of five monkey species in the Kibale forest, Uganda. In D. J. Chivers and J. Herbert, eds., Recent Advances in Primatology, vol. 1, Behaviour, Academic Press, London, pp. 225–48.

———. 1981. Polyspecific associations among tropical rain-forest primates. Z. Tierpsychol. 57:268–304.

———. 1997. Ecology of an African Rain Forest: Logging in Kibale and the Conflict between Conservation and Exploitation, Univ. Press of Florida, Gainesville.

Struhsaker, T. T., and L. Leland. 1979. Socioecology of five sympatric monkey species in the Kibale forest, Uganda. Advances in the Study of Behavior, Academic Press, New York, vol. 4, pp. 159–227.

Struhsaker, T. T., and J. F. Oates. 1975. Comparison of the behavior and ecology of red colobus and black-and-white colobus monkeys in Uganda: a summary. In R. H. Tuttle, ed., Socioecology and Psychology of Primates, Mouton, The Hague, pp. 103–24.

Sussman, R. W. 1978. Nectar-feeding by prosimians and its evolutionary and ecological implications. In Recent Advances in Primatology, vol. 3, Evolution. D. J. Chivers and K. A. Joysey, eds., Academic Press, London, pp. 119–25.

Sussman, R. W., and W. G. Kinzey. 1984. The ecological role of the Callitrichidae: a review. Amer. J. Phys. Anthropol. 64:419–49.

Tagliaro, C. H., M. P. C. Schneider, H. Schneider, I. C. Sampaio, and M. J. Stanhope. 1997. Marmoset phylogenetics, conservation perspectives, and evolution of the mtDNA control region. Mol. Biol. Evol. 14(6):674–84.

Tappen, N. 1964. Primate studies in Sierra Leone. Current Anthropol. 5(4):339–40.

Tattersall, I. 1982. The Primates of Madagascar, Columbia Univ. Press, New York.

———. 1988. Cathemeral activity in primates: a definition. Folia Primatol. 49:200–202.

Tattersall, I., and R. W. Sussman. 1998. "Little brown lemurs" of northern Madagascar. Folia Primatol. 69(suppl. 1):379–88.

Terborgh, J. 1983. Five New World Primates: A Study in Comparative Ecology, Princeton Univ. Press, Princeton.

——. 1990. An overview of research at Cocha Cashu Biological Station. *In* A. H. Gentry, ed., Four Neotropical Rainforests, Yale Univ. Press, New York, pp. 48–59.

Terborgh, J., and C. P. Van Schaik. 1987. Convergence vs. nonconvergence in primate communities. *In* J. H. R. Gee and P. S. Giller, eds., Organization of Communities: Past and Present, Blackwell Scientific Publications, Oxford, pp. 205–26.

Van Lawick-Goodall, J. 1971. In the Shadow of Man, Houghton Mifflin Company, Boston.

van Roosmalen, M. G. M. 1985. Habitat preferences, diet, feeding strategy and social organization of the black spider monkey (*Ateles paniscus paniscus* Linnaeus 1758) in Surinam. Acta Amazonica 15(3/4, suppl.):1–238.

van Roosmalen, M. G. M., and L. L. Klein. 1988. The spider monkeys, genus *Ateles. In* R. A. Mittermeier, A. B. Rylands, A. F. Coimbra-Filho, and G. A. B. da Fonseca, eds., Ecology and Behavior of Neotropical Primates, World Wildlife Fund, Washington, D.C., vol. 2, pp. 455–538.

van Roosmalen, M. G. M., T. van Roosmalen, R. A. Mittermeier, and G. A. B. da Fonseca. 1998. A new and distinctive species of marmoset (Callitrichidae, Primates) from the lower Rio Aripuanã, state of Amazonas, central Brazilian Amazonia. Goeldiana Zoologia (22):1–27.

Van Schaik, C. P., P. R. Assink, and N. Salafsky. 1992. Territorial behavioral in Southeast Asian langurs: resource defense or mate defense? Amer. J. Primatol. 26:233–42.

Van Schaik, C. P., and E. Mirmanto. 1985. Spatial variation in the structure and litterfall of a Sumatran rain forest. Biotropica 17:196–205.

Walker, S. E. 1996. The evolution of positional behavior in the saki-uakaris (*Pithecia, Chiropotes,* and *Cacajao*). *In* M. A. Norconk, A. L. Rosenberger, and P. A. Garber, eds., Adaptive Radiations of Neotropical Primates, Plenum Press, New York, pp. 335–76.

Walters, J. R., and Seyfarth, R. M. 1987. *In* B. B. Smuts, D. L. Cheney, R. M. Seyfarth, R. W. Wrangham, and T. T. Struhsaker, eds., Primate Societies, Univ. Chicago Press, pp. 306–17.

Waser, P. M. 1980. Polyspecific association of *Cercocebus albigena:* geographic variation and ecological correlates. Folia Primatol. 57:268–304.

——. 1987. Interactions among primate species. *In* B. B. Smuts, D. L. Cheney, R. M. Seyfarth, R. W. Wrangham, and T. T. Struhsaker, eds., Primate Societies, Univ. Chicago Press, pp. 210–26.

Watanabe, K. 1981. Variations in group composition and population density of the two sympatric Mentawaian leaf-monkeys. Primates 22:145–60.

Watts, D. P., and A. E. Pusey. 1993. Behavior of juvenile and adolescent great apes. *In* M. E. Pereira and L. A. Fairbanks, eds., Juvenile Primates: Life History, Development and Behavior, Oxford Univ. Press, New York, pp. 148–72.

Wilson, J. M., L. R. Godfrey, E. L. Simons, P. D. Stewart, and M. Vuillaume-Randriamanantena. 1995. Past and present lemur fauna at Ankarana, north Madagascar. Primate Conservation (16):47–52.

Wilson, J. M., P. D. Stewart, G. S. Ramangason, A. M. Denning, and M. S. Hutchings. 1989. Ecology and conservation of the crowned lemur, *Lemur coronatus,* at Ankarana, N. Madagascar. Folia Primatol. 52:21–26.

Wright, P. C. 1981. The night monkeys, genus *Aotus. In* A. F. Coimbra-Filho and R. A. Mittermeier, eds., Ecology and Behavior of Neotropical Primates, Academia Brasileira de Ciências, Rio de Janeiro, Brazil, vol. 1, pp. 211–41.

——. 1984. Biparental care in *Aotus trivirgatus* and *Callicebus moloch. In* M. F. Small. ed., Female Primates: Studies by Women Primatologists, Alan R. Liss, New York, pp. 59–75.

——. 1986. Ecological correlates of monogamy in *Aotus* and *Callicebus. In* J. G. Else and P. C. Lee, eds., Primate Ecology and Conservation, Cambridge Univ. Press, Cambridge, pp. 159–68.

——. 1996a. The future of biodiversity in Madagascar: a view from the Ranomafana

National Park. *In* B. D. Patterson and S. M. Goodman, eds., Environmental Change in Madagascar, Smiths. Inst. Press, Washington, D.C., pp. 381–405.

———. 1996*b*. The Neotropical primate adaptation to nocturnality: feeding in the night (*Aotus nigriceps* and *A. azarae*). *In* M. A. Norconk, A. L. Rosenberger, and P. A. Garber, eds., Adaptive Radiations of Neotropical Primates, Plenum Press, New York, pp. 369–82.

———. 1997. Behavioral and ecological comparisons of Neotropical and Malagasy primates. *In* W. G. Kinzey, ed., New World Primates: Ecology, Evolution and Behavior, Aldine de Gruyter, New York, pp. 127–41.

Wright, P. C., and M. Randrimanantena. 1989. Comparative ecology of three sympatric bamboo lemurs in Madagascar. Amer. J. Phys. Anthropol. 78:327.

Wu, H. Y., and Y. S. Lin. 1994. Study on population ecology of the Formosan macaque (*Macaca cyclopis*), Yushan National Park, Taiwan. Paper presented at the 15th Congress of the International Primatological Society (IPS), Kuta, Bali, Indonesia.

Yamamoto, M. E., and Box, H. O. 1997. The role of non-reproductive helpers in infant care in captive *Callithrix jacchus*. Ethology 103:760–71.

Zeller, A. C. 1987. Communication by sight and smell. *In* B. B. Smuts, D. L. Cheney, R. M. Seyfarth, R. W. Wrangham, and T. T. Struhsaker, eds., Primate Societies, Univ. Chicago Press, pp. 433–39.

Zimmerman, E., S. Cepok, N. Rakotoarison, V. Zietemann and U. Radepsiel. 1998. Sympatric mouse lemurs in north-west Madagascar: A new rufous mouse lemur species (*Microcebus ravelobensis*). Folia Primatol. 69:106–114.

Primates

Lorises, Pottos, and Galagos

This family of 9 genera and 18 species is found in Africa south of the Sahara, southern India, Sri Lanka, southeastern Asia, and the East Indies. Petter and Petter-Rousseaux (1979) divided the family into two subfamilies: the Lorisinae for the lorises and pottos *(Arctocebus, Loris, Nycticebus, Perodicticus)* and the Galaginae for the galagos *(Euoticus, Galago, Otolemur, Galagoides)*. This sequence of genera is followed here, with the newly described *Pseudopotto* (see account thereof) being added first to reflect its primitive nature. Some authorities (e.g., Goodman 1975; Schwartz and Tattersall 1985) consider both subfamilies to be full families. Schwartz (1987) not only gave familial rank to the Lorisinae and the Galaginae but also placed *Nycticebus* and *Perodicticus* in a new subfamily, the Nycticebinae. Groves (1989) did not at first accept such a familial division, but subsequently (*in* Wilson and Reeder 1993) he did, and he also followed Jenkins (1987) in using the names Loridae (in place of Lorisidae) and Galagonidae (in place of the more commonly used Galagidae). Schwartz (1996) argued for maintenance of Lorisidae.

In the lorises and the pottos head and body length is 170–390 mm and the tail is short or absent. The head and eyes are round, and the small, rounded ears are nearly hidden by the surrounding fur. The forelimbs and hind limbs are nearly equal in length. The hands and feet and the digits, especially the first, are relatively stronger than in the galagos. The first digit of each limb is more widely opposable to the other digits than in the galagos, and the index finger is more reduced. All the digits of lorises and pottos have nails except the second digit of the foot, which has a claw. The wrist and ankle joints have more freedom of movement than in galagos. The grip by either hands or feet is powerful and can be maintained for long periods. Male lorises and pottos do not have a baculum, and females have two or three pairs of mammae.

The forms of galagos are very different from those of lorises and pottos. The bodies of galagos are slender; the tail is usually longer than the head and body and is almost bushy. The arms and legs are comparatively longer and more slender with rather long fingers and toes. Head and body length is 105–465 mm. The ears are large and mobile. The hind limbs are considerably longer than the forelimbs. Males have a baculum, and females usually have two pairs of mammae.

The dental formula for the family is: (i 2/2, c 1/1, pm 3/3, m 3/3) × 2 = 36. In lorises and galagos the lower incisors project forward and slightly upward, the canines are long and sharp, and the molars have three or four cusps. The orbits in lorises and pottos face more directly forward than in the galagos.

Lorises and pottos are essentially arboreal and nocturnal, sheltering by day in hollow trees, in crevices of trees, or on branches. They generally sleep in a flexed position, with the head tucked between the arms. They usually move with slow, deliberate, hand-over-hand movements. They move quickly at times but do not leap or jump. The forward progress of lorises and pottos is accompanied by a rhythmic sinuous movement of the spine, produced by placing a hand and then a foot in consecutive order on the surface they are traversing near the midline of the body. They travel along the underside of limbs or poles as readily as on the

Pygmy slow loris (*Nycticebus pygmaeus*). Photo by David Haring, Duke University Primate Center.

top. The powerful grasp is facilitated by the great development and wide angle of divergence of the first digit on the hand and foot, the mobility of the wrist and ankle joints, and the presence of storage channels *(reta mirabilia)* for blood in the vessels of the hands and feet. These channels enable the muscles to remain contracted over long periods without fatigue by aiding the exchange of oxygen and carbon dioxide in the muscle fibers.

The geological range of this family is early Miocene to Recent (Thorington and Anderson 1984).

PRIMATES; LORISIDAE; Genus PSEUDOPOTTO
Schwartz, 1996

The single species, *P. martini,* is known only by two specimens taken years ago and now in the Anthropological Institute and Museum, University of Zurich-Irchel. One, a nearly complete skeleton, represents an adult female that was captured in "the Cameroons" and subsequently lived at the Zurich Zoo. The other, consisting only of a skull and mandible, represents a subadult male collected somewhere in "Equatorial Africa." These geographic terms, in use until about 1960, indicate that the first specimen originated either in the current nation of Cameroon or in extreme eastern Nigeria and that the second came from either

Cameroon or an immediately neighboring nation. The specimens had been cataloged as *Perodicticus potto* but recently were reexamined by Schwartz (1996) and described as a new genus and species.

The phylogenetic position of *Pseudopotto* is uncertain, and Schwartz did not even assign it definitely to any family. However, as noted above in the account of the Lorisidae, Schwartz (1987) treated the Galaginae as a full family (Galagidae). Since Schwartz (1996) indicated that *Pseudopotto* appears to be more closely related to what he had termed the Lorisidae than to what he had termed the Galagidae, and since those two groups are here considered only subfamilies of the family Lorisidae, it seems reasonable to assign *Pseudopotto* to that family here. Also, in keeping with Schwartz's indication that *Pseudopotto* is more primitive than any of the other living lorisids, it is here placed first among the group.

Pseudopotto bears a superficial resemblance to *Perodicticus,* and its cranial and humeral measurements fall at the lower end of the size range of the latter genus. It is immediately distinguished from *Perodicticus* and all other living lorisids, however, by its much longer tail (obvious even though at least one caudal vertebra is missing from the specimen with postcranial elements) and in lacking a distinctly hooked ulnar styloid process.

Pseudopotto is dentally primitive relative to *Perodicticus, Arctocebus,* and *Nycticebus* in retaining a more bucal-

Slender loris *(Loris tardigradus)*, photo by Bernhard Grzimek.

ly emplaced cristid obliqua and lacking deep hypoflexid notches on the lower molars, as well as in having relatively longer lower middle and last premolars. *Pseudopotto* is similar to *Perodicticus* and *Nycticebus* in having a bulky skull and mandible, expanding suprameatal shelves, and a broad snout, but differs in having a more hooked mandibular coronoid process and goneal region. It differs from *Perodicticus* and *Arctocebus* in having taller cheektooth cusps; more crisply defined cusp apices, crests, and margins; and an entepicondylar foramen. *Pseudopotto* is distinguished from most of the Lorisidae and Cheirogaleidae (but not *Phaner*) in having a very diminutive third upper premolar, from most extant prosimians (but not *Nycticebus*) in having the lacrimal fossa incorporated into the inferior orbital margin, and from all extant prosimians in having a diminutive third upper molar.

PRIMATES; LORISIDAE; Genus ARCTOCEBUS
Gray, 1863

Golden Potto, or Angwantibo

Petter and Petter-Rousseaux (1979) recognized a single species, *A. calabarensis*, occurring in the tropical forest belt from southeastern Nigeria to southern Congo and western Zaire. Groves (1989 and *in* Wilson and Reeder 1993) accepted the distinction of a second, allopatric species, *A. aureus*, occupying that part of the range of the genus to the south of the Sanaga River, which runs through central Cameroon.

Head and body length is 229–305 mm and the tail is only a slight protuberance. Napier and Napier (1967) gave the weight as 266–465 grams. The fur has a golden sheen and is thick, long, and wool-like over the body, but the hands and feet are thinly haired. The species or subspecies *aureus* is generally bright golden red above and grayish beneath; the species or subspecies *calabarensis* is yellowish brown to fawn above and paler, almost whitish, on the underparts. The face is darker than the back, and the sides of the face are light. There is a white line from the brow to the nose. *Arctocebus* resembles *Perodicticus*, and the two are frequently confused. *Arctocebus* is slightly smaller and has a relatively longer face. The index finger is reduced to a mere stub. The feet of *Arctocebus* are well suited for grasping for long periods of time with little or no change of position.

Although found within the overall tropical forest region, the golden potto seems to prefer clearings where trees have been felled by storms or areas of young secondary growth with dense and intricate vegetation. In primary forest it has never been observed at a height above 15 meters, and it is generally found below 5 meters in the undergrowth. It is nocturnal, sleeping by day in thick foliage. It moves with a stealthy, slow climb in which three extremities are involved in gripping supports at any one time. If threatened, it will remain firmly attached to a branch with its body rolled into a ball, and it may reach out under an arm to bite. The diet consists largely of insects, mostly caterpillars in some areas, and also includes fruits. Insects may be located under leaves by smell, and this primate also may rear on its hind legs to catch moths in flight. Population density has been estimated at 2/sq km in primary forest and 7/sq km in clearings and secondary forest (Charles-Dominique 1977; Charles-Dominique and Bearder 1979; Hladik 1979; Walker 1979).

The golden potto appears to be a generally solitary species. Several vocalizations have been reported, including a deep, hoarse growl in agonistic encounters and a fairly

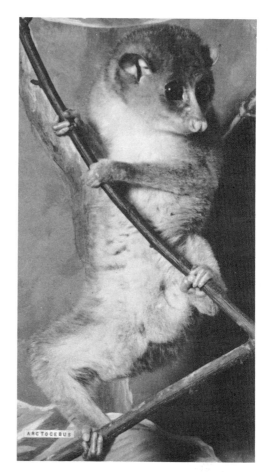

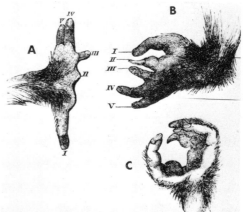

Golden potto *(Arctocebus calabarensis)*, photo from Zoological Society of London. A. Left hand spread; B. Left foot in natural position as seen from above; C. The same as seen from below, photos from *Proc. Zool. Soc. London.*

powerful call for distant communication (Petter and Charles-Dominique 1979). Estrous cycles of 36–45 days have been recorded in a captive female (Van Horn and Eaton 1979). Females apparently can breed more than once a year, and in Gabon births have been observed throughout the year, but with a minimum number from June to August. The gestation period is 131–36 days, and weight at birth is 24–30 grams. There is one young per birth. It clings to the belly of the female for about 3–4 months, after which time it is weaned and begins to follow the mother or ride dorsally. Adult weight and sexual maturity are attained at 8–10 months (Charles-Dominique 1977). Record longevity is 13 years (Jones 1982).

Arctocebus is threatened by loss of forest habitat to logging and cultivation and by being hunted for its meat. Its ability to inhabit secondary forests, however, may give it a chance to survive (Lee, Thornback, and Bennett 1988). It is on appendix 2 of the CITES. The IUCN distinguishes *A. aureus* and *A. calabarensis* as separate species and designates both as near threatened.

PRIMATES; LORISIDAE; **Genus LORIS**
E. Geoffroy St.-Hilaire, 1796

Slender Loris

Petter and Petter-Rousseaux (1979) and Corbet and Hill (1992) recognized a single species, *L. tardigradus,* occurring in southern India below about 15° N and in Sri Lanka. Groves (1989 and *in* Wilson and Reeder 1993) suggested that the three subspecies in Sri Lanka are sharply distinct and might represent full species. Goonan (1993) noted an apparent loss of reproductive viability when individuals from two of the Sri Lankan subspecies were joined in a captive group.

Head and body length is 175–264 mm, there is no ex-

ternal tail, and weight is 85–348 grams. The fur is soft, thick, and woolly and varies in coloration from yellowish gray to dark brown above, while the underparts are silvery gray to buff. Superficially this animal resembles the better-known slow loris *(Nycticebus),* but it is smaller and much more slenderly built. The great toes and the thumbs are opposable to the other toes and fingers, thereby producing highly efficient grasping organs, and the limbs are very slender. The eyes are round and large, and the prominent ears are thin, rounded, and naked at the edges.

In Sri Lanka this species occurs both in montane forest, where temperatures are relatively low, and in deciduous forest in thick vegetation (Hladik 1979). The slender loris is arboreal and nocturnal, spending daylight hours sleeping on branches. It moves with a slow, stealthy climb, at least three limbs always grasping a support. *Loris* is almost always found on the top side of branches, whereas *Arctocebus* sometimes rests by hanging below (Walker 1979). According to Roonwal and Mohnot (1977), *Loris* is capable of running at a fair pace on the ground but cannot swim. The diet is chiefly insectivorous but also includes shoots, young leaves, fruits with hard rinds, birds' eggs, and small vertebrates. When trying to catch prey, the method of attack is usually a cautious stalk followed by a quick grab with both hands. In Sri Lanka each individual utilizes approximately 1 ha. of forest (Hladik 1979).

According to Roonwal and Mohnot (1977), the slender loris usually lives alone or with a mate. Most reports indicate that it is aggressive, and when several individuals are kept together there initially is constant squealing and fighting, sometimes resulting in death. Eventually the

Slender loris *(Loris tardigradus)*, photo from New York Zoological Society.

animals become more tolerant of one another. Reported vocalizations are a "growl" indicating disturbance, a "screech" heard at night, a "whistle" probably expressing satisfaction upon meeting a mate, and loud distress and warning calls. Females apparently enter estrus twice a year and are receptive for about a week. In southern India mating reportedly takes place in April–May and October–November, and in Sri Lanka births have been recorded in May and December. There may be one or two young per birth. Studies of a captive colony (Izard and Rasmussen 1985) indicated no reproductive seasonality and the following parameters: estrous cycle, 29–40 days; gestation, 166–69 days; interbirth interval, 9.5 months; litter size, 1 young; lactation, 6–7 months; and sexual maturity, 10 months in females and about 18 months in males. According to Marvin L. Jones (Zoological Society of San Diego, pers. comm., 1995), one individual lived in captivity for 15 years and 5 months and another was still living after 14 years and 5 months.

Schulze and Meier (1994) indicated that *Loris* may be endangered in India because its forest habitat has been reduced to isolated fragments and it is hunted for body parts that have alleged medicinal use. Much of its habitat in Sri Lanka also has been destroyed and it is absent from most areas even where suitable habitat seems to remain. *Loris* currently is classified as vulnerable by the IUCN and is on appendix 2 of the CITES.

PRIMATES; LORISIDAE; **Genus NYCTICEBUS**
E. Geoffroy St.-Hilaire, 1812

Slow Lorises, or Cu Lan

There are two species (Corbet and Hill 1992; Eudey 1987; Groves 1971*b*; Petter and Petter-Rousseaux 1979; Wolfheim 1983; Zhang, Chen, and Shi 1993):

N. pygmaeus, east of the Mekong River in southern Yunnan, Laos, Viet Nam, and Cambodia;

N. coucang, eastern Bangladesh, Assam, Burma, southern Yunnan and Guangxi, Thailand, Indochina, Malay Peninsula and the nearby islands of Penang and Tioman, Riau Archipelago, Sumatra, Bangka, Borneo, Natuna Islands, southwestern part of the Sulu Archipelago in the southern Philippines.

Groves (1971*b*) stated that *N. coucang* also occurred on Mindanao, but perhaps as a result of introduction. Fooden (1991*a*) concluded that records from Mindanao were completely erroneous; both he and Timm and Birney (1992) showed that the species is found no farther east than the small islands of Tawitawi, Sanga Sanga, Bongao, and Simunul, which politically are part of the Philippines though they are located less than 100 km off northeastern

Slow loris *(Nycticebus coucang),* photo by Lim Boo Liat. Inset photo by Ernest P. Walker.

Slow loris (Nycticebus coucang), two-day-old baby, photo by Ernest P. Walker.

Borneo. Petter and Petter-Rousseaux (1979) considered N. pygmaeus a subspecies of N. coucang, but the two were regarded as specifically distinct by Corbet and Hill (1991), Eudey (1987), Groves (1971b and in Wilson and Reeder 1993), Napier and Napier (1985), Wolfheim (1983), and Zhang, Chen, and Shi (1993).

In N. coucang head and body length is 265–380 mm and weight is 375–2,000 grams (Roonwal and Mohnot 1977). In N. pygmaeus head and body length is about 180–210 mm. The tail is vestigial in this genus. The fur is short, thick, and woolly. The coloration above ranges from light brownish gray to deep reddish brown, sometimes with a hoary effect produced by the light tips of individual hairs. The coloration beneath is usually somewhat lighter, varying from almost white to buffy or grayish. There is usually a dark midline along the neck and back and a light streak between the dark orbital rings.

The thumb is practically perpendicular to the other fingers, and the great toe is perpendicular or even points slightly backward. The muscular arrangement enables an effortless grasp, with the hands and feet rigidly clenched.

The remainder of this account, except as noted, is taken from Lekagul and McNeely (1977) and applies to N. coucang. The slow loris is found in primary or secondary forest or in groves of bamboo. It is nocturnal and arboreal, seldom descending to the ground. By day it sleeps curled up in the fork of a tree or in a clump of bamboo. It walks hand over hand along branches, alternately placing a hand, then bringing the foot of the same side up to the hand, then extending the other hand, then bringing the other foot up to the hand, and repeating the process. It generally climbs slowly and deliberately but climbs considerably faster when the day is windy. The diet consists of large mollusks and insects, lizards, birds, small mammals, and fruits. Although it appears to be a slow creature, Nycticebus can strike with amazing speed. It grips a branch with both hind feet, stands erect, throws its body forward, and seizes its prey with both hands.

Little is known of social structure, but pairs and family groups have been collected and successfully maintained in captivity. Births evidently sometimes occur in the open rather than in a nest. The infant is carried by the mother and perhaps also by the father but may be left clinging to a branch for brief periods. Adult males appear to be strongly territorial and will not tolerate the presence of another adult male. The territory is marked with urine. Reported vocalizations include a low buzzing hiss or growl when disturbed, a single high-pitched rising tone for making contact, and a high pure whistle produced by the female when in estrus. There also are whistles for making contact and emitted by either sex during courtship (Daschbach, Schein, and Haines 1981; Zimmermann 1985).

Examination of specimens indicates that breeding in the wild continues throughout the year. Such also occurred in one captive colony in North Carolina (Izard, Weisenseel, and Ange 1988), but there was a peak in estrous activity from August to December and a birth peak from March to May. In a captive group in Germany (Zimmermann 1989) mating occurred mainly from June to mid-September, and parturition from December to March. Observations of these captives indicated that the estrous cycle is 29–45 days, estrus lasts 1–5 days, and gestation is 185–97 days. Litter size was always one in these two colonies, but occasional twins have been reported elsewhere. The young weigh 30–60 grams at birth and are weaned after 5–7 months. Females reach sexual maturity at 18–24 months. Males may be physiologically capable of reproduction at 17 months. Fathers become increasingly hostile when their male offspring are 12–14 months old and eventually seek to chase them out of the group. A number of individuals have lived in captivity for more than 20 years and one was still living in December 1994 at an estimated age of more than 26 years (Marvin L. Jones, Zoological Society of San Diego, pers. comm., 1995).

MacKinnon (1986) estimated that N. coucang once occupied 610,570 sq km of habitat in Indonesia, that this area had been reduced to 227,883 sq km through logging and clearing for agriculture, and that the number of animals remaining was 1,139,415. Hunting and deforestation have reduced the number in southeastern China to a few hundred (Tan 1985). The lesser slow loris, N. pygmaeus, is listed as vulnerable by the IUCN and as threatened by the USDI and was regarded as highly vulnerable by Eudey (1987). The restricted habitat of this species is said to have been subject to severe environmental disruption through military activity. However, Duckworth (1994) suggested that it still is common in Laos. Both species of Nycticebus are on appendix 2 of the CITES.

PRIMATES; LORISIDAE; Genus PERODICTICUS
Bennett, 1831

Potto

The single species, P. potto, is found in the tropical forest belt from Guinea to western Kenya and central Zaire (Petter and Petter-Rousseaux 1979).

Head and body length is 305–90 mm and tail length is 37–100 mm. Charles-Dominique (1977) gave the weight as 850–1,600 grams. The dense, woolly coat of adults varies from a brownish gray to a rich, dark reddish brown, slightly lighter on the underparts. The index finger is a mere rudiment, and the thumb is readily opposable to the three remaining fingers, thus producing an excellent grasping organ. The great toe is so placed that it readily opposes the other toes, making the foot equally efficient for grasping.

Potto *(Perodicticus potto)*, photo by Ernest P. Walker.

The second toe is slightly shortened and has a long, sharp claw. The nails of all the other toes and fingers are flattened like those of humans. The modification of the hand is similar to, but not as extreme as, that of *Arctocebus*. The presence of a moderate tail distinguishes *Perodicticus* from *Arctocebus*.

The spinal processes of the neck vertebrae are very long and project above the general contour of the flesh of the neck. They are hidden by thick fur but can be felt on either living or dead specimens. Walker (1970) stated that it was not true, as sometimes claimed, that the spinal processes actually penetrate through the skin. Instead the spines are

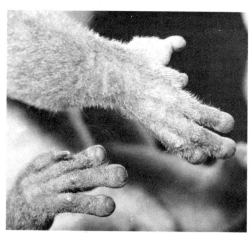

Potto *(Perodicticus potto)*, hand and foot, photo by Bernhard Grzimek.

covered by sensitive hairless tubercles. This nuchal region was postulated to function in tactile stimulation rather than for defense, aggression, abrasion, or muscular support. Charles-Dominique (1977), however, thought this region was used for defense. He stated that the potto lowers its head between its forelegs when threatened, thereby presenting an attacker with a scapular shield, and that it may also thrust forward with this shield to knock a predator off balance. Oates (1984) suggested that the scapular shield may provide defense while the potto forages along a branch with its head down to locate food by olfaction.

The remainder of this account, except as noted, is taken from Charles-Dominique (1977). The potto is an inhabitant of dense forest. It is nocturnal and arboreal, generally being found in trees at a height of 5–30 meters. It spends the day sleeping in foliage. It is generally slow, moving strictly by climbing rather than leaping, but it can make quick grabs with its hands and mouth. The diet consists largely of fruits but includes some gums and insects. The latter are located by smell, and ants are lapped up with the tongue. Small vertebrates, such as birds and bats, also may be killed and eaten.

In studies in Gabon population density was estimated to be 8–10/sq km in primary forest and 28/sq km in flooded forest. Home ranges were 6–9 ha. for females and 9–40 ha. for males. Both sexes are territorial. The females defend an area large enough to support themselves and their young. The older and larger males establish individual home ranges, from which other males are excluded but which overlap the home ranges of one or more females. A relationship between male and female is maintained throughout the year, but contact is made only at night, and the sexes do not sleep together. Vocalizations include a high-pitched "tsic" for communication between mother and young, a whistling call emitted by females in estrus, a groan used in threats, and a high-pitched distress call. The potto also leaves urine trails about 1 meter long on branches, apparently as a means of passing on information to conspecifics. Although studies in the wild indicate a solitary nature for this genus, long-term observations of captives (Cowgill, States, and States 1989) show that individuals, even adult males, form attachments and benefit from sociable activity.

The estrous cycle averages 39 days (Van Horn and Eaton 1979), estrus lasts less than 2 days (Hayssen, Van Tienhoven, and Van Tienhoven 1993), and the gestation period is 180–205 days (Cowgill, States, and States 1989; Müller *in* Grzimek 1990). In Gabon all births occurred from August to January, and weaning took place 4–5 months later, from January to March, when fruit was most abundant. Birth weight is 30–52 grams, and there is normally a single young, but twins were reported once in captivity. The young clings to the belly of the female for the first few days of life but subsequently is left suspended from a hidden branch during the night and is collected by the mother and returned to the nest in the morning. At 3–4 months the young begins to follow the mother or ride dorsally. Adult weight is attained at 8–14 months, and sexual maturity at 18 months. Young males leave the mother's home range when they are only 6 months old. Young females are independent after 8 months but still share the mother's range. Captive pottos have lived to be more than 26 years old (Cowgill, States, and States 1989). *Perodicticus* is on appendix 2 of the CITES.

PRIMATES; LORISIDAE; Genus EUOTICUS
Gray, 1863

Needle-clawed Bushbabies

There are three species (Groves 1989; Hill and Meester *in* Meester and Setzer 1977):

E. pallidus, north of the Sanaga River in Cameroon and southeastern Nigeria, island of Bioko (Fernando Poo) in Equatorial Guinea;
E. elegantulus, south of the Sanaga River in Cameroon, Rio Muni (mainland Equatorial Guinea), Gabon, Congo;
E. matschiei, northeastern Zaire, western Uganda.

Euoticus was regarded only as a subgenus of *Galago* by Meester et al. (1986), Olson (*in* Nash, Bearder, and Olson 1989), and Petter and Petter-Rousseaux (1979) but was considered a full genus by Corbet and Hill (1991), Groves (1974b, 1989, and *in* Wilson and Reeder 1993), and Napier and Napier (1985). The species *E. matschiei* (sometimes called *E. inustus*) was transferred to *Galago* by Groves but was retained in *Euoticus* by Corbet and Hill, Napier and Napier, Olson, and Petter and Petter-Rousseaux. The species *E. pallidus* was distinguished from *E. elegantulus* by Groves but not by any of the other authorities cited herein.

In *E. matschiei* head and body length is 147–95 mm, tail length is 195–279 mm, and weight is 170–250 grams (Kingdon 1974a). In *E. pallidus* and *E. elegantulus* head and body length is 182–210 mm, tail length is 280–310 mm, and weight is 270–360 grams (Happold 1987). The upper parts are orange-cinnamon to dull cinnamon-gray. The fur is thick, woolly, soft, and without luster. From *Galago*, *Euoti-*

cus is distinguished by the presence of pointed and keeled nails on all digits except the first, which has flat nails.

Needle-clawed bushbabies are found in the closed canopy of tropical rainforest and are completely arboreal (Napier and Napier 1985). In studies in Gabon, Charles-Dominique (1977) found that *E. elegantulus* lived at heights of 5–50 meters above the ground, hardly ever descended, and slept by day rolled up in a ball in foliage. It was seen to make horizontal leaps of 2.5 meters and to cover up to 8 meters with some loss of height. This species spends the entire night searching for the gums of trees, its primary food. The nails on the feet of this species allow it to investigate large tree trunks and other areas inaccessible to most galagines. It locates gums by smell and visits 500–1,000 collection points per night. Insects and fruits also form part of its diet.

According to Charles-Dominique (1977), *E. elegantulus* has a population density of 15–20/sq km in favorable habitat. It sleeps in tight clusters of two to seven individuals, which break up at night. It does not urine wash, as some other bushbabies do, but deposits urine directly on branches. In Gabon births occur mostly from January to March, when fruits and insects are most abundant, and there are no births from June to August, when resources are minimal. Litter size is normally one, and it is cared for by the mother, as described below in the account of *Galagoides*. Kingdon (1974a) indicated that a breeding peak in *E. matschiei* may occur in November–December. Müller (*in* Grzimek 1990) listed a longevity of more than 15 years for captive *Euoticus*.

The IUCN classifies all three species as near threatened. They occupy relatively small ranges that are subject to severe disturbance by human activity. All species are on appendix 2 of the CITES.

PRIMATES; LORISIDAE; Genus GALAGO
E. Geoffroy St.-Hilaire, 1796

Galagos, or Bushbabies

There are three species (Hill and Meester *in* Meester and Setzer 1977; Meester et al. 1986; Nash, Bearder, and Olson 1989; Petter and Petter-Rousseaux 1979):

G. senegalensis, savannah and forest savannah zones from Senegal to Somalia and Tanzania;
G. gallarum, southern Ethiopia, southern Somalia, northern Kenya;
G. moholi, Zaire and Tanzania to northern Namibia and Transvaal.

Euoticus, *Otolemur*, and *Galagoides* (see accounts thereof) sometimes are considered subgenera or synonyms of *Galago*. Meester et al. (1986) summarized the different views relative to the status of these genera and the assignment of species thereto. Groves (*in* Wilson and Reeder 1993) included within *Galago* what are here treated as the species *Euoticus matschiei* and *Galagoides alleni*.

Head and body length is 88–210 mm, tail length is 180–303 mm, and weight is about 95–300 grams (Nash, Bearder, and Olson 1989). The fur is dense, woolly, rather long, slightly wavy, and without luster. Coloration is silvery gray to brown, being slightly lighter on the underparts. The large ears have four transverse ridges and can be independently or simultaneously bent back and wrinkled downward from the tips well toward the base. This furling

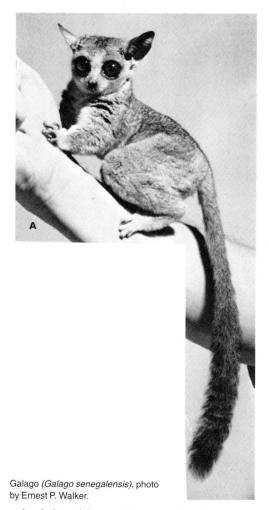

Galago *(Galago senegalensis)*, photo by Ernest P. Walker.

sity of *G. moholi* was 95–200/sq km in favorable habitat. This species often occurs in small family groups of about two to seven individuals. These groups may consist of an adult pair with or without young, two adult females plus infants, or an adult female with young. Such groups spend the day sleeping at the same site, but the adults split up at night to forage. Adult males are territorial and may fight viciously for control of a home range that overlaps the ranges of several females. Harcourt and Bearder (1989) reported home range sizes of 1.5–22.9 ha. for males and 4.4–11.7 ha. for females.

Vocalizations identified for *Galago* include a clicking sound by which the young calls its mother, a more powerful version of the same used by adults of some species to assemble at the sleeping site, a loud cry or bark used for distant communication and territorial encounters, and a high-pitched alarm call (Petter and Charles-Dominique 1979). Galagos also may communicate by olfactory methods, especially through urination. Individuals, especially dominant males, frequently "urine wash" their hands and feet, so that scent is spread over the entire three-dimensional space through which the animals move. This process apparently allows dissemination of information regarding the presence and condition of individuals. However, studies by C. Harcourt (1981) suggest that at least in some in-

and unfurling of the ears is frequent and produces a most quizzical expression. On the ends of all digits are flat disks of thickened skin that help in grasping limbs and slippery surfaces. The long front digits and flattened hind digits have flattened nails, except the second digit of the hind foot, which is armed with a curved grooming claw. Females have two pairs of mammae.

In contrast to most other galagines, which are restricted largely to dense forests, *Galago* occurs in open woodlands, scrub, wooded savannahs, and grasslands with thickets (Doyle and Bearder 1977; Happold 1987). Like the others, *Galago* is arboreal and nocturnal. In studies in South Africa Doyle and Bearder (1977) observed *G. moholi* to have a midnight rest period and to make use of two or three favorite sleeping sites during the day. The females made nests to give birth and shelter the family. According to Smithers (1983), this species may construct an open-topped, platformlike nest or shelter in an unused bird nest, a hollow tree, or a clump of dense foliage. Bushbabies are alert, sprightly, and agile, making long leaps from branch to branch. Doyle and Bearder (1977) found *G. moholi* to progress mainly by hops and leaps and only rarely to walk quadrupedally. On the average night it covered about 2.1 km. It fed exclusively on acacia gums and insects, catching its prey with rapid strikes of one hand.

According to Doyle and Bearder (1977), population den-

Galago *(Galago moholi)*, photo by David Haring.

dividuals urine washing functions primarily to facilitate grip.

Studies indicate that females experience at least two restricted periods of mating and two pregnancies per year, the resulting birth seasons being September–October and January–February in southern Africa, December–February and June in Uganda, and March and July in Sudan. The estrous cycle averages about 32 days, estrus lasts 2–7 days, and the gestation period is 120–42 days. Litter size is commonly one or two, occasionally three. Birth weight is about 9.5 grams in *G. moholi* and about 12 grams in *G. senegalensis*. The young are fully furred and have their eyes open at birth. They first leave the nest at 10–11 days, and they are able to catch insects at 4 weeks. Weaning was reported at 11 weeks of age in a laboratory study of *G. senegalensis*. Young males leave their mother when they reach physiological sexual maturity, at about 10 months, but young females may remain longer, even though they also mature at this time (Charles-Dominique 1977; Doyle and Bearder 1977; Hayssen, Van Tienhoven, and Van Tienhoven 1993; Kingdon 1974*a*; Nash, Bearder, and Olson 1989; Smithers 1983; Van Horn and Eaton 1979). Record longevity for galagines is held by a *G. senegalensis* that was still living after 18 years and 10 months in captivity (Marvin L. Jones, Zoological Society of San Diego, pers. comm., 1995).

Lee, Thornback, and Bennett (1988) did not regard *Galago* as threatened. Nonetheless, the newly distinguished species *G. gallarum* occupies a restricted area that may be undergoing severe habitat disruption because of human activity. It is designated as near threatened by the IUCN. All species of *Galago* are on appendix 2 of the CITES.

PRIMATES; LORISIDAE; **Genus OTOLEMUR**
Coquerel, 1859

Greater Bushbabies

There are two species (Hill and Meester *in* Meester and Setzer 1977; Masters 1986, 1988; Meester et al. 1986; Nash, Bearder, and Olson 1989):

O. garnettii, southern Somalia to southeastern Tanzania, and on the nearby islands of Pemba, Mafia, and Zanzibar;
O. crassicaudatus, eastern Zaire and southwestern Kenya to Angola and eastern South Africa.

Otolemur sometimes has been regarded as a synonym or subgenus of *Galago* but was recognized as a distinct genus by Corbet and Hill (1991), Groves (1989 and *in* Wilson and Reeder 1993), Meester et al. (1986), and Napier and Napier (1985).

Head and body length is 230–465 mm, tail length is 300–550 mm, and weight is about 600–2,000 grams (Charles-Dominique 1977; Kingdon 1974*a*; Nash, Bearder, and Olson 1989; Nash and Harcourt 1986). The upper parts are pale gray tinged with buff or brown and sometimes penciled with black; the underparts are paler. The tail is usually lighter and buffier than the upper parts, and the head is darker around the eyes (Smithers 1983). From *Galago*, *Otolemur* is distinguished by larger size, a pronounced postorbital constriction of the skull, a long and robust muzzle, and a foramen magnum that is directed backwards rather than downwards (Meester et al. 1986). Characteristics of the ears and digits are about the same as described above for *Galago*.

Greater bushbabies are found in forests, thickets, and

Greater bush baby (*Otolemur* sp.), photo from Zoological Society of London.

well-developed woodlands. They may occupy urban areas if there are sufficient trees for shelter and food sources (Smithers 1983). In studies in South Africa Doyle and Bearder (1977) observed *O. crassicaudatus* to sleep in a nest 5–12 meters high during the day. Its nocturnal activity lasted from 9.5 hours in summer to about 12 hours in winter, with a rest period in the middle of the night. It generally walked quadrupedally when undisturbed and could make leaps of up to 2 meters without losing height. It moved over a distance of about 1 km per night. Although some fruits and insects were eaten, 62 percent of the diet throughout the year consisted of gums (tree exudates). Population density in favorable habitat was 72–125/sq km, and one maternal group had a home range of 7 ha. In coastal Kenya, Nash and Harcourt (1986) found *O. garnettii* to move an average of 1.6 km per night. Population density in this species was 31–38/sq km, and home range averaged 17 ha. for males and 12 ha. for females.

O. crassicaudatus often occurs in small, stable family groups of about two to six individuals. These groups may comprise an adult pair with or without young, two adult females plus infants, or an adult female with young. Such groups spend the day sleeping at the same site, but the adults split up at night to forage. Adults may be territorial; the ranges of adjacent groups overlap only to a small extent, but the ranges of males overlap extensively with those

Greater bush baby *(Otolemur crassicaudatus)*, photo from San Diego Zoological Society.

of females. The males evidently fight one another to secure control of such areas (Doyle and Bearder 1977). *O. garnettii* has a similar social structure but more pronounced territorial behavior. Fully adult animals, even of opposite sexes, rarely share a nest. There is little overlap between the ranges of adults of the same age but extensive overlap between the ranges of individuals of different ages. Young females mature in their natal ranges, but males disperse to other areas (Nash and Harcourt 1986).

According to Kingdon (1974a), *Otolemur* announces its presence for several months of the year by a loud, croaking wail repeated at frequent intervals during the early part of the night. This vocalization appears to correspond to the breeding season and to have a territorial function. Its resemblance to the cry of a child is said to be the basis for the common name "bushbaby." Urine washing such as described above in the account of *Galago* also is characteristic of *Otolemur*.

Except as noted, the following information on natural history was taken from Doyle and Bearder (1977), Izard (1987), Kingdon (1974a), Nash and Harcourt (1986), Smithers (1983), and Van Horn and Eaton (1979). In the northern Transvaal, mating occurs mainly in June or July and births are restricted largely to November. Births have been reported in March in Somalia and between August and November in Tanzania and Zambia, at the end of the dry season. Pregnancies seem to peak in August on Zanzibar and Pemba. *O. garnettii* breeds once a year, from August to October, in coastal Kenya. The estrous cycle of *O. crassicaudatus* averages about 44 days, and gestation about 133 days. Litter size is commonly two young and occasionally three in *O. crassicaudatus* but usually only one in *O. garnettii*. The young weigh about 40 grams, have their eyes open at birth, and can crawl after about 30 minutes. They begin feeding on their own at around 1 month, though lactation may last for nearly 5 months. Both sexes of *O. gar-*

nettii are able to breed by 20 months of age. Captives of both species have lived for about 18 years or more (Marvin L. Jones, Zoological Society of San Diego, pers. comm., 1995).

Lee, Thornback, and Bennett (1988) did not consider either species of *Otolemur* to be threatened. Nonetheless, *O. garnettii*, at least, could be of long-term concern, as it occupies a restricted area undergoing increasing human activity and habitat disruption. It is designated as near threatened by the IUCN. Both species are on appendix 2 of the CITES.

PRIMATES; LORISIDAE; Genus GALAGOIDES
A. Smith, 1833

Dwarf Galagos

There are two subgenera and four species (Hill and Meester *in* Meester and Setzer 1977; Meester et al. 1986; Nash, Bearder, and Olson 1989):

subgenus *Sciurocheirus* Gray, 1873

G. alleni, southeastern Nigeria to Congo, Bioko (Fernando Poo);

subgenus *Galagoides* A. Smith, 1833

G. demidoff, forests from Senegal to Uganda and western Tanzania, Bioko (Fernando Poo);
G. thomasi, an extremely disjunctive range including the vicinity of Mount Cameroon in southwestern Cameroon, possibly Mount Marsabit in northern Kenya, the Kivu and Ituri regions of eastern Zaire and

B GALAGO

Dwarf galago *(Galagoides demidoff)*, photo from Paignton Zoo, Great Britain.

southwestern Uganda, the region between Dilolo and Kolwezi in southern Zaire, and the Loanda Highlands of central Angola;

G. zanzibaricus, southern Somalia, coastal Kenya and Tanzania, Zanzibar, Mozambique, Malawi, eastern Zimbabwe.

Galagoides sometimes is regarded as a subgenus or synonym of *Galago* but was given generic rank by Corbet and Hill (1991), Groves (1989 and *in* Wilson and Reeder 1993), Meester et al. (1986), Napier and Napier (1985), and Olson (*in* Nash, Bearder, and Olson 1989). However, Corbet and Hill, Groves, and Napier and Napier included *G. alleni* in *Galago* rather than in *Galagoides. G. granti* of Mozambique was listed as a separate species by Corbet and Hill but was included within *G. zanzibaricus* by Groves, Meester et al., and Nash, Bearder, and Olson. *G. thomasi* was distinguished from *G. demidoff* by Nash, Bearder, and Olson but not by Groves or any of the other authorities cited above. Based on karyological analysis, Stanyon et al. (1992) suggested that *G. demidoff* may actually comprise several species; like some other authorities, they used the name *G. demidovii* for that taxon.

The following specific descriptions were taken largely from Nash, Bearder, and Olson (1989). In *G. alleni* head and

body length is 155–240 mm, tail length is 205–300 mm, and weight is 200–445 grams. The upper parts are very dark brown to almost black, the underparts are paler, the flanks of the limbs are brightly colored, and the face has dark circumocular rings. The head is long and narrow, the muzzle is pointed, and the tail is bushy (Nash, Bearder, and Olson 1989). In *G. demidoff* head and body length is 73–155 mm, tail length is 110–215 mm, and weight is 44–97 grams. The upper parts vary in color from rufus to reddish brown, the underparts and flanks of the limbs are paler, and the circumocular markings are variable. The head is narrow, the muzzle is pointed and upturned, and the tail is not bushy. In *G. thomasi* head and body length is 123–66 mm, tail length is 150–233 mm, and weight is 55–149 grams. The upper parts are blackish brown, the underparts and flanks of the limbs are yellowish, and the circumocular markings are prominent. The head is narrow, the muzzle is pointed, and the tail is not bushy. In *G. zanzibaricus* head and body length is 120–90 mm, tail length is 170–265 mm, and weight is 104–203 grams. The upper parts are cinnamon reddish brown, the underparts are bright yellow, the flanks of the limbs are close in color to the upper parts, and the circumocular markings are prominent. The muzzle is long, pointed, and concave dorsally. The tail is moderate to bushy.

Galagoides resembles *Galago* but differs in that the upper parts are generally brownish or cinnamon rather than predominantly gray and in having a relatively longer muzzle. The nails of the feet are not pointed. Except in *G. alleni* there is a pronounced elongation of the premaxillae of the skull in front of the incisor teeth.

Dwarf galagos live in primary or secondary forests, usually where there is dense foliage, and are nocturnal and arboreal (Napier and Napier 1985). In studies in Gabon, Charles-Dominique (1977) found *G. alleni* only 1–2 meters above the ground; it spent each day sleeping in one of several favored tree hollows within its home range. *G. demidoff* occurred 5–40 meters above the ground and slept in leaf nests or thick vegetation. The latter species was seen to make horizontal leaps of 1.5–2 meters. *G. demidoff* fed largely on insects, mostly small beetles and nocturnal moths, and also ate some fruits and gums. Population densities in favorable habitat were 15–20/sq km for *G. alleni* and 50–117/sq km for *G. demidoff.* Harcourt and Nash (1986) determined that in the coastal forests of Kenya *G. zanzibaricus* moved about 1.5–2.0 km per night, had a population density of 170–80/sq km, and had a group home range of 1.6–2.8 ha. Charles-Dominique and Bearder (1979) listed the following home range sizes: *G. alleni,* females, 3.9–16.6 ha., and males, 17–50 ha.; *G. demidoff,* females, 0.6–1.4 ha., and males, 0.5–2.7 ha.

In *G. alleni* and *G. demidoff* usually a single animal occupies a sleeping site, but sometimes two or three adult females plus young are found together (Charles-Dominique 1977). Adult males are aggressive toward one another and apparently are territorial. Adults of opposite sexes, however, may share the same area, sometimes sleep together by day, and have some contact at night even though foraging separately. Each male seeks to control a home range that overlaps those of several females, and intense competition may result. Charles-Dominique (1977) distinguished four categories of male *G. demidoff.* The heaviest males (averaging 75 grams) have large home ranges that include at least one female's home range and often are in a central position overlapping a number of female ranges. The home ranges of several such males converge at a common point of slight overlap, where interaction occurs. The lightest males (averaging 56 grams) are tolerated within the ranges of the heavy males and have small home ranges of their own. Medium-sized males (averaging 61 grams) occupy

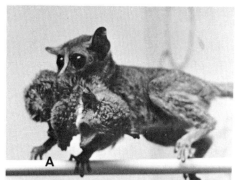

A. Mother dwarf galago *(Galagoides demidoff)* carrying one of twins, aged 30 days; B. Baby, aged two days. Photos by A. S. Woodhall.

relatively large home ranges, but on the periphery of female ranges. These males may associate with other peripheral males and eventually may gain weight and shift into a central position. The last category, nomadic males, includes mostly young animals that do not remain long in any one area.

In *G. zanzibaricus* in coastal Kenya there is a closer regular association between the sexes, and both are probably territorial. Adult males usually have nonoverlapping ranges that are shared with one or two females and their offspring. The male usually sleeps with these other animals, but a mother will sleep alone for about three weeks after giving birth. Females can breed twice a year, the peak birth seasons in this area being February–March and August–October. Gestation is 120 days, and there usually is a single young, occasionally two. Young males disperse from their natal ranges, but females remain in the area and initially give birth there at about 12 months (C. Harcourt 1986; Harcourt and Nash 1986). One *G. zanzibaricus* lived at the San Antonio Zoo for 16 years and 6 months (Marvin L. Jones, Zoological Society of San Diego, pers. comm., 1995).

The following additional information on natural history was taken from Charles-Dominique (1977), Hayssen, Van Tienhoven, and Van Tienhoven (1993), and Van Horn and Eaton (1979). In *G. demidoff* there usually is one pregnancy per female per year; mating takes place in Congo in September–October and January–February; and births occur all year in Gabon, with a peak in January–April, when fruits and insects are most abundant. Gestation has been reported to be about 133 days in *G. alleni* and 111–14 days in *G. demidoff*. Birth weight is about 5–10 grams in *G. demidoff*. In *G. alleni* and *G. demidoff* the female takes the young out of the nest when it is a few days old, leaves it hidden in vegetation while she forages during the night, and carries it back to the nest in the morning. After about a month the young is able to follow the female, but it still is carried on occasion. It has been observed that if the young is unable to leap across a gap after its mother, it will emit a call, upon which the female will return, pick up the young, and make the jump. Weaning occurs after about 6 weeks and sexual maturity at 8–10 months. Wild *G. alleni* are known to have lived at least eight years.

The IUCN classifies *G. alleni* and *G. zanzibaricus* as near threatened (placing both in the genus *Galago*). The indigenous forests on which the species depend are being cleared for development purposes or cut and replaced with exotic conifers (Lee, Thornback, and Bennett 1988). All species are on appendix 2 of the CITES.

PRIMATES; **Family CHEIROGALEIDAE**

Dwarf Lemurs and Mouse Lemurs

This family of five genera and eight species is found only in Madagascar. Petter and Petter-Rousseaux (1979) divided the family into two subfamilies: the Cheirogaleinae, for *Microcebus, Mirza, Cheirogaleus,* and *Allocebus;* and the Phanerinae, for *Phaner.* Tattersall (1982, 1986), however, did not consider *Phaner* to warrant subfamilial separation. In any case, all of these genera once were placed within the family Lemuridae but now are thought to represent a distinct family on the basis of new anatomical, behavioral, ecological, and cytogenetic information. Moreover, Tattersall (1982) united the Cheirogaleidae with the Lorisidae in the superfamily Lorisoidea of the strepsirhine infraorder Lorisiformes, while he assigned all other Malagasy primates to the superfamily Lemuroidea of the infraorder Lemuriformes. Dene, Goodman, and Prychodko (1980) considered the Cheirogaleidae to represent a distinct superfamily, Cheirogaleoidea, within the Lemuriformes. Groves (1989 and *in* Wilson and Reeder 1993) did not accept a close affinity between the Cheirogaleidae and the Lorisidae.

The members of the Cheirogaleidae are smaller than those of the Lemuridae. Head and body length is about 125–275 mm and tail length is about 125–350 mm. The hind legs are substantially longer than the forelegs. The fur is soft, thick, and woolly. In this family the eyes are large and set closely together, whereas in the Lemuridae the eyes are smaller and placed more laterally. Female cheirogaleids have three pairs of mammae located pectorally, abdominally, and inguinally, while female lemurids, except in *Varecia,* have only a single pair located pectorally. The dental formula of the Cheirogaleidae is: (i 2/2, c 1/1, pm 3/3, m 3/3) × 2 = 36. The incisors are well developed, not small and peglike as in the Lemuridae or absent as in the Megaladapidae.

The Cheirogaleidae are arboreal and nocturnal. *Cheirogaleus* and *Microcebus* are known to accumulate fat reserves on the hind legs and base of the tail and then to become torpid during the dry season.

Mouse lemur *(Microcebus murinus)*, photo by Howard E. Uible.

PRIMATES; CHEIROGALEIDAE; **Genus MICROCEBUS**
E. Geoffroy St.-Hilaire, 1828

Mouse Lemurs

There are three species (Mittermeier et al. 1994; Petter and Petter-Rousseaux 1979; Schmid and Kappeler 1994):

M. murinus, western and southern Madagascar;
M. myoxinus, west-central Madagascar;
M. rufus, northern and eastern Madagascar.

The distinction of *M. myoxinus* from *M. murinus* was recognized only recently. Still another species, *M. coquereli,* is here assigned to the separate genus *Mirza* (see account thereof) but has been retained in *Microcebus* by Groves (*in* Wilson and Reeder 1993) and some other authorities.

In *M. murinus* and *M. rufus* head and body length is usually 125–50 mm, tail length is about the same, and weight is 39–98 grams. *M. myoxinus,* with a total length of 178–219 mm and a weight of 24–38 grams, is evidently the smallest of primates. All species usually have soft fur, a short snout, a rounded skull, prominent eyes and ears, long hind limbs, and a long tail. The upper parts are generally grayish in *M. murinus,* rufous brown with an orange tinge in *M. myoxinus,* and brownish in *M. rufus.* There is a dorsal stripe along the middle of the back, which is not always distinct, and a distinct white median nasal stripe. The underparts are white.

Mouse lemurs live in forests and are associated with trees. They have been found to be most common in secondary forest and to construct spherical leaf nests in foliage or nest in hollow trees (Martin 1973). Mouse lemurs are agile and active at night, usually traveling along branches on all four legs, leaping at times, and using the tail as a balancing organ. On the ground *M. murinus* moves with froglike hops (Martin 1973). *Microcebus* does not undergo true hibernation, but there is a decline in activity during the winter (June–September). From July to December fat accumulates in the tail, and then in the dry season the animals experience short periods of torpor with a decline in body temperature. The diet consists of insects, spiders, occasionally small frogs and lizards, a large amount of fruit, flowers, nectar, some gums and insect secretions, and leaves (Hladik 1979).

Petter (1978) reported population densities of about 300–400/sq km for *M. murinus.* Pollock (1979) listed densities of up to 262/sq km for *M. rufus* and 1,300–2,600/sq km for *M. murinus.* Martin (1973) stated that the individual home range of both species is not more than 50 meters in diameter. Hladik, Charles-Dominique, and Petter (1980) reported home range not to exceed 2 ha. in dry forest and to be smaller in humid areas. Females form relatively stable groups, usually comprising two to nine individuals, that share a sleeping nest but forage alone by night. A male occasionally may be found with a group of females even outside of the breeding season. Males usually nest alone or in pairs (Martin 1973; Pollock 1979). A male's home range may overlap those of several females; ranges often are scent-marked with urine and feces (Mittermeier et al. 1994). The vocalizations of *M. murinus* are very high-pitched but are known to include a variety of calls for seeking contact, mating, distant communication, alarm, and distress (Petter and Charles-Dominique 1979).

Reproduction seems to be well known only for *M. murinus* (Doyle 1979; Hayssen, Van Tienhoven, and Van Tienhoven 1993; Van Horn and Eaton 1979). There is a restricted breeding season from August to March in Madagascar, but it shifts to the other half of the year among captive animals in the Northern Hemisphere. The estrous cycle usually lasts about 45–55 days, estrus 1–5 days. Gestation averages 60 (54–68) days. Litter size is normally two or three, the newborn weigh 5 grams each, and weaning occurs after 25 days. Sexual maturity is attained at 10–29 months by females and at 7–19 months by males. A captive *M. murinus* reportedly lived to an age of 15 years and 5 months (Marvin L. Jones, Zoological Society of San Diego, pers. comm., 1995).

Because of the general concern about the pace of habitat destruction in Madagascar, all species of *Microcebus* are listed as endangered by the USDI and are on appendix 1 of the CITES. The IUCN classifies *M. myoxinus* as vulnerable. Although *M. murinus* and *M. rufus* are still relatively common and may be the only Malagasy lemurs not undergoing a general decline, they are apparently losing habitat in some areas through logging and excessive grazing by cattle and goats (Richard and Sussman 1975). Information compiled by Mittermeier et al. (1992, 1994) suggests that *M. murinus* and *M. rufus* may still number in the hundreds of thousands, if not millions, of individuals but that *M. myoxinus,* which occupies a much smaller known range, may be vulnerable to extinction.

PRIMATES; CHEIROGALEIDAE; **Genus MIRZA**
Gray, 1870

Coquerel's Dwarf Lemur

The single species, *M. coquereli,* occurs on the northwest and west-central coasts of Madagascar. Until recently this species generally was assigned to *Microcebus,* and Groves (*in* Wilson and Reeder 1993) continued to do so, but *Mirza* was recognized as a distinct genus by Corbet and Hill (1991), Mittermeier et al. (1994), and Tattersall (1982).

Head and body length is about 250 mm, tail length is about 280 mm, and weight is 280–335 grams. The upper parts are dark gray washed with rufous, being darkest along the midline; many of the hairs have golden tips, producing an olive brown effect. The underparts are yellowish gray. Facial markings are indistinct or absent. The ears are long and hairless. Females have two pairs of mammae.

Mirza is much larger than *Microcebus.* The two genera are both characterized by a relatively small second upper premolar, a small pericone on the first and second upper molars, the reduction or loss of the paraconid on the posterior lower premolar, and a complete buccal cingulum on all lower molars (Tattersall 1982). *Mirza,* however, lacks the greater bullar development, obliteration of the interincisal gap, and toothrow curvature of *Microcebus* (Groves 1989).

Mirza generally lives near rivers or ponds in thick forest, where it is usually observed 1–6 meters above the ground (Hladik 1979; Pages 1978). A spherical nest 50 cm in diameter is constructed of interlaced vines, branches, and leaves, usually in a tree fork (Pages 1980). *Mirza* is nocturnal, is active all year except on very cold nights, and is not known to accumulate fat reserves (Petter 1978). During the wet season it has a varied diet consisting of fruits, flowers, gums, insects, spiders, frogs, lizards, small birds, and eggs. In the dry season it feeds largely on a sweet liquid secreted onto branches by the larvae of the homopteran insect family Flatidae and also eats the insects themselves (Pages 1980; Petter 1978).

Reported population densities have varied from about 30/sq km to as high as 385/sq km (Mittermeier et al. 1994). Pages (1980) reported that each individual has a home range with up to 12 nests in the central portion. Male

Greater mouse lemur *(Mirza coquereli)*, photo by David Haring.

ranges consist of a core area of about 1.5 ha. and a peripheral, less used area of up to 4.0 ha. Females have a core area of 2.5–3.0 ha. and a peripheral zone of up to 4.5 ha. Males are solitary and seem to defend their core areas from other males. However, a male's core area may overlap that of a female. Marking is accomplished through urination, salivation, and anogenital dragging. Vocalizations include a "hum" that accompanies all movements and meetings between individuals.

Observations of a captive colony at the Duke University Primate Center in North Carolina (Stanger, Coffman, and Izard 1995) indicate that reproduction occurs throughout the year. Females there usually had one litter per year but sometimes had a second if the first did not survive or was removed. The estrous cycle averaged 22.1 days, estrus lasted no longer than a day, and the gestation period averaged 89.2 days. Of 51 litters, 21 contained a single infant and 60 had twins. Mean birth weight was 17.5 grams. Females first gave birth at an average age of 22.5 months. According to Hayssen, Van Tienhoven, and Van Tienhoven (1993), lactation lasts 134–38 days. Jones (1982) listed a record longevity of 15 years and 3 months.

Mirza is designated as vulnerable by the IUCN and as endangered by the USDI and is on appendix 1 of the CITES. It occupies a restricted area of humid forest and is threatened both by a long-term drying trend in the climate and by human destruction of the forests. Mittermeier et al.

Dwarf lemur *(Cheirogaleus medius)*, photo by David Haring.

(1992, 1994) suggested that a reasonable estimate of numbers would be 10,000–100,000 individuals.

PRIMATES; CHEIROGALEIDAE; **Genus**
CHEIROGALEUS
E. Geoffroy St.-Hilaire, 1812

Dwarf Lemurs

There are two species (Petter and Petter *in* Meester and Setzer 1977; Petter and Petter-Rousseaux 1979):

C. major, northern and eastern Madagascar, possibly one restricted area in west-central Madagascar;
C. medius, western and extreme southern Madagascar.

Groves (1989) also recognized *C. crossleyi* of northeastern Madagascar as a separate species, but subsequently (*in* Wilson and Reeder 1993) he included it within *C. major.*

Head and body length is 167–264 mm, tail length is 195–310 mm, and weight is 177–600 grams (Tattersall 1982). The fur is soft, woolly, and sometimes silky. The upper parts vary from buffy or gray to reddish brown; the underparts are whitish, often tinged with yellow. The ears are thin, membranous, and naked. The eyes are large, lustrous, and surrounded by dark rings. Females have two pairs of mammae.

These dwarf lemurs live in forests. They are nocturnal, spending the daylight hours in globular nests that they construct of twigs and leaves in hollow tree trunks or in the tops of trees. They move about their arboreal heights on all fours, but they sit up to eat, holding the food in their hands. The diet consists mainly of fruits, flowers, nectar, and insects (Hladik 1979). Those living in areas subject to periodic dry seasons store up quantities of fat in the basal portion of the tail during the wet season and estivate when food is scarce. In a study in western Madagascar Petter (1978) found that *C. medius* gained weight from November to

March, went into torpor within hollow tree trunks in April, and emerged in October or November, at the start of the rainy season. Individuals could lose 100 grams of body weight during estivation.

Pollock (1979) reported a population density of 75–110/sq km for *C. major.* Hladik, Charles-Dominique, and Petter (1980) found densities of 200/sq km and 350/sq km for *C. medius* and determined that most individuals did not have ranges of more than 200 meters in diameter. There was much overlap of these areas and no evidence of territorial-

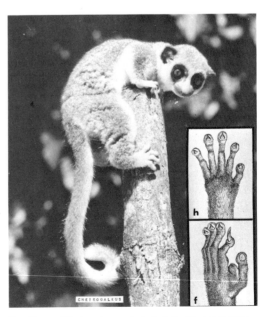

Dwarf lemur *(Cheirogaleus major),* photo by Howard E. Uible. Insets: hand and foot of dwarf lemur, photos from *Proc. Zool. Soc. London.*

ity; groups of 3–5 estivated together. Nonetheless, these animals are usually solitary, and captives are generally intolerant of others of the same sex. They are relatively quiet but have a number of weak calls for contact and a louder cry in agonistic situations (Petter and Charles-Dominique 1979). Captive individuals reportedly are easy to tame, and at least one became quite affectionate, enjoyed being handled, and would come when called by name. Limited data indicate that in the wild mating occurs in October or November, births take place in January or February, and litter size is 2–3 (Klopfer and Boskoff 1979; Mittermeier et al. 1994; Van Horn and Eaton 1979). Estrus lasts 2–3 days in *C. major*, gestation 70 days, and lactation about 45 days (Hayssen, Van Tienhoven, and Van Tienhoven 1993). In a captive colony of *C. medius* in North Carolina mating takes place from April to early June, births occur from June to August, the estrous cycle averages 19.7 days, and males compete fiercely for the estrous females. Gestation lasts 61 days, and litter size is one to four. The newborn are fully furred and have their eyes open. Females reach sexual maturity at 10–14 months (Foerg 1982). One captive specimen was still living at 19 years and 2 months (Marvin L. Jones, Zoological Society of San Diego, pers. comm., 1995).

According to Richard and Sussman (1975), both species of *Cheirogaleus* are declining because of loss of forest habi-

tat. Both are listed as endangered by the USDI and are on appendix 1 of the CITES. Mittermeier et al. (1992, 1994) suggested that each species probably numbers more than 100,000 individuals.

PRIMATES; CHEIROGALEIDAE; **Genus ALLOCEBUS**
Petter-Rousseaux and Petter, 1967

Hairy-eared Dwarf Lemur

The single species, *A. trichotis*, is known with certainty only from a few specimens collected in lowland rainforest in the vicinity of Mananara in northeastern Madagascar, though there have also been recent reports from the Masoala Peninsula farther north (Mittermeier et al. 1994).

Petter-Rousseaux and Petter (1967) described a specimen with a head and body length of 133 mm and a tail length of 170 mm. Gunther, who originally described *trichotis* as a species of *Cheirogaleus*, gave the measurements of the type specimen as: head and body length, 152 mm, and tail length, 149 mm. He emphasized that the tail was shorter than the head and body, but Petter-Rousseaux and Petter (1967) thought that Gunther's specimen may have been

Hairy-eared dwarf lemur *(Allocebus trichotis)*, photos by Bernhard Meier.

damaged. In four recently collected living adults head and body length was 125–45 mm, tail length was 150–95 mm, and weight was 75–98 grams (Meier and Albignac 1991). The upper parts are brownish gray, the underparts are whitish gray, and the tail is reddish brown. The ears are short with tufts of long hair in front and on the internal side of the lobe.

This genus resembles *Microcebus* in size and color but has a noticeably longer tail and better-developed hind limbs. Its muzzle is elongated and rounded off at the end, whereas that of *Microcebus* is short and pointed. The hands and feet are larger than in *Microcebus*. According to Groves (1989), *Allocebus* differs from *Microcebus*, *Mirza*, and *Cheirogaleus* but resembles *Phaner* in lacking enlarged molars, convergent toothrows, and a molariform fourth upper premolar. *Allocebus* is unique in having caniniform second and third upper premolars, a greatly enlarged first upper incisor, and a deflated tympanic bullar region, hardly defined from the mastoid inflation.

This genus is extremely rare, and little is known of its habits in the wild. Meier and Albignac (1991) found *Allocebus* only in primary lowland rainforest. The specimen described by Petter-Rousseaux and Petter (1967) was taken from a hole in a tree. Walker (1979) stated that *Allocebus* apparently progresses in a quadrupedal fashion, like a large *Microcebus murinus*. Coimbra-Filho and Mittermeier (1977*b*) observed that *Allocebus* has large upper incisors like those of *Phaner* and may also use these teeth to scrape tree bark to obtain exudates.

Allocebus had not been seen in more than 20 years when, in March 1989, it was rediscovered in the vicinity of the Mananara River in northeastern Madagascar. In a survey for the World Wildlife Fund, Yoder (1989) obtained considerable new information through interviews with native people in the Mananara area. *Allocebus* is found only in primary forest, probably at low densities. It is nocturnal and arboreal. It makes nests of fresh leaves in small holes in either living or dead trees, usually 3–5 meters above the ground. The people encounter it only from October to March, when they cut down nest trees in the course of slash-and-burn agriculture. During the entire cold season, from early May to mid-October, *Allocebus* evidently hibernates deep within tree holes. The diet includes new leaves and small fruits. *Allocebus* is found only in pairs consisting of one male and one female or in pairs with a single infant.

Observations of four captive individuals have confirmed part of Yoder's report and provided some new information (Meier and Albignac 1991). The animals take a wide variety of foods but seem to prefer large insects, which they capture with both hands. They are less active during the dry season, June–September, but fat reserves that would facilitate dormancy are less obvious than in *Cheirogaleus* and are distributed across the body, not accumulated in the tail. The captive *Allocebus* use straw to construct nests in holes, where they sleep together as a group during the day. Various evidence suggests that estrus occurs at the beginning of the wet season, in November–December, with births in January–February.

Yoder (1989) indicated that *Allocebus* is being killed and eaten regularly by people, and this factor, along with rapid deforestation, places its survival in doubt. Meier and Albignac (1991) stated that it is certainly declining rapidly. Mittermeier et al. (1992) gave the main threat as slash-and-burn agriculture, estimated the total population at only 100–1,000 individuals, and considered the genus of the highest conservation priority. *Allocebus* is classified as critically endangered by the IUCN and the USDI and is on appendix 1 of the CITES.

Fork-marked Dwarf Lemur

The single species, *P. furcifer*, occupies scattered areas across northern Madagascar, a section of the west-central part of the island, and a small area in the far southeast; four distinctive subspecies now are recognized (Groves and Tattersall 1991; Mittermeier et al. 1994).

Head and body length is about 227–85 mm, tail length is 285–370 mm, and weight is 300–500 grams. The thickly furred body is reddish gray or brownish gray; the color is brightest on the head and neck. Black streaks extend from the eyes to the top of the head, where they converge in a middorsal stripe to about the hips. This particular color pattern on the top of the head results in the vernacular name. The throat and underparts are pale rufous or yellowish. The hands and feet are dark brown, and the bushy tail is dark reddish brown with a black or white tip. *Phaner* differs from other members of the Cheirogaleidae in chromosomal features, cranial and dental characters, the presence of distinctive dermatoglyphs and a specific marking gland on the surface of the neck, and its unique behavior (Petter and Petter-Rousseaux 1979). The second upper premolar tooth is uniquely long, resembling a second canine, and the first upper incisor is long, procumbent, and separated from the second by a gap (Groves 1989).

Phaner is primarily a forest inhabitant but also frequents narrow lines of trees that project into or through savannahs. During the day some individuals use nests made by *Mirza coquereli*, but most rest in holes in tree trunks or branches. *Phaner* is active from evening to early morning. It runs quadrupedally along branches and jumps from one to another. It is capable of horizontal leaps of 4–5 meters but can cover 10 meters with some loss of height (Petter, Schilling, and Pariente 1975). In a study in an area of dry forest it was found to forage at heights of 8–10 meters and often in the treetops (Hladik, Charles-Dominique, and Petter 1980). Apparently, it neither accumulates fat reserves nor estivates (Petter 1978). It is a specialized gum eater. In October–November it sometimes feeds on nectar, but its diet consists primarily of vegetable resins and insect secretions. Its keeled nails allow it to descend along smooth tree trunks to favored feeding places; there it scrapes off the bark with its incisor teeth and laps up sap (Hladik 1979; Petter, Schilling, and Pariente 1975).

Population densities of 40–870/sq km have been reported (Hladik, Charles-Dominique, and Petter 1980; Pollock 1979). According to Petter and Charles-Dominique (1979), females occupy well-defined territories that may overlap to some extent. The males have individual territories, separate from those of other males but including the territories of one or more females. An associated male and female often move around together by night, with the female taking the lead. Territorial encounters are purely vocal and do not involve urine marking. Small groups of animals regularly assemble in encounter zones, all calling at once, and then return to normal foraging activity. Identified vocalizations include a mild "hong" sound emitted every few seconds by an adult in motion; a series of "tia" calls exchanged by an associated pair, which may stimulate a territorial encounter with other individuals; a powerful call emitted by males for territorial purposes; and a series of staccato calls and grunts expressing agitation.

Charles-Dominique and Petter (1980) stated that territory size averages 4 ha. and that breeding occurs during the austral spring. Petter, Schilling, and Pariente (1975) wrote

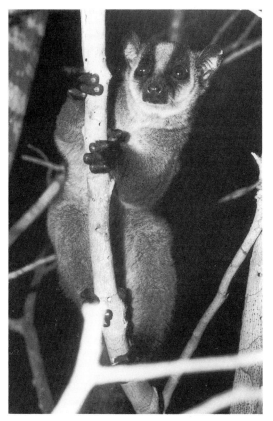

Fork-marked mouse lemur *(Phaner furcifer)*, photos by Russell A. Mittermeier.

that according to local informants, *Phaner* gives birth about 15 November, around the time of the first rains. The single young is initially sheltered in a tree hole, then clings to the belly of the mother, and finally rides on her back. Several individuals reportedly have lived about 12 years in captivity (Marvin L. Jones, Zoological Society of San Diego, pers. comm., 1995).

The fork-marked dwarf lemur is declining because of loss of its forest habitat to human logging and agricultural activity. It is listed as endangered by the USDI and is on appendix 1 of the CITES. Each of the four recognized subspecies of *P. furcifer* was given a high-priority rating for conservation action by Mittermeier et al. (1992), and each was estimated to number 1,000–10,000 individuals. Of the four, the IUCN classifies *P. f. furcifer*, of northeastern Madagascar, as near threatened and *P. f. electromontis*, of the extreme north, *P. f. parienti*, of the northwest (south of Ambanja), and *P. f. pallescens*, of the west and possibly the southeast, as vulnerable.

PRIMATES; Family LEMURIDAE

Lemurs

This family of 4 living genera, one Recently extinct genus *(Pachylemur)*, and 11 species is confined to Madagascar and (probably through introduction) the Comoro Islands. Several genera that once were usually placed in this family are now considered to belong to the families Cheirogaleidae

and Megaladapidae (see accounts thereof). With regard to the remaining genera there is disagreement about relationships. Some authorities consider *Varecia, Pachylemur, Lemur,* and *Eulemur* to be closely related or even congeneric but place *Hapalemur* in a separate subfamily (Schwartz and Tattersall 1985; Tattersall and Schwartz 1991). Others see close affinity of *Lemur* and *Hapalemur,* with *Varecia, Pachylemur,* and *Lemur* being more distantly related (Groves 1989; Groves and Eaglen 1988; Simons and Rumpler 1988).

In the four living genera head and body length is 280–458 mm and tail length is 280–600 mm. Lemurs have long, heavily furred tails and slender bodies and limbs. The hind limbs are longer than the forelimbs. The pelage is usually soft, thick, woolly, and solid-colored, but sometimes there are color patterns. In *Lemur catta,* the ring-tailed lemur, the tail is ringed with black and white; in all other species it is unicolored. The face is elongated and foxlike in *Lemur, Eulemur,* and *Varecia* but shortened in *Hapalemur.* The ears, which are at least partially haired, are short or of only moderate length. A clawlike nail is present on the second toe of the hind foot of some forms, and the naked areas of the palms and soles are generally marked by large, ridged pads. In *Lemur, Eulemur,* and *Hapalemur* there is only one pair of mammae, located pectorally, but *Varecia* has one pectoral pair and two abdominal pairs.

The dental formula is: (i 2/2, c 1/1, pm 3/3, m 3/3) × 2 = 36. The upper incisors are small and peglike. The first upper incisors are separated from each other by a wide space. The canines are somewhat elongate, sharp, and separated from the premolars by a space. The lower incisors and canines project forward and somewhat upward. The first lower premolar resembles a canine tooth. The molars have three tubercles.

Although the geological range of the lemurs often is said to extend back into the Pleistocene, the oldest known remains are actually subfossils dating from about 2,850 B.P. The subfossil material evidently represents several of the species that are still extant, as well as the extinct genus *Pachylemur*.

PRIMATES; LEMURIDAE; **Genus HAPALEMUR**
I. Geoffroy St.-Hilaire, 1851

Bamboo Lemurs, or Gentle Lemurs

There are three species (Meier et al. 1987; Petter and Petter-Rousseaux 1979; Rumpler and Dutrillaux 1978; Vuillaume-Randriamanantena, Godfrey, and Sutherland 1985; Wilson et al. 1989):

H. griseus (bamboo lemur), eastern, northwestern, and west-central Madagascar;
H. simus (broad-nosed gentle lemur), found in much of eastern Madagascar until at least 100 years ago, also known by subfossil material from central and far northern Madagascar;
H. aureus (golden bamboo lemur), southeastern Madagascar.

Head and body length is 260–458 mm, tail length is 240–560 mm, and weight is about 1.0–2.5 kg. The fur is soft and of moderate length. Coloration is brownish gray, reddish gray, orangish brown, grayish green, or reddish green above; it is darkest on top of the head. The underparts are whitish, buffy, gray, or yellowish. The head is globose, the muzzle is short, and the ears are short and hairy. The hands and feet are short and broad with large pads under the tips of the toes and fingers. All the teeth except the molars have a serrated cutting edge adapted for tough foods such as bamboo and other coarse vegetation.

On the inner side of the wrist of *H. griseus* is a rough tract of skin over a gland. In the male this area is covered with spinelike processes, but in the female the processes are hairlike. *H. simus* does not possess such glandular areas.

Bamboo lemurs seem to be restricted to humid forests and marshes where bamboos and reeds are abundant. They have been observed at all hours of the day but are most active in the evening and early morning. They can run quickly on the ground and jump considerable distances from branch to branch in the trees. *H. simus* is more terrestrial than *H. griseus* and often runs across the ground in the forest. The subspecies *H. griseus alaotrensis*, found at Lake Alaotra, apparently is semiaquatic and can swim well. The natural diet reportedly consists almost entirely of bamboo shoots and leaves (J.-J. Petter 1975; Petter and Peyrieras 1975).

Population densities of 47–120/sq km and defended group territories of 6–15 ha. have been reported for *H. griseus*, while a group of *H. aureus* maintained an exclusive territory of about 80 ha. (Harcourt and Thornback 1990). These animals are seen most frequently in groups of 3–5 individuals thought to be family units, comprising an adult male, 1 or 2 adult females, and 1 or 2 juveniles (Petter and Peyrieras 1975). *H. griseus alaotrensis* has been reported to form aggregations of up to 40 individuals when Lake Alaotra is at high water, though permanent group size

Bamboo lemur *(Hapalemur griseus)*, photo by David Haring.

may be no greater than that of other populations (Feistner and Rakotoarinosy 1993). Group sizes of 2–6 individuals and a mean group home range of 2 ha. for that subspecies were observed by Mutschler and Feistner (1995). There are numerous vocalizations, including a weak grunt to maintain group cohesion, a strong call for distant communication, and a very powerful "creee" in threat situations (Petter and Charles-Dominique 1979).

Births of *H. griseus* in Madagascar have been reported to occur from late October to February and to always be of a single young (Harcourt and Thornback 1990; Petter and Peyrieras 1975). Records from Duke University, however, show that of four births in captivity one was of twins; gestation lasted 135–50 days (Klopfer and Boskoff 1979). One newborn weighed 32 grams (Hayssen, Van Tienhoven, and Van Tienhoven 1993). The young initially may be carried ventrally, but by 3 weeks of age they are riding on the mother's back. One captive infant was carried until 11 weeks and was weaned at 20 weeks (Steyn and Feistner 1994). A captive specimen of *H. griseus* was still living at 17 years (Marvin L. Jones, Zoological Society of San Diego, pers. comm., 1995).

All species of *Hapalemur* are listed as endangered by the USDI and are on appendix 1 of the CITES. The IUCN classifies *H. simus*, *H. aureus*, and *H. griseus alaotrensis* as critically endangered and *H. griseus occidentalis*, found in a small part of west-central Madagascar, as vulnerable. Mittermeier et al. (1994) estimated a population of more than 10,000 individuals for *H. g. occidentalis*, which is jeopardized by human destruction of forest habitat. The population of *H. g. alaotrensis*, consisting of about 7,500 animals, is threatened by burning of the reed beds in which it lives, draining of parts of Lake Alaotra for agricultural purposes, and hunting (Harcourt and Thornback 1990; Mutschler

Gentle lemur *(Hapalemur griseus)*, photo by David Haring.

Gentle lemur *(Hapalemur griseus)*, photo by David Haring.

and Feistner 1995). *H. g. alaotrensis* also seems to be a popular pet (Feistner and Rakotoarinosy 1993). *H. g. griseus*, which occurs throughout the eastern forests of Madagascar, numbers more than 100,000; it is generally declining through loss of forest habitat and excessive hunting by people but may now occur at higher than original densities in certain areas where bamboo has replaced the original forest. A fourth subspecies of *H. griseus (H. g. meridionalis)* was recently reported to occur at one locality in the extreme southeast but was not accepted as a distinct form by Mittermeier et al. (1992, 1994) or Harcourt and Thornback (1990).

Skeletal remains indicate that *H. simus* may still have been present in the Ankarana region at the northern tip of Madagascar until less than 50 years ago, and there are some indications that it still exists there (Wilson et al. 1989). Otherwise *H. simus* and *H. aureus* are known to survive only in a few small areas east of Fianarantsoa in southeastern Madagascar, where they are critically endangered by logging, clearing of forest habitat for agriculture, and direct hunting by people (Meier and Rumpler 1987). Both species are estimated to number only about 1,000 individuals and are considered to be of the highest priority for conservation attention (Mittermeier et al. 1992).

PRIMATES; LEMURIDAE; **Genus LEMUR**
Linnaeus, 1758

Ring-tailed Lemur

The single species, *L. catta,* is found in southern Madagascar south and west of a line running approximately from the vicinity of Morondava on the west coast, east to Fianarantsoa, and then south to Taolanaro (Harcourt and Thornback 1990; Tattersall 1982). Several other species usually placed in *Lemur* are here assigned to the separate genera *Eulemur* and *Varecia* (see accounts thereof). Simons and Rumpler (1988) considered *L. catta* to be more closely related to *Hapalemur* than to *Eulemur.*

Head and body length is 385–455 mm, tail length is 560–624 mm, and weight is about 2.3–3.5 kg (Tattersall 1982). The upper parts are brownish gray, the underparts are whitish, and the tail is ringed with black and white. The palms and soles are long, smooth, and leatherlike, affording a firm footing on slippery rocks, and the great toe is considerably smaller than in the various species formerly assigned to *Lemur,* which are more arboreal.

According to Simons and Rumpler (1988), *Lemur* resembles *Hapalemur* but differs from *Eulemur* in having an antebrachial (carpal) gland on the forearm as well as brachial glands. Males rub fatty secretions from these various glands onto the tail in order to disperse a scent during agonistic interaction. The karyotype of *Lemur* also resembles that of *Hapalemur* and differs from that of *Eulemur.* Tattersall and Schwartz (1991) suggested that the resemblance of *Lemur* and *Hapalemur* was the result of parallel evolution and that numerous cranial and dental characters indicate that *Lemur* is more closely related to *Eulemur.* Among the characters shared by *Lemur, Eulemur,* and *Varecia* are a laterally swollen, prenasopalatine portion of the palatine bone and at least some posterior expansion of this element; a paranasal sinus; at least some obscuring of the maxilla in the lateral part of the orbital floor; an infraorbital foramen lying anterior to the lacrimal foramen; the maxilla unexposed in the medial orbital wall; and some cingular development on the first and second upper molars.

The ring-tailed lemur is capable of arboreal activity but

Ring-tailed lemur *(Lemur catta)*, photo by Bernhard Grzimek.

is partly terrestrial and sometimes is found in thinly wooded country. It ranges farther into the interior highlands of Madagascar than any other lemur. In a study in southwestern Madagascar, Sussman (1975, 1977) observed *L. catta* to spend more time on the ground than within any one level of the trees. Groups foraged intensively, covering 900–1,000 meters a day, and took only a short rest at midday. Jolly (1966) reported that *L. catta* may be moderately active at any time of the night, though like most species of *Lemur*, it is primarily diurnal. The diet includes mostly fruits, some leaves and other plant parts, and only rarely insects (Hladik 1979).

Most reported population densities are from about 150/sq km to 350/sq km, though there is one estimate of only 17.4/sq km in an area of disturbed forest (Harcourt and Thornback 1990; Pollock 1979; Sussman 1975, 1977; Sussman and Richard 1974). Jolly (1966) found a group of 20 to have a home range of 5.7 ha. In southwestern Madagascar groups used home ranges of 6.0–8.8 ha. (Sussman 1977; Sussman and Richard 1974). Budnitz and Dainis (1975) reported that troops inhabiting closed canopy and open forest had home ranges of 6 ha. and 8.1 ha., whereas a troop occupying an area of brush, scrub, and open forest utilized 23.1 ha. Sauther and Sussman (1993) stated that known home range size is 6–35 ha., with those in wet habitats averaging about twice the size of those in drier areas. Group home ranges overlap, with few or no areas of exclusive use.

In her study of *L. catta* Jolly (1966) found group sizes of 12–24, no consistent group leadership, and considerable agonistic activity, including some fighting. The sexes had separate dominance hierarchies, and the females dominated the males. Budnitz and Dainis (1975) found group size

to range from 5 to 22, averaging 12.8 before the birth season and 17.3 after. They reported the basic troop to be organized around a core group of adult females and their infants, young juveniles, and sometimes 1 or more dominant males. The average troop comprised 6 adult females, 4 adult males, and 4 young. There was no constant leader of the troop, but adult females dominated others and seemed responsible for territorial defense, while males were generally peripheral to group activity. Females remained in the troop of their birth, but males moved among troops. Sauther and Sussman (1993) reported that males emigrate from their natal group upon reaching adulthood, that adult males usually change groups every 3–5 years, and that each group appears to have a single "central male" that interacts with the females at a greater rate than do other males and that is the first to mate.

Jolly (1966) reported that troops of *L. catta* have well-defined, nonoverlapping territories within their overall home ranges. However, Sauther and Sussman (1993) questioned whether the species is truly territorial; there is almost total overlap of the ranges used by different groups, though certainly there is vigorous defense of areas being used at a given time. Resulting disputes generally involve two opposing groups of females running at each other and vocalizing, but direct physical contact is rare (Budnitz and Dainis 1975). Jolly (1966) identified 15 different vocalizations, including a howl audible to humans at a distance of 1,000 meters.

Like other lemurs, *L. catta* mates from about April to June and gives birth from August to October, just before or at the beginning of the rainy season (Petter 1965; Pollock 1979). In any one area, however, mating and births may be synchronized to occur within a period of a few days (Jolly 1966). Most females give birth annually in the wild (Sauther and Sussman 1993). The estrous cycle averages 40 days, estrus lasts 1 day, and weight at birth is 50–80 grams (Hayssen, Van Tienhoven, and Van Tienhoven 1993). Doyle (1979) listed a precise gestation period of 136 days. According to Van Horn and Eaton (1979), females usually produce a single young, though twins are not rare. For the first two weeks of life the young clings to the underside of the mother and rides longitudinally; subsequently it rides on her back (Tattersall 1982). Infants may suckle until they are 5 months old, but they begin to take some solid food during the second month of life. Female *L. catta* first conceive at an average age of 19.56 months (Van Horn and Eaton 1979). Males are sexually mature at 2.5 years but may not be allowed to mate then by the older males (Budnitz and Dainis 1975). A captive specimen was still living at about 33 years (Marvin L. Jones, Zoological Society of San Diego, pers. comm., 1995).

According to Mittermeier et al. (1994), *L. catta* generally has been considered common, but its preferred habitats—gallery forests along rivers and *Euphorbia* bush—are disappearing rapidly because of fires, overgrazing by livestock, and cutting of trees for charcoal production. The species can only survive in primary vegetation. Recent surveys indicate that suitable habitat has been diminished alarmingly through human activity, that hunting pressure is severe, and that the species may be more threatened than was thought previously (Harcourt and Thornback 1990). Mittermeier et al. (1992) considered its conservation of high priority and estimated its total numbers at only 10,000–100,000 individuals. *L. catta* now is classified as vulnerable by the IUCN and as endangered by the USDI and is on appendix 1 of the CITES.

PRIMATES; LEMURIDAE; **Genus VARECIA**
Gray, 1863

Variegated Lemur, or Ruffed Lemur

The single species, *V. variegata*, is found in the forests of eastern Madagascar. There are two distinctive subspecies, *V. v. ruber* (red ruffed lemur) on the Masoala Peninsula east of the Antainambalana River and *V. v. variegata* (black and white ruffed lemur) from the Antainambalana River in the north to the Mananara River in the south (Harcourt and Thornback 1990; Mittermeier et al. 1994; Tattersall 1982). Until recently this species usually was placed in the genus *Lemur,* but because of various distinctive anatomical and behavioral characteristics it now is considered to represent a separate genus (Groves and Eaglen 1988; Petter and Petter-Rousseaux 1979; Rumpler 1974). The extinct genus *Pachylemur* (see account thereof) often has been included in *Varecia.*

V. variegata is the largest living member of the family Lemuridae. Head and body length is 510–60 mm, tail length is 560–650 mm, and weight is about 3.2–4.5 kg (Tattersall 1982). The fur is long and soft, and the ears are hidden by a ruff of hair. The color pattern varies and may be different on the right and left sides of a specimen. In the subspecies *V. v. variegata* much of the pelage is black, but there are large white areas on the limbs, back, and head. In the subspecies *V. v. ruber* the pelage is mostly red, there may be white markings on the limbs, and the tail and belly are black.

Varecia differs from *Lemur* and *Eulemur* in dermatoglyphic pattern and the presence of a marking gland on the neck (Rumpler 1974). Whereas the other genera have only a single pair of mammae, *Varecia* has three pairs. According to Tattersall and Schwartz (1991), *Varecia* shares numerous cranial and dental characters with *Lemur* (see account thereof) and *Eulemur.* From the others, however, *Varecia* can be distinguished by its elongate talonid basins in the lower molars, the absence of entoconids on the lower molars, the protocone fold of the first upper molar, the posterolingual opening of the talonid basin of the second lower molar, and the anterior expansion of the lingual cingulum of the first and second upper molars.

The ruffed lemur is an arboreal forest dweller. It normally progresses by walking or running on larger branches and makes leaps from tree to tree, but its locomotion is more labored and cautious than that of *Lemur* (Walker 1979). It is crepuscular, being most active from 1700 to 1900 hours, when its peculiar calls are commonly heard (Petter and Charles-Dominique 1979; Pollock 1979). The diet consists largely of fruits. On the Masoala Peninsula, animals

Ruffed lemur *(Varecia variegata),* photo from San Diego Zoological Society.

Black and white ruffed lemur *(Varecia variegata variegata)*, photo by David Haring.

Cooper, and Benirschke 1977; Boskoff 1977; Harcourt and Thornback 1990; Hayssen, Van Tienhoven, and Van Tienhoven 1993; Klopfer and Boskoff 1979; Mittermeier et al. 1994; Pollock 1979). The reported gestation period of *Varecia* is only 90–102 days, considerably shorter than that of any species of the other genera. The estrous cycle lasts about 30 days, with estrus averaging 6.25 days. On Nosy Mangabe Island most mating occurs in June and July, births in September and October. More than half of the births are of twins, and the remainder are mostly single or of triplets, but litters with as many as six young have been reported in captivity. Birth weight is 80–100 grams. The young are initially carried in the mouth of the female and are often deposited in a convenient place while she forages. By 5 weeks the young can climb to the tops of trees. Weaning occurs at around 135 days. Females may become pregnant at 20 months, but average age at first reproduction in captivity is 3.4 years. According to Marvin L. Jones (Zoological Society of San Diego, pers. comm., 1995), a number of specimens have lived in captivity for more than 25 years and one was still living at about 33 years.

The ruffed lemur is listed as endangered by the USDI and is on appendix 1 of the CITES. It reportedly is declining because of human destruction of its forest habitat, hunting for use as food, and commercial exportation (Richard and Sussman 1975; Wolfheim 1983). Although the species occupies a relatively large range, there are few well-protected areas therein (Mittermeier et al. 1992). Each of the two subspecies

moved an average of 436 meters per day but generally remained in one area with large fruit trees for several weeks before shifting to another area (Rigamonti 1993). Population densities of 20–30/sq km have been reported on Nosy Mangabe Island, a protected reserve, but apparently are much lower elsewhere (Harcourt and Thornback 1990).

There apparently is some variation in social organization. In southeastern Madagascar an adult male and female formed a cohesive pair and foraged through a home range of 197 ha. (Harcourt and Thornback 1990). Two groups on the Masoala Peninsula consisted of 5 and 6 members, used home ranges of about 25 ha., and during the cool wet season (May–August) fragmented into subgroups that used different core areas (Rigamonti 1993). A study on Nosy Mangabe Island indicated that groups consist of 8–16 individuals, all members use a common home range, groups are aggressive toward one another, and females form the core of the group and defend its home range; subgroups of 2–5 individuals formed during the cool season, possibly representing mated pairs and offspring (Morland 1991). There are a variety of vocalizations, the most characteristic being an intense roar of alarm and a powerful plaintive call for territorial expression (Petter and Charles-Dominique 1979). Both sexes scent-mark, females only with the anal-genital region but males mainly with a unique process of rubbing the chest, chin, and neck onto the substrate (Pereira, Seeligson, and Macedonia 1988).

The reproductive pattern of this genus differs from those of *Lemur* and *Eulemur* in a number of ways (Bogart,

Red ruffed lemur *(Varecia variegata ruber)*, photo by David Haring.

is thought to number only 1,000–10,000 individuals in the wild (Mittermeier et al. 1994), and there are another 859 in captivity (Olney, Ellis, and Fisken 1994). The IUCN now classifies *V. v. ruber* as critically endangered, forecasting a decline of at least 80 percent over the next decade, and *V. v. variegata* as endangered.

PRIMATES; LEMURIDAE; Genus PACHYLEMUR
Lamberton, 1948

The single species, *P. insignis*, is known only by subfossil material from various sites in northern, central, and southern Madagascar (Jenkins 1987; Simons et al. 1990; Tattersall 1982). *P. jullyi*, named on the basis of subfossil material from central Madagascar, sometimes is treated as a separate species but is here considered at most subspecifically distinct from *P. insignis*. These species commonly have been referred to the genus *Varecia*, together with the living species *V. variegata*. *Pachylemur*, originally described as a subgenus for *Varecia insignis* and *V. jullyi*, was accepted as a full genus by Mittermeier et al. (1994), Ravosa (1992), and Simons et al. (1990).

Based on skeletal remains, *Pachylemur* resembles *Varecia* but is somewhat larger, having a cranial length of 114.5–126.0 mm, compared with 97.2–110.7 mm in the latter genus. According to Mittermeier et al. (1994), the skull structure of *Pachylemur* is very close to that of *Varecia*, but its postcranial bones have a heavier build. This condition may suggest a more terrestrial way of life than that of the highly arboreal *Varecia*. Tattersall (1982) noted that the skull of *Pachylemur* also can be distinguished by the presence of sagittal and nuchal cresting and by the more forward orientation of its orbits. In a statistical analysis of cranial and dental measurements, Ravosa (1992) found substantial differences between the two genera. The skull of *Pachylemur* is relatively broader, the jaws more massive, and the molar teeth larger. These adaptations may indicate that the diet of *Pachylemur* was more obdurate or fibrous than that of *Varecia*.

Pachylemur has been found at sites dated at about 2,000–1,000 years ago, but there is little information about when or why it finally disappeared. Except for *Mesopropithecus* (see account thereof), it was the smallest of the subfossil Malagasy prosimians and it may have survived until about the same time. If it was terrestrial, it probably was more susceptible than *Varecia* to hunting by people.

PRIMATES; LEMURIDAE; Genus EULEMUR
Simons and Rumpler, 1988

Lemurs

There are five species (Harcourt and Thornback 1990; Mittermeier et al. 1994; Petter and Petter-Rousseaux 1979; Tattersall 1982, 1993):

E. coronatus (crowned lemur), extreme northern Madagascar;

E. rubriventer (red-bellied lemur), eastern rainforest zone of Madagascar;

E. macaco (black lemur), northwestern Madagascar north of Narinda Bay;

E. mongoz (mongoose lemur), northwestern Madagascar south of Narinda Bay, Moheli and Anjouan islands in the Comoros (probably introduced);

Black lemurs *(Eulemur macaco)*, photo by Bernhard Grzimek.

E. fulvus (brown lemur), Madagascar, Mayotte Island in the Comoros (probably introduced).

These species usually have been placed in the genus *Lemur*, but Simons and Rumpler (1988) concluded that they should be put in a genus different from the type species of *Lemur, L. catta*, and proposed the name *Eulemur* for such a genus. Almost simultaneously, Groves and Eaglen (1988) independently came to a similar conclusion, proposing the name *Petterus* for the same group of species. Most subsequent authorities, including Groves (*in* Wilson and Reeder 1993), have accepted *Eulemur* as the appropriate name, though Corbet and Hill (1991) used *Petterus*. Based on detailed analyses of craniodental characters, Tattersall (1993) and Tattersall and Schwartz (1991) recommended that the species of *Eulemur* be restored to the genus *Lemur*. However, *Eulemur* was maintained as a separate genus by Crovella, Montagnon, and Rumpler (1993), Shedd and Macedonia (1991), and Mittermeier et al. (1992, 1994). Matings between *E. fulvus* and some other species of *Eulemur* have

Crowned lemur *(Eulemur coronatus)*, photo by David Haring.

resulted in fertile offspring (Ratomponirina, Andrianivo, and Rumpler 1982; Tattersall 1993).

The species are allopatric, except for *E. fulvus*, which overlaps all the others and is well differentiated from them. According to Tattersall (1982), its head and body length is 380–500 mm, tail length is 465–600 mm, and weight is 2.1–4.2 kg. There is much variation in color, and in some subspecies there is sexual dichromatism. The upper parts are usually gray or brown, the underparts are paler, and the tail often darkens distally. The face is usually dark, though there may be light patches above the eyes. In some subspecies the head is mostly white or pale gray, and in some the ears are tufted. All subspecies have mystacial, submental, superciliary, and carpal vibrissae, and in all the face is clothed in short hair except at the tip of the muzzle. The circumanal region is distinguished by an area of naked, wrinkled, glandular skin. Females have one or two pairs of mammae, though only the anterior pair are functional.

In the other four species head and body length is 300–450 mm, tail length is 400–640 mm, and weight is about 2–3 kg. The tail is slightly shorter than the head and body in some forms but longer in others. The fur is soft and relatively long, and there is a pronounced ruff about the neck and ears. The coloration varies considerably. Some forms are speckled reddish brown or gray; others are more or less reddish, brownish, or blackish throughout. In *E. macaco* males are usually black, and females, brown.

According to Simons and Rumpler (1988), *Eulemur* differs from *Lemur* and *Varecia* in having a hairy, not naked, scrotum. *Eulemur* lacks the antebrachial and brachial glands of *Lemur* and *Hapalemur* but possesses perianal glands not seen in *Varecia, Lemur,* or *Hapalemur*. It is further distinguished from *Lemur* by a suite of dental characters that include smaller third upper and lower molars, distinct protostyle development on the first and second molars, a more developed anterior basin in the posterior lower premolars, and more continuous crests in the trigonids and talonids of the molars. Groves and Eaglen (1988)

added that *Eulemur* is unique among the Lemuridae in having the paranasal air sinuses much dilated and the interorbital region projecting above the plane of the remainder of the skull roof, creating a bubblelike effect in its cranial profile. *Eulemur* resembles *Varecia* in the reduction of the anterior upper premolar and in lacking any lingual outlet for the talonid basin. Tattersall and Schwartz (1991) pointed out that *Eulemur* shares numerous other cranial and dental characters with *Varecia* and *Lemur* (see account thereof).

All species are arboreal forest dwellers. *L. coronatus* occurs mainly in dry forest and *L. rubriventer* in rainforest, but all species show some adaptability to various habitats, including moderately disturbed areas (Harcourt and Thornback 1990). Lemurs are active, quadrupedal animals that run and walk on horizontal and diagonal branches and are capable of leaping to and from vertical and horizontal supports. Resting postures range from sitting upright to lying sprawled on a horizontal branch. Movement on the ground is normally by quadrupedal walking, running, or galloping, but short bouts of bipedal running have been observed (Walker 1979). Most activity is diurnal, but all species are sometimes active by night. *E. mongoz* has been found to be exclusively nocturnal in some places in Madagascar and the Comoros but largely diurnal in others; its shift from day to night activity may coincide with the transition from the rainy to the dry season (Harcourt and Thornback 1990; Harrington 1978; Tattersall 1978a). The diet of *Eulemur* consists largely of flowers, fruits, and leaves (Tattersall 1982). *E. macaco* also is known to eat bark, and *E. mongoz* feeds extensively on nectar (Harcourt and Thornback 1990). Captive *E. macaco* will accept meat (Kolar *in* Grzimek 1990).

In a study in southwestern Madagascar, Sussman (1975, 1977) found *E. fulvus* to contrast sharply with *Lemur catta* in that it spent 95 percent of its time in the tops of trees, rarely descending to the ground. It was relatively sedentary, staying within only a few trees and covering about

Red-bellied lemur *(Eulemur rubriventer)*, photo by Russell A. Mittermeier.

125–50 meters in its daily foraging. It rested most of the afternoon and had peaks of feeding activity from 0600 to 0930 and 1700 to 1825 hours. In a study on Mayotte Island in the Comoros, Tattersall (1977a) also found the brown lemur to prefer the upper levels of the forest but found it to be active both day and night and to range 450–1,150 meters per day. The diet consisted predominantly of kily leaves in southwestern Madagascar and fruits on Mayotte Island.

Population densities as great as 900–1,000/sq km have been reported for *E. fulvus* (Sussman 1975, 1977), though such figures would apply only in small remaining areas of quality habitat, and densities as low as 40–60/sq km have been reported for the species in some areas (Harcourt and Thornback 1990). Other recorded densities are 58–200/sq km for *E. macaco* (Colquhoun 1993; Tattersall 1982), 350/sq km for *E. mongoz* (Wolfheim 1983), and 50–500/sq km for *E. coronatus* (Harcourt and Thornback 1990; Mittermeier et al. 1994). Lemurs appear to be territorial. A group of 12 *E. fulvus* utilized an area of 7 ha. in northwestern Madagascar (Harrington 1975). Other studies of that species have indicated home ranges as large as 14 ha. (Mittermeier et al. 1994). Groups of 5–14 *E. macaco* occupied ranges of 3.5–7.0 ha. (Colquhoun 1993). In *E. fulvus* and *E. mongoz* home ranges overlap extensively, but troops avoid one another or engage in disputes over feeding areas through vocalization and gestures (Harrington 1975; Sussman 1975; Tattersall 1978a). There is a rich repertoire of calls for contact, greeting, territorial expression, threats, and alarm (Petter and Charles-Dominique 1979).

Social organization in *Eulemur* is variable. In southwestern Madagascar 18 groups of *E. fulvus* contained an average of 9.4 (4–17) individuals. These groups appeared cohesive and peaceful, agonistic activity was rare, and no

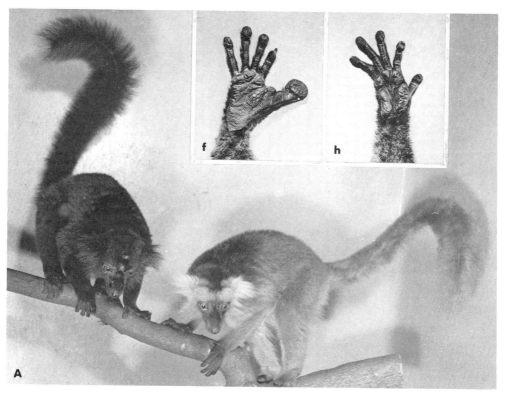

Black lemurs *(Eulemur macaco)*, photo by Ernest P. Walker. Insets: *E. mongoz*, photos from *Histoire physique, naturelle et politique de Madagascar*, Alfred Grandidier.

Mongoose lemur *(Eulemur mongoz)*, photo by David Haring.

dominance hierarchy was observed (Harrington 1975; Sussman 1975, 1977). On Mayotte Island, Tattersall (1977a) found group size to average 9.1 (2–29) individuals, but the groups changed in composition from day to day. Investigation of *E. macaco* indicates a group size of 4–15 in-

Brown lemur *(Eulemur fulvus)*, photo by David Haring.

dividuals, leadership by a female, and some exchange of adults between troops (Pollock 1979). At least in some areas, *E. mongoz* has been found to live in small family groups comprising a permanently bonded adult pair and not more than 2 young (Pollock 1979; Tattersall 1978a). *E. rubriventer* has been observed in groups of 5 or fewer individuals (Pollock 1979). *L. coronatus* and *L. rubriventer* generally occur in groups of 2–10 animals, including several adults of both sexes (Harcourt and Thornback 1990).

It seems likely that all species of *Eulemur* throughout Madagascar mate from about April to June, have a gestation period of about 4.5 months, and give birth from August to October, just before or at the beginning of the rainy season (Petter 1965; Pollock 1979). In any one area, however, mating and births may be synchronized to occur within a period of a few days (Jolly 1966). Doyle (1979) listed the following precise gestation periods: *E. fulvus,* 117 days; *E. macaco,* 127 days; *E. mongoz,* 128 days; and *E. rubriventer,* 127 days. Hayssen, Van Tienhoven, and Van Tienhoven (1993) listed birth weights of 60–70 grams, estrous cycles of about 1 month, and estrus lengths of 1–2 days. There usually is a single young, though twins are not rare. For the first four weeks of life the young grips the mother's fur and rides ventrally in a transverse position; subsequently it rides on her back (Tattersall 1982). Young *E. rubriventer* frequently are carried by the father (Mittermeier et al. 1994). Doyle (1979) reported weaning to occur at 135 days and sexual maturity to be attained at 548 days in *E. fulvus* and *E. macaco.* In *L. coronatus,* which seems to have a few unusual characteristics, gestation is 125 days, some births evidently occur in January and May, twins appear to be as common as singletons, and both males and females reach sexual maturity at 20 months (Harcourt and Thornback 1990). Lemurs thrive in captivity and are often exhibited in zoos. A specimen of *E. fulvus* lived to more than 36 years and a captive hybrid *E. macaco* × *E. fulvus* survived for 39 years (Marvin L. Jones, Zoological Society of San Diego, pers. comm., 1992).

All species of *Eulemur* are listed as endangered by the USDI and are on appendix 1 of the CITES. The IUCN classifies most species and subspecies of the genus. Those classifications, together with population estimates by Mittermeier et al. (1992, 1994), are: *E. fulvus albifrons* (found in northeast Madagascar), more than 100,000; *E. fulvus albo-*

Red-bellied lemur *(Eulemur rubriventer)*, photo by David Haring.

ably were introduced through human agency long ago, face the same threats as those faced by the lemurs in Madagascar (Tattersall 1977b). In addition, there appears to have been a recent increase in the frequency of cyclones striking the islands, which have a devastating effect on the remaining natural vegetation (Harcourt and Thornback 1990). The estimated number of *E. fulvus* on Mayotte fell from about 50,000 in 1975 to less than half that in 1991 (Mittermeier et al. 1994).

PRIMATES; Family MEGALADAPIDAE

Sportive Lemurs, or Weasel Lemurs, and Koala Lemurs

This family of one living and one Recently extinct genus and 10 species is found only in Madagascar. The living genus, *Lepilemur,* long was placed in the family Lemuridae. Recent systematic investigations involving especially cytogenetics indicated that it represents a separate family that took the name Lepilemuridae (Petter and Petter-Rousseaux 1979; Rumpler 1975; Rumpler and Albignac 1975); however, it now seems that *Lepilemur* belongs in the same family as the subfossil *Megaladapis* and that the correct name for the resulting group is Megaladapidae (Buettner-Janusch and Tattersall 1985; Tattersall 1982, 1986). Some authorities, including Corbet and Hill (1991), have continued to place *Lepilemur* in the Lemuridae, but new studies by Groves and Eaglen (1988) indicate that *Lepilemur* is more closely related to the Indriidae than to the Lemuridae. Other authorities, including Jenkins (1987), while acknowledging affinity between *Lepilemur* and *Megaladapis,* consider the latter to warrant placement in a separate family. Tattersall (1982) recognized three subfamilies: the Lepilemurinae, with *Lepilemur;* the Hapalemurinae, with *Hapalemur;* and the Megaladapinae, with *Megaladapis.* However, Schwartz and Tattersall (1985) and Groves and Eaglen (1988) regarded *Hapalemur* to be a member of the Lemuridae, to which it also has been assigned by most other authorities.

Lepilemur and *Megaladapis* differ markedly in size and certain other morphological characters, and the latter genus probably lived in a manner unlike that of modern lemurs. In the skull of both genera there is a median vertical articular area in addition to the transverse articular surface of the mandible (Jenkins 1987). The close relationship between the two genera also is indicated by the dentition (Tattersall 1982). Each has the same dental formula as that of the Lemuridae, but in both *Lepilemur* and *Megaladapis* the adults lose the upper incisors, so that there are only 32 teeth. The cheek teeth of both genera show much resemblance. The premolars broaden posteriorly, and the upper molars display buccal cingula and parastyles, lingual buttressing of the paracones and metacones, and distal displacement of their lingual moieties. Differences between *Lepilemur* and the Lemuridae have been found in studies of dermatoglyphics, hematology, parasitology, behavior, and cytogenetics (Petter and Petter-Rousseaux 1979).

Geologically the Megaladapidae are known only from the Recent of Madagascar, subfossil remains having been found at sites dated nearly as far back as 3,000 years ago. However, Tattersall (1982) indicated close affinity between the Megaladapidae and the prosimian family Adapidae of the Eocene of Europe.

collaris (southeast), endangered, 1,000–10,000; *E. fulvus collaris* (southeast), vulnerable, 10,000–100,000; *E. fulvus fulvus* (northwest and east-central), more than 100,000; *E. fulvus mayottensis* (Mayotte Island, not recognized as a valid taxon by Mittermeier et al.), fewer than 25,000; *E. fulvus rufus* (west and southeast), more than 100,000; *E. fulvus sanfordi* (extreme north), vulnerable, 10,000–100,000; *E. macaco flavifrons,* critically endangered, 100–1,000; *E. macaco macaco,* vulnerable, 10,000–100,000; *E. mongoz,* vulnerable, 1,000–10,000; *E. coronatus,* vulnerable, 10,000–100,000; *E. rubriventer,* vulnerable, 10,000–100,000.

All of these populations are considered to be declining through loss of forest habitat to industrial activity, logging for local use and export, plantations, and slash-and-burn agriculture (Mittermeier et al. 1992, 1994; Richard and Sussman 1975; Wolfheim 1983). The relatively high population densities reported in field studies may be misleading in that they generally apply only to small, protected reserves. In addition to habitat destruction, certain forms reportedly are threatened by intensive hunting and trapping for use as food and because of their alleged raids on crops. Harcourt and Thornback (1990) reported the following for some of the most critically endangered forms: *L. coronatus* has less than 1,300 sq km of suitable habitat remaining, the area is rapidly being reduced and fragmented by logging, burning, and grazing, and poaching is rampant; *L. macaco flavifrons* has an even smaller area of natural habitat, most of which has been converted to agriculture; the mainland population of *L. mongoz* is declining as forests are cleared for pastures and charcoal production. The populations of *E. mongoz* and *E. fulvus* on the Comoro Islands, which prob-

PRIMATES; MEGALADAPIDAE; **Genus LEPILEMUR**
I. Geoffroy St.-Hilaire, 1851

Sportive Lemurs, or Weasel Lemurs, and Koala Lemurs

There are seven species (Petter and Petter *in* Meester and Setzer 1977; Petter and Petter-Rousseaux 1979; Rumpler 1975; Rumpler and Albignac 1975, 1977):

L. dorsalis, extreme northwestern Madagascar and Nosy-Be Island;
L. ruficaudatus, southwestern Madagascar;
L. edwardsi, west-central Madagascar;
L. leucopus, extreme southern Madagascar;
L. mustelinus, northern part of eastern forests of Madagascar;
L. microdon, southern part of eastern forests of Madagascar;
L. septentrionalis, extreme northern tip of Madagascar.

Compiled from the sources cited above, this list of seven species is based to a large extent on studies involving cytogenetics. Tattersall (1982) questioned the validity of such work and tentatively followed the more traditional view that there is only a single species, *L. mustelinus.* That position also was taken by Corbet and Hill (1991), but the above seven species of *Lepilemur* were accepted by Groves (1989 and *in* Wilson and Reeder 1993), Harcourt and Thornback (1990), and Mittermeier et al. (1992, 1994).

Head and body length is 300–350 mm, tail length is 255–305 mm, and weight is about 500–900 grams. The upper parts are rufous, brown, or gray, and the underparts are white or yellowish. The head is conical and short; the ears are large, round, and membranous; the feet are only slightly elongated, and the fourth and fifth toes are the longest. Except for the nail of the great toe, which is large and flat, the nails are keeled. Females have a single pair of mammae located pectorally.

These lemurs are arboreal forest dwellers and strictly nocturnal (Pollock 1979). They normally move by rapid leaps from one vertical support to another using a powerful extension of the hind limbs. On the ground they usually progress with bipedal hops, as do kangaroos (Walker 1979). During the day they sleep rolled up in a ball in a hollow tree or thick foliage. Their diet is folivorous, consisting mostly of leaves and flowers (Hladik 1979; Klopfer and Boskoff 1979). The species *L. leucopus* feeds mainly on thick, juicy leaves and reingests part of its fecal material, as do some lagomorphs (Hladik 1978).

Most species are not well known, but Harcourt and Thornback (1990) compiled population density estimates of 57/sq km for *L. edwardsi,* 13–100/sq km for *L. microdon,* 60–564/sq km for *L. septentrionalis,* and 180–350/sq km for *L. ruficaudatus.* Several species have been found to occupy defended home ranges of about 1 ha. in size. Two or three individual *L. edwardsi* may sleep together in a tree hole by day, but they move about separately by night.

Field studies in southern Madagascar have provided substantial information on *L. leucopus* (Charles-Dominique 1974; Hladik 1978; Hladik and Charles-Dominique 1974; Petter and Charles-Dominique 1979; Pollock 1979). Population densities vary from about 200/sq km to 810/sq km. Home ranges are small and apparently coincide with stable, well-defined territories; those of adult females average 0.18 (0.15–0.32) ha. and those of males average 0.30 (0.20–0.46) ha. The territory of a large male may

Lemur (*Lepilemur* sp.), photo by Jean-Jacques Petter.

overlap those of up to five females, but that of a small male overlaps those of only one or two females. The animals are basically solitary, except that there may be some association between related females. Territorial defense against members of the same sex is very pronounced and occupies the greater part of an individual's time during the night. Since a territory is relatively small, the entire area can be surveyed from a high branch by the resident, and animals may spend hours observing one another. Defense involves visual displays, vocalization, chases, and sometimes severe fighting. A variety of calls ranging from weak squeals to powerful, high-pitched sounds function in distant communication and territoriality.

The mating season is May–August. A single young is born in the period from mid-September to December and is raised in a nest within a hollow tree. The gestation period has been calculated as 120–50 days (Klopfer and Boskoff 1979). Birth weight is about 50 grams. The female often leaves the young clinging to a branch when she forages. Weaning occurs at about 4 months, but the young may follow its mother until it is more than 1 year old. Sexual maturity is attained at around 1.5 years. Captives have lived for 12 years (Doyle 1979; Kolar *in* Grzimek 1990).

Like the other Malagasy prosimians, all species of *Lepilemur* are listed as endangered by the USDI and are on appendix 1 of the CITES. The IUCN classifies *L. dorsalis* and

L. septentrionalis as vulnerable. These animals are threatened by loss of forest habitat to slash-and-burn agriculture, clearing for settlement, burning to encourage growth of pastures, and overgrazing by cattle and goats (Harcourt and Thornback 1990). Most species also are subject to hunting for use as food (Mittermeier et al. 1994). Total populations have been estimated to number more than 100,000 individuals for *L. edwardsi* and *L. leucopus* and 10,000–100,000 for each of the other five species (Mittermeier et al. 1992).

PRIMATES; MEGALADAPIDAE; **Genus MEGALADAPIS**
Forsyth Major, 1894

Koala Lemurs

Vuillaume-Randriamanantena et al. (1992) recognized two subgenera and three species, all extinct:

subgenus *Peloriadapis* Grandidier, 1899

M. edwardsi, known by subfossil material from
    southwestern and extreme southeastern Madagascar;

subgenus *Megaladapis* Forsyth Major, 1894

M. madagascariensis, known by subfossil material from
    southwestern and perhaps northern Madagascar;
M. grandidieri, known by subfossil material from central
    and perhaps northern Madagascar.

Remains referable to the subgenus *Megaladapis* have been found at four sites in northern Madagascar but have not yet been distinguished by species.

The three species are estimated to have ranged in weight from about 40 kg to 80 kg (Mittermeier et al. 1994). *M. edwardsi*, with a cranial length of 277–317 mm, is among the largest of the known prosimians. It also is distinguished by its huge molar teeth, far larger than would be expected in an animal of its overall body size. Cranial length is 235–44 mm in *M. madagascariensis* and 273–300 mm in *M. grandidieri*. The cranium is relatively narrow and greatly elongated. The facial region is long, the orbits divergent, the braincase remarkably small, and the auditory bullae flat. The nasal bones are long and project well beyond the anterior end of the palate, possibly indicating the presence of a mobile snout in life. The zygomatic arches are massive, and there are strong nuchal and sagittal crests. The foramen magnum is backward-facing, and the occipital condyles are oriented perpendicularly to the cranial base. Adults have no upper incisor teeth; in their place are bony ridges suggesting the presence in life of a horny pad such as is found in some ruminants. The molars have complex cusps and increase strikingly in size from front to rear. The skull is disproportionately large in relation to the short, stocky postcranial skeleton. The forelimbs are longer than the hind limbs, and all four extremities are long and somewhat curved, clearly having been powerful grasping organs (Jenkins 1987; Tattersall 1978*b*, 1982; Tattersall and Schwartz 1975; Vuillaume-Randriamanantena et al. 1992).

Both the cranial and the postcranial morphology of *Megaladapis* indicate that its locomotion and lifestyle paralleled that of *Phascolarctos*, the living koala of Australia. It evidently clung to tree trunks and branches with all four limbs, moved upward by a series of short hops, and crossed to nearby trees by short leaps. It presumably fed by crop-

Koala lemur *(Megaladapis edwardsi)*, reconstruction by Stephen D. Nash.

ping leaves pulled by the forelimbs within reach of the mouth (Jenkins 1987; Preuschoft 1971; Tattersall 1982). The species *M. grandidieri* and *M. madagascariensis* had postcranial specializations suggesting greater flexibility of limbs and perhaps more pronounced arboreal adaptation than in *M. edwardsi*. Newly discovered pedal remains suggest also that hind limb suspension may have been an important behavior of the former two species (Vuillaume-Randriamanantena et al. 1992).

Megaladapis is known from sites with radiocarbon dates of about 2,850 to 600 years ago (Culotta 1995). Humans invaded Madagascar around the middle of this period, probably then moved across the island with flocks of domestic livestock, and also introduced various suids that became feral (Dewar 1984). The resulting hunting pressure and environmental disruption, especially the elimination of natural forest habitat, would have been disastrous for a huge, arboreal, and slow lemur that had evolved without predators. There have been recent suggestions that severe drought may already have reduced the lemur population, which thus would have been especially vulnerable to human disturbances, and that people also may have brought a lethal disease to Madagascar (Culotta 1995). *Megaladapis* and the other Malagasy primates known only as subfossils probably became extinct prior to the arrival of Europeans about 500 years ago. There is, however, a seventeenth-century description of a living animal that could fit any of several of the larger extinct genera (Tattersall 1982).

PRIMATES; **Family INDRIIDAE**

Avahi, Sifakas, and Indri

This family of three genera and five species is confined to forests and scrublands of Madagascar. Groves (*in* Wilson and Reeder 1993) followed Jenkins (1987) in spelling the name of this family Indridae. The sequence of genera presented here follows that suggested by Tattersall (1982), who, however, considered the extinct families Palaeopropithecidae and Archaeolemuridae (see accounts thereof) to be subfamilies of the Indriidae.

The three genera may be distinguished as follows: *Avahi*, size small, tail long, pelage uniformly brown and woolly; *Propithecus*, size moderate, tail long, pelage largely white and more silky than woolly; and *Indri*, size large, tail short, and pelage silky. The muzzle is shortened and bare, so that these primates have a somewhat monkeylike appearance rather than the foxlike appearance of certain members of the family Lemuridae. The eyes are large; the orbits of the skull are large and well separated, being directed more forward than laterally in *Avahi* but about midway between forward and laterally in the other genera. As in some genera of Lemuridae, the external ears are largely concealed by the pelage. The hand is more elongated and narrowed than in the Lemuridae, and all fingers have pointed nails. The thumb is short and only slightly opposable. A fold of skin along the arm that extends to the side of the chest represents a vestigial parachutelike membrane. The legs are about one-third longer than the arms, and the foot is larger than the hand but also elongated and narrow, differing in the relatively greater development of the first toe. The other four toes are united at the base by a web of skin, and they function as a single unit in opposition to the first toe, providing good grasping ability for the feet. The palms and soles are padded. Females have a single pair of mammae in the chest region, and the males have a penis bone.

The salivary glands are greatly enlarged, as in the langurs and colobus monkeys. The stomach is large, and the intestine is relatively long, due partly to the elongated caecum.

The members of this family have fewer teeth than other primates, the dental formula being: (i 2/1, c 1/1, pm 2/2, m 3/3) $\times$ 2 = 30. Some workers interpret the arrangement of the teeth as: (i 2/2, c 1/0, pm 2/2, m 3/3) $\times$ 2 = 30. The upper incisors are often large, and the upper canines are elongate and sharp. The lower front teeth project forward and slightly upward as in the Lemuridae. The cusps and ridges of the molars are arranged in alternating V's or crescents.

Avahis are nocturnal, whereas the other members of this family are diurnal. These animals are sometimes solitary, but they usually associate in pairs or in groups of 10 or more individuals. They are arboreal, though they also often descend to the ground. In both arboreal and terrestrial locomotion they differ from the Lemuridae. When climbing they utilize a slow, deliberate hand-over-hand movement. Generally they leap from one upright to another, gaining footholds by means of the powerful grasping action of the hind limbs. In trees they usually cling in an erect position to vertical branches. They descend awkwardly, tail first. On the ground they stand upright with their arms held out in front of their body and usually progress by a series of short leaps or hops. Common resting positions are clinging to an upright branch, sprawling horizontally on a lateral branch with limbs dangling over the sides, and sitting upright in a crotch, sometimes with their arms held outward so as to expose their underparts to the sun. This latter resting position has given rise to a Malagasy superstition that these animals worship the sun.

These primates are strictly vegetarian in the wild, feeding on leaves, buds, fruits, nuts, bark, and flowers. They thus occupy a dietary position in the tropical forests of Madagascar similar to that of the leaf-eating monkeys of Africa and Asia and the howler monkeys (*Alouatta*) of tropical America. When feeding on the ground they often pick up food directly with the mouth; however, they often use their hands to convey food to the mouth and then lick their palms.

Geologically the Indriidae are known only from the Recent of Madagascar. Subfossil remains of some living species, together with those of extinct members of related families, have been found at various sites dated between about 2,000 and 1,000 years ago.

PRIMATES; INDRIIDAE; **Genus INDRI**
E. Geoffroy St.-Hilaire and G. Cuvier, 1796

Indri

The single species, *I. indri*, now occurs in northeastern Madagascar from the vicinity of Sambava to the Mangoro River. Subfossil remains indicate that the range of this species once extended as far west as the Itasy Massif in central Madagascar and north to the Ankarana Range (Mittermeier et al. 1994).

This is the largest living prosimian. Head and body length is 610–900 mm, tail length is 50–64 mm, and weight is about 6–10 kg. The body is thickly covered with long, silky fur above, which becomes shorter beneath. Coloration is usually patterns of grays, browns, and blacks, but there is variation. Some individuals are black, while others are almost or entirely white. The head, ears, shoulders, back, and arms are usually black. On the rump there is a large triangular patch of white, usually surrounded by black. The great toe is long and opposable, and the remaining toes are joined by a web as far as the ends of the first joint. The muzzle is small and nearly naked, the eyes are large, and the skin is black. Females have two mammae.

The indri inhabits coastal and montane rainforest from sea level to about 1,800 meters. It is diurnal and arboreal. It moves mainly by powerful leaps between large vertical stems and trunks (Walker 1979). On the ground it stands erect and usually progresses by jumps, holding its relatively short arms above its body. It is generally found in trees at heights of 2–40 meters, and 30–60 percent of its activity involves feeding. Activity begins two to three hours after dawn and continues until two to three hours before dusk, the period of daily activity being shorter in winter than in summer. Groups move about 300–700 meters in a day. The diet consists of leaves, flowers, and fruits (Pollock 1977, 1979).

Population densities have been estimated at 9–16/sq km but are thought to be lower in some areas. The ranging areas of two groups were 17.7 ha. and 18.0 ha. A large central part of each ranging area constitutes a defended territory from which other groups are excluded. Defense is by adult males, which mark territories with urine and secretions from glands on the muzzle. The most characteristic vocalization of the indri is a melodious song that can be heard by humans up to 2 km away. Often there are loud singing sessions by several members of a group, with each song lasting 40–240 seconds and consisting of a series of cries or howls. These calls probably function to unite groups, express territoriality, and convey information relative to age,

Indris *(Indri indri)*, photos by Russell A. Mittermeier.

sex, and reproductive availability of individuals (Pollock 1975, 1977, 1979, 1986).

According to Pollock (1977), groups comprise two to five individuals, apparently small families with a mated adult pair and their offspring. There is frequent intragroup agitation and competition for food, and it has been observed that the female and young are dominant in such situations, easily displacing the adult male. Nonetheless, it is the male that is responsible for territorial defense, and during encounters with intruders the female moves to a safe location. A single young is born to the female at two- to three-year intervals. It is carried ventrally by the mother for a month or more and then rides on her back until about eight months. Adult size, and presumably full reproductive maturity, is attained at seven to nine years. Thalmann et al. (1993) indicated that births occur in May in the southern part of the range but in December and onward at one site near the northern edge of the range. Doyle (1979) listed the gestation period as 137 days and the time of weaning as 180 days after the birth.

The indri is classified as endangered by the IUCN and the USDI and is on appendix 1 of the CITES. Its range has been reduced and fragmented through the spread of slash-and-burn agriculture and the commercial logging of the forests on which it depends. Like other prosimians, it is protected by law in the Malagasy Republic, but unlike most others, it is also still protected through local custom in some areas (Harcourt and Thornback 1990). With a total wild population estimated at only 1,000–10,000 individu-

als and no captives being maintained, the indri was considered to be of the highest priority for conservation attention by Mittermeier et al. (1992).

PRIMATES; INDRIIDAE; **Genus AVAHI**
Jourdan, 1834

Avahi, or Woolly Lemur

Groves (1989 and *in* Wilson and Reeder 1993), Harcourt and Thornback (1990), Mittermeier et al. (1992), Petter and Petter-Rousseaux (1979), and Tattersall (1982) recognized a single species, *A. laniger,* with two subspecies: *A. l. laniger,* in the eastern forests of Madagascar from near the northern tip of the island to the extreme southeast and, based on subfossil remains, formerly in central Madagascar at least as far west as Analavory; and *A. l. occidentalis,* now in northwestern Madagascar to the north and east of the Betsiboka River but possibly once found as far south as Morondava. Based on karyological analysis by Rumpler et al. (1990), Mittermeier et al. (1994) recognized these two taxa as separate species. Rakotoarison, Mutschler, and Thalmann (1993) reported the discovery of an isolated population of *Avahi,* probably referable to *occidentalis,* still living in west-central Madagascar, not far north of Morondava.

Head and body length is 300–450 mm, tail length is about 325–400 mm, and weight is about 600–1,200 grams.

Avahi (*Avahi laniger*); Top, photo by R. D Martin; Bottom, photo by Chris Raxworthy.

The fur is thick and woolly, not silky as in *Indri* and *Propithecus*. The face is covered with short hairs, and the small ears are concealed by the fur on the head. The most common coloration is brown gray with a lighter rump. The legs, forearms, hands, and feet are white and the tail is reddish orange. There is considerable color variation in this genus, however, some individuals being almost white and others reddish. The head is almost spherical in shape, the snout is short, and the eyes are large.

The avahi is found in rainforests. It spends most of its time in trees but sometimes descends to the ground. In trees it maintains a nearly vertical position and clings to upright limbs and trunks. It is nocturnal, spending the day-

light hours sleeping in hollow trees or thick vegetation. On the ground it stands upright and leaps like an indri or sifaka, in a vertical position with arms held upward and legs close together. The diet in the wild seems to be exclusively vegetarian, consisting chiefly of leaves and also including, buds, bark, and fruit.

Harcourt and Thornback (1990) cited estimated population densities of 72–100/sq km and home range sizes of up to 4 ha. They noted also that the ranges of the subspecies *A. l. occidentalis* are larger than those of *A. l. laniger* and overlap more and that *occidentalis* shows less territorial aggression. Ganzhorn, Abraham, and Razanahoera-Rakotomalala (1985) reported that 10 groups of *A. l. laniger* occupied nonoverlapping home ranges of 1–2 ha. each and that a baby was born to four of these groups in August or September. There is some question regarding social structure. Groups consist of up to five individuals, and it has been suggested that these represent either a mated pair and young or an adult female with several generations of offspring (Klopfer and Boskoff 1979; Pollock 1979). Based on a radio-tracking study and other observations, Harcourt (1991) concluded that *Avahi* is monogamous, being usually found in male-female pairs that sleep and forage together. The various vocalizations include a high-pitched whistle for distant communication and perhaps territorial expression (Petter and Charles-Dominique 1979). The gestation period is four to five months, and there is normally a single young per birth. The young are born in the dry season, August–November (Hayssen, Van Tienhoven, and Van Tienhoven 1993), and are transported on the back of the mother for several months.

The avahi is classified as endangered by the USDI and is on appendix 1 of the CITES. It is declining through loss of forest habitat, especially because of clearing and burning to create pastures and cropland. Unlike many primates, the avahi is extremely difficult to keep in captivity, none currently known to be held. The number of individuals in the wild may be more than 100,000 for *A. l. laniger* and 10,000–100,000 for *A. l. occidentalis* (Mittermeier et al. 1994). The IUCN recognizes the latter as a separate species and classifies it as vulnerable.

PRIMATES; INDRIIDAE; **Genus PROPITHECUS**
Bennett, 1832

Sifakas

There are three species (Petter and Petter-Rousseaux 1979; Simons 1988):

P. verreauxi, western and southern Madagascar;
P. diadema, eastern Madagascar;
P. tattersalli, extreme northeastern Madagascar.

Head and body length is 450–550 mm, tail length is 432–560 mm, and weight is 3–7 kg. The fur is rather long and soft with woolly hair above but is sparse below. Coloration varies greatly within each species, from white, often tinged with yellowish, to black, gray, or reddish brown, arranged in various patterns. The face is hairless and black. The short arms are limited in their movement by small gliding membranes; the thumbs are scarcely opposable to the fingers; and the hind limbs are large and strong. *Propithecus* is distinguished from *Indri* by its long tail and smaller ears, which are largely concealed by hair.

Sifakas inhabit deciduous and evergreen forests and are diurnal. They stay mainly in large trees and sleep at a

Sifaka *(Propithecus verreauxi)*, photo from San Diego Zoological Society.

which the troop moved in a 7- to 10-day period. Estimated home ranges of 20 ha. to more than 250 ha. have been reported for *P. diadema* (Harcourt and Thornback 1990) and of 9–12 ha. for *P. tattersalli* (Mittermeier et al. 1994).

Harcourt and Thornback (1990) compiled additional available information on *P. diadema* and *P. tattersalli* but suggested that their social behavior may be much like that of *P. verreauxi*. Groups of the latter species are territorial in at least some areas. Jolly (1966) reported that there was an area of 0.6–1.8 ha. within the home range of each troop that other troops did not enter. Richard (1977) stated that each group had exclusive use of 24–51 percent of its home range. Adult males mark territories with urine and with a scent gland located on the front of the throat. Territorial confrontations between two troops may involve growling, scent marking, and ritualistic leaping toward the enemy territory but not direct physical contact. In addition to growls at other troops, vocalizations include barks at aerial predators, a resonant bark for distant communication and spacing, and the sound "sifaka," which is made when there are intruders on the ground (Jolly 1966; Petter and Charles-Dominique 1979).

Groups include 2–13 individuals, there usually being 2–3 adult males, 2–3 adult females, and several young (Pollock 1979). Groups are compact, with the various individuals generally keeping in sight of one another (Jolly 1972). According to Jolly (1966), intragroup social life is peaceful, with the young solicitously cared for and the adults playing together and grooming one another. During the breeding season, however, there are fights and sometimes serious injuries. Richard (1974) observed that at this time there is a breakdown in group structure, with adult males undergoing "roaming" behavior and there being competition for

height of about 13 meters. They can make leaps of about 10 meters from tree to tree using the strength of the hind legs. On the ground they stand on their hind feet and progress by bipedal leaps, throwing their arms above their heads for balance. When picking up food from the ground they usually stoop and seize the food in the mouth, seldom using their hands. The diet consists mainly of leaves, flowers, bark, and fruits. In a study of *P. verreauxi*, Richard (1977) found that in the wet season there was an increase in the take of flowers and fruits, whereas in the dry season there was an increase in the amount of leaves eaten. Reported estimates of population density are up to 20/sq km for *P. diadema*, 50–500/sq km for *P. verreauxi*, and 60–70/sq km for *P. tattersalli* (Harcourt and Thornback 1990; Mittermeier et al. 1994).

Comparatively little is known of *P. diadema* and *P. tattersalli*, but *P. verreauxi* is one of the most intensively investigated of the Malagasy prosimians, and the remainder of this account, except as noted, deals with the latter species. Jolly (1966) reported that *P. verreauxi* woke up at about dawn, fed until 0800–0900 hours, moved to a sunning site for a while, and then fed again. In warm weather the sifakas slept at noon, had a second period of activity, and then settled for sleep at dusk. Richard (1977, 1978) found peak feeding activity during the wet season at 0700–0900 and 1300–1400 hours. She studied sifakas in both the northern and the southern parts of their range. The mean distance moved in a day in the north was 1,100 meters in the wet season and 750 meters in the dry season. In the south the respective distances were 1,000 and 550 meters. Group home range in these studies was 6.75–8.50 ha., and the entire range was visited by the troop about every 10–20 days. Jolly (1966) reported home ranges of only 2.2–2.6 ha., through

Sifaka *(Propithecus tattersalli)*, photo by David Haring.

Sifaka *(Propithecus verreauxi)*, photo by Martin E. Nicoll.

and are on appendix 1 of the CITES. The IUCN also classifies all species and subspecies. Those classifications, together with population estimates by Mittermeier et al. (1992), are: *P. diadema diadema* (found in eastern Madagascar from the Antanambalana to the Mangoro River), endangered, 1,000–10,000; *P. d. edwardsi* (in the east, south of the Mangoro River), endangered, 1,000–10,000; *P. d. candidus* (to the north and east of the Antanambalana River), critically endangered, 100–1,000; *P. d. perrieri* (extreme north), critically endangered, 2,000; *P. verreauxi verreauxi* (in the south and west, as far north as the Tsiribihina River), vulnerable, more than 100,000; *P. v. coquereli* (in the northwest, to the east and north of the Betsiboka River), endangered, 1,000–10,000; *P. v. coronatus* (in the northwest, between the Betsiboka and the Mahavavy du Sud River), critically endangered, 100–1,000; *P. v. deckeni* (in the west, from the Mahavavy du Sud to the Tsiribihina River), vulnerable, 1,000–10,000; and *P. tattersalli*, critically endangered, 8,000. The primary threat to all of these taxa is destruction and fragmentation of natural forest habitat because of slash-and-burn agriculture, commercial logging, charcoal production, fires to stimulate growth of pasture, and overgrazing by domestic livestock. Enforcement of protective laws is often ineffective, even in parks and reserves, and hunting for use as food also jeopardizes certain populations (Harcourt and Thornback 1990). All of these problems are especially severe for the very narrowly restricted *P. diadema perrieri*, *P. d. candidus*, *P. verreauxi*, and *P. tattersalli*, which are considered to be among the most critically endangered primates. Although *P. verreauxi* has been successfully maintained and bred in captivity, most attempts to keep *P. diadema* have failed (J.-J. Petter 1975); a few individuals of *P. d. diadema* and *P. tattersalli* are cur-

Sifaka *(Propithecus diadema)*, photo by Gustav Peters.

females, food, and resting places. The dominance hierarchy that existed in the nonmating season may change. A formerly subordinate male may take over his own group or leave his group and attain dominant status in another group. Females permit mating only by males that retain or achieve dominance during the mating season. Based on such behavior, Richard (1985) suggested that sifaka groups are primarily foraging units of closely related females and a varying number of males.

The mating season extends from January to March. Females in a given group enter estrus only once a year and for a relatively short time (Van Horn and Eaton 1979). Gestation periods of 130 and 141 days have been reported (Doyle 1979; Jolly 1966). The single young usually is born during June–September and weighs about 40 grams (Kolar *in* Grzimek 1990; Mittermeier et al. 1994). Jolly (1966) wrote that young born in July were carried on the belly of the mother until October, rode on her back until December–January, and reached full size at 21 months. Doyle (1979) reported that weaning occurred at 180 days, and sexual maturity at 913 days. A captive *P. verreauxi* was still living at 23 years and 4 months (Marvin L. Jones, Zoological Society of San Diego, pers. comm., 1995).

All three species are listed as endangered by the USDI

rently maintained at the Duke University Primate Center (Mittermeier et al. 1994).

PRIMATES; **Family PALAEOPROPITHECIDAE**

Sloth Lemurs

This family of four Recently extinct genera and five species is known only by subfossil remains found in Madagascar. The Palaeopropithecidae were treated as a subfamily of the Indriidae by Groves (1989) and Tattersall (1982) but as a full family by Jenkins (1987), Jungers et al. (1991), Mittermeier et al. (1994), and Simons et al. (1992).

This family usually had been thought to comprise only the genera *Palaeopropithecus* and *Archaeoindris*. However, substantial postcranial remains of *Mesopropithecus*, only recently discovered, suggest that the genus is allied with the Palaeopropithecidae rather than the Indriidae, to which it was assigned previously. Moreover, the newly described genus *Babakotia* is a somewhat intermediate form with affinity to both *Mesopropithecus* and *Palaeopropithecus*. In all genera of the family the skull is heavily built, the dental formula is the same as in the Indriidae, and the forelimbs are substantially longer than the hind limbs. According to Tattersall (1982), *Palaeopropithecus* and *Archaeoindris* share unusual conditions of the nasal and auditory regions. The premaxillae are enlarged bilaterally in their superior portions to produce paired protuberances projecting from the nasal aperture. Auditory bullae, which are prominent in the Indriidae and Archaeolemuridae, are entirely lacking. *Mesopropithecus* and *Babakotia* do not have the nasal projections and do retain inflated bullae. However, Simons et al. (1992) noted that all four genera of palaeopropithecids are characterized by relatively small orbits, robust zygomatic arches with convex superior margins, and a bony palate more or less rectangular in outline, its posterior rim bearing prominent tubercles for the attachment of the muscles of the soft palate.

Although most of the extinct Malagasy lemurs were described about a century ago, much information on their likely overall appearance, natural history, and systematic affinity has become available only in the last decade or is still unpublished. Simons et al. (1992) used the term "sloth lemurs" for the Palaeopropithecidae because of their many extraordinary convergences with the South American sloths. Mittermeier et al. (1994) suggested that the Palaeopropithecidae evolved from and are closely related to the Indriidae. They were medium-sized to large primates, apparently slow-moving, and at least to some extent slothlike in suspensory habit. *Mesopropithecus*, the smallest and most primitive genus, probably weighed about 10 kg. It and *Babakotia* seem to represent stages, in size and adaptation for arboreal suspension, that culminate in *Palaeopropithecus*. *Archaeoindris*, which weighed up to 200 kg, probably was largely terrestrial and somewhat analogous to a giant ground sloth. The prominent shearing crests on the molar and premolar teeth of the Palaeopropithecidae suggest that all members were highly folivorous (Simons et al. 1992).

Geologically the family is known only from Recent sites dated as about 8,000–1,000 years old, though *Mesopropithecus*, at least, could have survived to within 500 years ago. All genera probably disappeared as a direct result of the human invasion of Madagascar (Tattersall 1971, 1978b; Tattersall and Schwartz 1975). Simons et al. (1992) stated that they were probably diurnal, noisy, slow to reproduce, and slow-moving and would have been easy to hunt.

PRIMATES; PALAEOPROPITHECIDAE; **Genus MESOPROPITHECUS**
Standing, 1905

There are two species, both extinct (Jenkins 1987; Mittermeier et al. 1994; Simons et al. 1990, 1992; Tattersall 1982):

M. globiceps, known only by subfossil remains from the type locality in central Madagascar;

M. pithecoides, known by subfossil remains from northern, central, western, and southwestern Madagascar.

M. globiceps sometimes has been placed in a separate genus, *Neopropithecus* Lamberton, 1936.

Until recently *Mesopropithecus* was known mainly from cranial material, on the basis of which it usually was assigned to the family Indriidae. Cranial length is 83.3–94.4 mm in *M. globiceps* and 94.0–103.5 mm in *M. pithecoides*. The skull is similar in size to that of *Indri* (cranial length 97.1–117.7 mm) but closely resembles that of *Propithecus* in proportions and dentition. It differs from that of *Propithecus* in having a more robust build, slightly smaller and more convergent orbits, more massive zygomatic arches, a broader snout, a more pronounced postorbital constriction, a steeper facial angle, a more rounded braincase, a sagittal crest or temporal lines that are anteriorly confluent, a distinct nuchal ridge confluent with the posterior root of the zygoma, and larger upper incisors and canines. *M. pithecoides* is the more divergent of the two species, being characterized by larger size, sagittal and nuchal crests, massive zygomatic arches, and a broad muzzle. *M. globiceps*, with its more gracile build and narrower snout, is more primitive and more like *Propithecus*.

Mesopropithecus is the smallest of the subfossil prosimians of Madagascar and may have been the last to become extinct. A recently discovered postcranial skeleton apparently referable to *M. pithecoides* shows that the species probably weighed about 10 kg and had elongated forelimbs. These and other characters indicate adaptation for arboreal suspension and affinity to the family Palaeopropithecidae rather than to the Indriidae. The dentition of *Mesopropithecus* suggests that the genus was folivorous in diet. It may have survived until about 500 years ago and then disappeared through direct hunting and/or environmental disruption by people.

PRIMATES; PALAEOPROPITHECIDAE; **Genus BABAKOTIA**
Godfrey, Simons, Chatrath, and Rakotosamimanana, 1990

The single species, *B. radofilai*, is known only by subfossil remains from Antsiroandoha and one other cave in the Ankarana Range in extreme northern Madagascar (Godfrey, Simons, et al. 1990; Jungers et al. 1991; Mittermeier et al. 1994; Simons et al. 1992).

Cranial length in one specimen is 114.5 mm. The skull has a superficial resemblance to that of *Indri* because of the elongation of the face, buccolingual compression and mesiodistal expansion of the maxillary and mandibular premolars, and the limited depth of the corpus of the mandible. *Babakotia* also resembles the indriids, as well as *Mesopropithecus*, in possessing inflated bullae, an unfused mandibular symphysis, and a relatively long and procumbent tooth comb. Details of the dental morphology, however, do not suggest a close relationship with any other genus. *Babakotia* has especially elongated upper premolars with shallow paracristas and metacristas; a mesial projection of the crown of the anterior upper premolar well beyond the mesial root of the tooth; fine enamel crenulations on the occlusal surfaces of the molars; incipient bilophodonty of the second maxillary molar; and a relatively large upper third molar with four distinct cusps.

A nearly complete postcranial skeleton indicates that *Babakotia* was a mid-sized primate weighing about 15–20 kg. It was adapted for climbing and hanging, as were the other palaeopropithecids, rather than leaping, a characteristic of the indriids. Its forelimbs are elongated, being about 20 percent longer than the hind limbs. The forefeet and hind feet are also long and adapted for strong grasping. The hind feet are reduced, like those of animals adapted for suspension and in contrast to those of leapers. These and other structural characters indicate specialized suspensory behavior involving frequent use of all four limbs and simultaneous feeding. The living arboreal sloths, *Bradypus* and *Choloepus*, possess similar features. *Babakotia* itself appears to have become extinct within the last 1,000 years, probably as a result of human hunting.

PRIMATES; PALAEOPROPITHECIDAE; **Genus PALAEOPROPITHECUS**
G. Grandidier, 1899

Jenkins (1987) and Tattersall (1982) recognized a single species, *P. ingens,* known only by subfossil remains from central, southwestern, and southern Madagascar. Mitter-

meier et al. (1994) reported that there is at least one other extinct species and that the range of the genus extends into northern Madagascar.

Weight probably was about 40–60 kg. Cranial length is 181–211 mm. The skull is long, relatively low, and robustly built. It resembles that of *Indri* but is much larger, and its braincase and orbits are relatively smaller. The orbits are forwardly and dorsally directed and are heavily ringed by bone. There is a pronounced postorbital constriction and a small degree of nuchal cresting. The structure of the face and the occipital condyles indicate that the head was habitually held at a considerable angle to the neck. In contrast to *Mesopropithecus* and *Babakotia*, as well as to the members of the Archaeolemuridae and the Indriidae, *Palaeopropithecus* has no auditory bullae. Much of the space they would have occupied is taken up by a massive structure consisting of the partially coalesced and highly developed styloid, mastoid/postglenoid, and paroccipital processes. The cheek teeth are similar to those of the Indriidae, and the dental formula is the same. The medial upper incisor is large and subcylindrical, while the lateral is greatly reduced.

The forelimbs are long relative to the hind limbs, and the feet of each are long and hooklike. These and other postcranial specializations indicate that *Palaeopropithecus* was exclusively arboreal and practiced a form of suspensory locomotion similar to that of *Pongo.* Recently collected postcranial remains suggest that of the extinct Malagasy lemurs, *Palaeopropithecus* was the most highly specialized for arboreal suspension and was slothlike in habit. The diet probably was like that of living indriids. *Palaeopropithecus* is known from sites dated at about 2,300 to 1,000 years ago, and it evidently became extinct not long after the end of this period. Humans arrived in Madagascar around the

Extinct Madagascar prosimians (with living *Indri* for comparison), reconstructions by Stephen D. Nash.

middle of this same period. They and their domestic stock probably eliminated much of the forest habitat on which *Palaeopropithecus* depended, and they also would have hunted this vulnerable giant lemur whenever possible.

PRIMATES; PALAEOPROPITHECIDAE; Genus ARCHAEOINDRIS
Standing, 1908

The single species, *A. fontoynonti,* is known only by sub-fossil remains found at Ampasambazimba in central Mada-gascar (Jenkins 1987; Mittermeier et al. 1994; Tattersall 1982; Vuillaume-Randriamanantena 1988).

Archaeoindris apparently was the largest of the Mala-gasy lemurs and, with an estimated weight of 160–200 kg, about as large as an adult male gorilla. The length of the single known skull is 269 mm. It is similar to that of *Palaeopropithecus* but larger and more robust, the facial region is shorter, the orbits are less dorsally oriented, and there is a broad sagittal crest. There are no auditory bullae. The few known postcranial bones have some resemblance to those of the genus *Megaladapis* and had suggested a similar form of locomotion. However, recent analysis indi-cates that in its postcranial morphology *Archaeoindris* ac-tually shows affinity to *Palaeopropithecus.* As in the latter genus, the forelimbs of *Archaeoindris* were longer than the hind limbs, but there was far less disparity between the humeral and femoral lengths.

Archaeoindris probably was largely terrestrial, having evolved from a more arboreal ancestor, and may have been ecologically comparable to the extinct ground sloths of the New World. The diet is thought to have been folivorous. There is some question about the period during which *Ar-chaeoindris* lived, but certain specimens were collected at a stratigraphic level recently radiocarbon dated at about 8,000 years B.P.. Considering also the apparent rarity of the genus, it is possible that it disappeared even before the hu-man invasion of Madagascar. In any event, it would cer-tainly have been extremely vulnerable to environmental disruption and hunting by people.

PRIMATES; Family ARCHAEOLEMURIDAE

Baboon Lemurs

This family of two Recently extinct genera and three species is known only by subfossil remains found in Mada-gascar. The Archaeolemuridae were treated as a subfamily of the Indriidae by Groves (1989) and Tattersall (1982), but as a full family by Jenkins (1987), Mittermeier et al. (1994), and Simons et al. (1992).

The members of this family were medium-sized pri-mates weighing about 15–25 kg. Their skulls resemble those of the Indriidae but are much more massive with es-pecially robust mandibles, fused and less oblique mandibu-lar symphyses, and sagittal and nuchal cresting. The dental formula is the same as that of the Indriidae and Palaeopro-pithecidae except that there are three, rather than two, up-per and lower premolars. The teeth are highly specialized, as described below in the generic accounts. Examination of the postcranial skeleton indicates that the archaeolemurids were short-limbed and powerfully built and that all four feet were strikingly short.

The Archaeolemuridae seem to have evolved from an ancestral indriid that still had three premolar teeth. The

Palaeopropithecidae represent an evolutionary branch that has closer affinity to the living indriids. Although the skulls of the indriids resemble those of the archaeolemurids, the latter show considerable divergence in their dentition and in a highly developed suite of locomotor adaptations for terrestrial living. The two genera have been called the eco-logical equivalents of the African baboons. Geologically, the Archaeolemuridae are known only from Recent sites dat-ed at about 3,000–1,000 years ago. Like the other extinct Malagasy prosimians, they probably disappeared as a direct result of the human invasion of Madagascar.

PRIMATES; ARCHAEOLEMURIDAE; Genus ARCHAEOLEMUR
Filhol, 1895

Jenkins (1987), Simons et al. (1990), and Tattersall (1982) recognized two species, both extinct:

A. majori, known by subfossil remains from southern Madagascar;
A. edwardsi, known by subfossil remains from northern and central Madagascar.

Godfrey and Petto (1981) suggested that there is only a sin-gle, geographically variable species. In contrast, Godfrey, Sutherland, et al. (1990), while finding extensive variation in *Archaeolemur,* thought the genus probably com-prises more than two species. Mittermeier et al. (1994) noted that *Archaeolemur* is perhaps the most widely dis-tributed of all the Malagasy subfossil lemurs, but they did not comment on specific differentiation. All of the sources cited above have been used in compiling the remainder of this account.

Archaeolemur is much larger than any living prosimi-an. Maximum weight probably was more than 22 kg. Cranial length is 122.3–135.0 mm in *A. majori* and 139.6–153.5 mm in *A. edwardsi.* Nonetheless, the skull is similar to that of living indriids, especially *Propithecus,* except for the presence of a sagittal crest. This crest is particularly pro-nounced in *A. edwardsi.* The skull also is similar to that of the extinct *Palaeopropithecus* and *Archaeoindris* except that the bullae are not reduced. It differs from the skull of these other genera in having more forwardly directed or-bits, as in monkeys. There are three upper and three lower premolar teeth, compared with only two of each in living indriids. These unique premolars are buccolingually com-pressed and together form a single longitudinal shearing blade. The molars are subsquare and bilophodont, the para-cone-protocone and metacone-hypocone pairs being unit-ed by transverse crests. The upper canines are reduced, the upper incisors are greatly expanded, and the lower canines and incisors are angled forward at only about 45°, though they apparently were derived from the procumbent condi-tion found in the Indriidae.

The postcranial skeleton is well known and bears some resemblance to that of living terrestrial cercopithecid mon-keys. *Archaeolemur* appears to have been a powerfully built and rather short-legged quadruped with reduced leap-ing power compared with that of some other lemurs. The foot seems to have been modified for grasping, and thus it is likely that *Archaeolemur* retained the ability to exploit arboreal food resources. Its general locomotor and postur-al characters evidently closely parallel those of *Papio.* Its cheek teeth also are similar to those of cercopithecids and may have been used for cropping, husking, and pulping a frugivorous or mixed diet.

Contrary to an earlier view that Madagascar was covered largely by dense forest prior to human modification, there now is evidence that the original vegetation over much of the island was woodland, savannah, bushland, or grassland. Such areas would have been suitable for baboonlike primates. *Archaeolemur* evidently occurred in a number of these habitats, there being a general pattern of larger individuals in wetter (mainly northern and central) zones and smaller animals in more arid (mainly southern and western) areas. Altogether the genus is known from more than 20 sites with radiocarbon dates ranging from about 2,850 to 1,035 years ago, though it may have survived several hundred years longer. People first arrived in Madagascar about 1,500–2,000 years ago. Although the disappearance of *Archaeolemur* and the other subfossil lemurs sometimes is attributed to a variety of factors (Dewar 1984), there is little doubt that human hunting and environmental disruption were directly responsible. A large and primarily terrestrial herbivore such as *Archaeolemur* would have been especially vulnerable to an expanding pastoral culture.

PRIMATES; ARCHAEOLEMURIDAE; Genus HADROPITHECUS
Lorenz Von Liburnau, 1899

The single species, *H. stenognathus*, is known only by subfossil remains from central, southwestern, and southern Madagascar (Jenkins 1987; Mittermeier et al. 1994; Tattersall 1982).

Hadropithecus is about the same size as *Archaeolemur*. Cranial length in two specimens is 128.2 and 141.8 mm. The skull differs from that of *Archaeolemur* in having a much shorter snout, a deeper facial region, more forwardly directed orbits, a more marked interorbital constriction, very robust and widely projecting zygomatic arches, larger auditory bullae, and even more specialized dentition. The incisor and canine teeth are greatly reduced. The posterior lower premolar is molariform, with a cruciform arrangement of rounded ridges, and the upper premolars increase in size and complexity posteriorly. The anterior molars, upper and lower, are large and subsquare, with high, rounded enamel folds replacing the transverse crests of *Archaeolemur*. This entire crowded battery of cheek teeth rapidly is worn flat, with thick enamel ridges enclosing shallow basins in the softer dentine, as occurs in various ungulates. The postcranial skeleton is not well known; identified bones are similar to those of *Archaeolemur* but more gracile.

Hadropithecus is thought to have been quadrupedal and terrestrial and more specialized in this regard than *Archaeolemur*. Its skull and dentition show many characters that are shared by the living gelada baboon *(Theropithecus)* of East Africa. Like the gelada, *Hadropithecus* apparently had an abrasive diet consisting of grass and other small, tough, low-growing items that require heavy grinding by the cheek teeth. This genus is known from sites dated at about 2,000–1,000 years ago and disappeared along with *Archaeolemur* in the face of the human invasion of Madagascar. As a large, terrestrial grazer, it would have been highly susceptible to the spread of people, their domestic flocks, and introduced suids (Dewar 1984).

PRIMATES; Family DAUBENTONIIDAE; Genus DAUBENTONIA
E. Geoffroy St.-Hilaire, 1795

Aye-ayes

The single genus, *Daubentonia*, contains two species (Harcourt and Thornback 1990; MacPhee and Raholimavo 1988; Petter and Petter *in* Meester and Setzer 1977; Tattersall 1982):

D. madagascariensis, formerly found in much of eastern, northern, and west-central Madagascar;

D. robusta, known only by subfossil remains found in southwestern Madagascar.

Groves (*in* Wilson and Reeder 1993) listed *robusta* as a synonym of *D. madagascariensis* but noted that it may be a distinct species.

In *D. madagascariensis* head and body length is 360–440 mm, tail length is 500–600 mm, and weight is about 2–3 kg. *D. robusta*, which is known only from postcranial material and a few teeth, is thought to have been about 30 percent larger. The pelage is coarse and straight, and the bushy tail has hairs up to 100 mm long. The coloration is dark brown to black, the pale bases of the individual hairs showing through to some extent. The nose, cheeks, chin, throat, and spots over the eyes are yellowish white. The hands and feet are black.

D. madagascariensis has a rounded head with a short face, large eyes, and large, naked, membranous ears. The body and limbs are slender. The fingers are long and nar-

Aye-aye *(Daubentonia madagascariensis)*, photo by David Haring.

Aye-aye *(Daubentonia madagascariensis)*, from *Bull. Acad. Malgache*. Insets: right hand and left foot photos from *Zoologie de Madagascar*, G. Grandidier and G. Petit; photo (right) by Jean-Jacques Petter.

row, the third finger extremely so. The thumb is flexible but not truly opposable; however, the first toe is opposable. All the digits have pointed, clawlike nails, except the first toe, which has a flat nail. Females have two mammae located abdominally, and males have a penis bone.

The incisors are large, curved, and similar to those in rodents, that is, chisel-like with enamel on only the front surface and ever growing. Canines are absent in the permanent dentition but present in the deciduous dentition. In the adult aye-aye there is a large space between the incisors and the premolars. The modified cheek teeth have flattened crowns with indistinct cusps. The formula for the permanent dentition is: (i 1/1, c 0/0, pm 1/0, m 3/3) × 2 = 18. The skull resembles that of a squirrel, but primate features are prevalent (see also above account of the order Primates).

The modern aye-aye lives mostly in forests but appears to be adaptable, having been found in secondary growth, mangroves, bamboo thickets, and cultivated areas, particularly coconut groves; it may, however, depend on large trees for nesting (Harcourt and Thornback 1990). It is arboreal and nocturnal. During the day it sleeps in a nest constructed amidst dense foliage in a strong fork of a large tree, about 10–15 meters above the ground. The nest, about 50 cm in diameter, probably requires 24 hours to build and is very complex. Made of twigs or interlaced leaves, it is closed at the top and has a lateral opening; the bottom is covered by a layer of shredded leaves. The 5-ha. home range of two animals contained 20 such nests. Each nest was used for several days in succession. The aye-aye leaves its nest at nightfall to forage and returns at first light. Its locomotion through the branches is much like that of a lemur, but it is less adept in horizontal movements. Vertical climbing is by

rapid successive leaps. It frequently descends to the ground and can make long trips there (J.-J. Petter 1977; Petter and Petter 1967). When moving about, it carries its tail in a curve. Occasionally the aye-aye, like a loris, will suspend itself by its hind feet, using its hands for feeding or cleaning. In the latter operation it uses the long third finger in combing, scratching, and cleansing; the other fingers are flexed during this performance.

On Nosy Mangabe Island, Sterling (1993) found individuals to move about 800–4,400 meters per night. The main foods there were seeds from the fruit of *Canarium*, cankerous growths on *Intsia*, nectar from *Ravenala*, and larvae from several families of insects. The aye-aye apparently listens carefully for the sound of larvae in decaying wood and often taps the surface of the wood with its long third finger. Smelling may also be involved in finding the larvae. Erickson (1991, 1994) found that the tapping serves to locate the galleries made by larvae, as well as adult ants and termites, within the wood. The aye-aye also usually can determine which cavities are actually occupied by prey. This capability could result from a combination of perceptual factors, including a cutaneous sense in the third finger and an echolocation system involving triangulation of the large ears of the aye-aye and the tones emitted in response to the tapping. When the prey is located, the aye-aye bites through the wood with its powerful incisors and inserts its third finger to crush and extract the larvae. The aye-aye also eats coconuts by gnawing a hole with its incisors and extracting the juice and pulp with the third finger (Hladik 1979; J.-J. Petter 1977; Petter and Petter 1967). Before biting into the coconut the aye-aye may tap the surface with its third finger, perhaps to evaluate the milk content (Winn 1989). Other studies on Nosy Mangabe Island, where the

Aye-aye *(Daubentonia madagascariensis)*, photo by David Haring.

aye-aye was introduced in 1967, indicate that the diet also includes bamboo shoots, tree exudates, large insects, and possibly small vertebrates (Pollock et al. 1985).

The aye-aye appears to be basically solitary, but six captive adults kept in the same cage did not show signs of aggression (J.-J. Petter 1977). The animals in a captive colony at the Jersey Wildlife Preservation Trust are kept in separate cages except when mixed for breeding purposes (Carroll and Beattie 1993). Some observations of captive animals suggest that females may be dominant to males (Rendall 1993). An adult female and a male not fully grown shared a home range of about 5 ha. (Petter and Petter 1967). A male, a female, and a juvenile ranged over an area about 5 km long (Hladik 1979). In a radio-tracking study on Nosy Mangabe Island, Sterling (1993) found home range size for two males to be 126 ha. and 215 ha. and that for two females to be 32 ha. and 40 ha. Male ranges overlapped greatly with one another and with those of several females, but female ranges were well separated from one another. Social interaction varied, with some animals avoiding one another but others forming foraging units consisting of two or three males or of a male and a female. Adults slept separately, though observations elsewhere indicate that two males sometimes share a nest. Reproductive activity was asynchronous and evidently occurred throughout the year. During a three-day period of estrus a female called repetitively and was subsequently surrounded by up to three males.

According to J.-J. Petter (1977), several vocalizations have been distinguished, including the sounds "rontsit" for alarm or danger and "creee" possibly as a contact signal. Both sexes frequently mark with urine. Reproduction in any one female may occur only once every two or three years, and apparently there is a single young, which is raised in a nest. A captured 1-year-old male and 2-year-old female were still with the mother and did not appear sexually mature. Limited observations in the wild suggest that

breeding may occur through much of the year (Sterling 1993). Beattie et al. (1992) reported that a female, apparently already pregnant when captured, gave birth at the Duke University Primate Center in April. The first fully captive-bred aye-aye was born at the Jersey Wildlife Preservation Trust in August after a gestation period of 158 days; it weighed about 140 grams. Another infant, subsequently born at Duke after a gestation period of 172 days, weighed only 103 grams. According to Winn (1989), a young male at the Paris Zoo was weaned at 7 months of age but still had not reached adult size by about 3 years. Although captive specimens are rare, one lived for 23 years and 3 months (Jones 1982).

D. robusta apparently disappeared less than 1,000 years ago (J.-J. Petter 1977). As with other large Malagasy prosimians, its extinction was probably brought on by human agency. Specimens of its teeth, evidently modified by people for ornamental purposes, have been recovered (MacPhee and Raholimavo 1988). *D. madagascariensis* still survives but is designated as endangered by the IUCN and the USDI and is on appendix 1 of the CITES. It has declined mainly through cutting of the large forest trees upon which it depends. It also is killed on sight by villagers, who believe it to be a harbinger of misfortune. Based on its rarity and systematic uniqueness, Mittermeier et al. (1986) considered its survival to be one of the highest primate conservation priorities, if not the highest, in the world. At that time only a few scattered individuals were thought to remain on the northeastern and possibly northwestern coasts of Madagascar. A number of animals also had been captured alive in 1967 and released on Nosy Mangabe Island, off northeastern Madagascar, which is a protected reserve, and evidently a population became established there (Constable et al. 1985).

Subsequently, considerable attention was devoted to locating and investigating the aye-aye. In 1985 a small population was discovered in a rainforest about 900–1,000 meters above sea level in east-central Madagascar (Ganzhorn and Rabesoa 1986). Several were observed and captured in the northeast in 1986 (Albignac 1987). In 1991 another population was discovered in the northwest to the east of Narinda Bay (Simons 1993). Mittermeier et al. (1994) thought the many new sightings over a large region to be an incredible turn of events since the aye-aye had been feared to be on the verge of extinction; nonetheless, they cautioned that the species remained rare and of much conservation concern. Only 1,000–10,000 individuals are estimated to survive in the wild (Mittermeier et al. 1992). There also are 17 in captivity, 3 of which were born in that state (Olney, Ellis, and Fisken 1994).

PRIMATES; Family TARSIIDAE; Genus TARSIUS
Storr, 1780

Tarsiers

The single Recent genus, *Tarsius*, contains five species (Groves 1976; Musser and Dagosto 1987; Niemitz 1984*b*; Niemitz et al. 1991; Petter and Petter-Rousseaux 1979):

T. spectrum, Sulawesi and the nearby islands of Sangihe, Peleng, and Salayar;

T. pumilus, central Sulawesi;

T. dianae, known only from the type locality in central Sulawesi;

T. bancanus, southern Sumatra and the nearby islands of

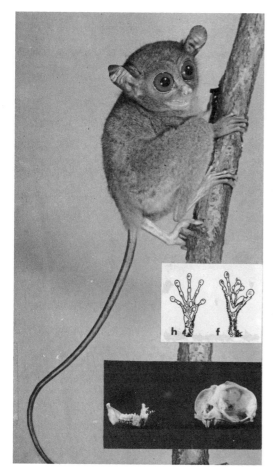

Mindanao tarsier *(Tarsius syrichta)*, photo by Ernest P. Walker. Insets: hand and foot (*Tarsius* sp.), photo from *Proc. Zool. Soc. London;* skull (*Tarsius* sp.), photo by P. F. Wright of specimen in U.S. National Museum of Natural History.

Bangka and Billiton, Borneo and the nearby islands of Karimata and Serasan, possibly Java;

T. syrichta, the islands of Samar, Leyte, Dinagat, Siargao, Bohol, Mindanao, and Basilan in the Philippines.

The forms *T. pelengensis,* of Peleng Island, and *T. sangirensis* of Sangihe Island, are sometimes regarded as species distinct from *T. spectrum,* but they were not regarded as such by Corbet and Hill (1992), Flannery (1995), or Groves (*in* Wilson and Reeder 1993).

Head and body length is 85–160 mm, tail length is 135–275 mm, and adult weight is 80–165 grams. The individual hairs have a wavy, silky texture, and certain parts of the body are sparsely haired. The coloration above ranges from buff or grayish brown to dark brown, and the underparts are buff, grayish, or slate. The tail is naked except for a few short hairs on the tip. The very large eyes are the most outstanding structural feature, the diameter of the eyeball being approximately 16 mm. The head is round with a reduced muzzle, and the neck is short. The ears are thin, membranous, and nearly naked. Members of this genus differ from all lemurs in that the nasal region is clothed with short hairs to the margins of the nostrils and there is a narrow strip of naked skin around the nostrils. Lemurs have a moist muzzle with a central prolongation

dividing the upper lip. The forelimbs of *Tarsius* are short, but the hind limbs are greatly elongated—the name "tarsier" refers to the elongated tarsal, or ankle, region. The digits are long and tipped with rounded pads that enable the animal to grip almost any surface. The thumb is not truly opposable, but the first toe is well developed and widely opposable. Except for the second and third toes, which have clawlike nails used in grooming, the digits have flattened nails.

The dental formula is: (i 2/1, c 1/1, pm 3/3, m 3/3) × 2 = 34. The upper incisors are large and pointed and the lower incisors point upward, not forward. The upper canines are relatively small. The cheek teeth are adapted for a diet of insects. The orbits of the skull are directed forward.

Tarsiers seem to prefer secondary forest, scrub, and clearings with thick vegetation, but they have also been found in primary forest and mangroves (IUCN 1978). They are nocturnal or crepuscular and mainly arboreal. They spend the day sleeping in dense vegetation on a vertical branch or, rarely, in a hollow tree. They are not believed to build nests. When clinging to an upright branch they rigidly apply their tail as a support, and when they are asleep the head may drop downward between the shoulders. If disturbed while resting, a tarsier moves up or down its support and faces the suspected enemy. Its mouth may be opened and the teeth bared at the same time.

A tarsier can rotate its head nearly 360°, giving the animal an extremely wide field of vision. Tarsiers are notable acrobats in trees and shrubs, making quick leaps of several meters with no apparent effort. MacKinnon and MacKinnon (1980) reported an average leap of 1.4 meters and a maximum leap of 5–6 meters for *T. spectrum.* The tail trails behind as they leap from one support to another. Their movements are much like those of a tree frog, except that in *Tarsius* grasping is also involved. On any flat surface they also leap froglike, with their tail arched over their back, but they can walk on all fours with the tail hanging down. Their leaps on the ground are 1,200–1,700 mm long and up to 600 mm high. The ears of tarsiers are in almost constant motion during the waking period, being furled or crinkled frequently.

Tarsiers prey mainly on insects and readily accept small lizards and crustaceans, such as shrimps, in captivity. A tarsier watches its moving prey, adjusts its position and focus, then suddenly leaps forward and seizes the prey with both hands. The prey is chewed with side-to-side movements of the jaw while the tarsier sits upright on its hindquarters. These primates drink water by lapping.

MacKinnon (1986) estimated that 9,912,500 individual *T. bancanus* occupied 198,250 sq km of suitable remaining habitat in Indonesia, at an average density of 50/sq km, and that 14,146,000 *T. spectrum* occupied 70,730 sq km at a density of 200/sq km. However, such figures may not account for widely varying habitat conditions and human disturbances and thus could be highly excessive.

Niemitz (1984*b*) reported that an adult pair of *T. bancanus* inhabits a home range of 1–2 ha. This area apparently is a territory, which is marked by urine and the scent of various glands. Fogden (1974), also working in Borneo, found home range to be 2–3 ha., larger for males than for females. Ranges of individuals of the same sex seemed mostly exclusive, but there was extensive overlap between the ranges of males and females. Despite reports that tarsiers are usually found in pairs, Fogden saw two together only eight times during his study. He recorded immature males mainly in primary forest, which is not considered good habitat, and suggested that these young animals were forced to disperse into marginal areas on attaining independence. Females, however, established ranges near those

Tarsier *(Tarsius syrichta)*, photo by Bernhard Grzimek.

of their parents. *T. syrichta* is usually seen in male-female pairs (IUCN 1978). Wright et al. (1987) reported maintaining captive *T. bancanus* in male-female pairs and *T. syrichta* in pairs or in groups of one male with two or three females.

In studies on northern Sulawesi, MacKinnon and MacKinnon (1980) found *T. spectrum* to occupy a wider range of habitats than that reported for *T. bancanus*. Family groups occupied an average home range of approximately 1 ha. and regularly slept together at the same sites each day. They were territorial, actively chasing others out of their range and marking the area by rubbing branches with urine and epigastric glands. Vocalizations included loud shrieks during territorial battles and loud duets, with the male and female of a pair having separate parts. Pairs formed a close and stable bond and were seen to remain together for more than 15 months. Young females remained

with their parents until adulthood, whereas young males departed as juveniles. One group had 3 males and 3 females but appeared to be splitting up. There were two breeding seasons 6 months apart, at the beginning and the end of the rainy season. Births occurred in May and November–December. The precocious young could travel with the group at 23 days and hunt alone at 26 days.

Fogden (1974) reported that *T. bancanus* has a sharply defined breeding season, with mating in October–December and births in January–March. Van Horn and Eaton (1979), however, questioned Fogden's data and pointed out that earlier investigation had found pregnant females in every month of the year on Bangka Island. These authorities also cited studies indicating that litter size in the genus is one and that the estrous cycle is about 25–28 days in *T. syrichta*, but they questioned a report that the gestation period in a specimen of *T. syrichta* lasted 6 months. Subsequent studies, however, have shown that gestation in *T. bancanus* is 178 days, an unusually long period for such a small mammal, and that the estrous cycle in this species is 18–27 days, with a 1- to 3-day estrus (Izard, Wright, and Simons 1985; Wright, Izard, and Simons 1986). The young is born in a fairly well-developed state—well furred, its eyes open, and capable of climbing and making short hops on a level surface. It is unable to leap until 1 month old. The weight at birth is approximately 20–31 grams. The young usually clings to the mother's abdomen but sometimes is carried in her mouth. It begins to capture prey at 42 days and is weaned shortly thereafter (Wright et al. 1987). Fogden (1974) stated that adult weight was attained at 15–18 months. A captive specimen of *T. syrichta* lived for 13 years and 5 months (Jones 1982).

T. syrichta is classified as threatened by the USDI and as conservation dependent by the IUCN. All species are on appendix 2 of the CITES. Tarsiers are jeopardized by destruction of forest habitat and capture by people (Wolfheim 1983).

Simons and Bown (1985) tentatively assigned fossils from the Oligocene of Egypt to the Tarsiidae, and Ginsburg and Mein (1987) described the species *Tarsius thailandica* from the lower Miocene of northwestern Thailand. Otherwise, fossils referable to the family have not been found, but the Omomyidae, a related family of the infraorder Tarsii (or Tarsiiformes), has a geological range of early Eocene to late Oligocene in North America and early Eocene to early Oligocene in Eurasia. This diverse group contains about 30 known genera, and there is increasing evidence that *Tarsius* is a living descendant (Beard and Wang 1991; Gingerich 1984).

PRIMATES; Family CEBIDAE

New World Monkeys

This family of 11 living genera, 3 Recently extinct genera (in the West Indies), and 65 species inhabits forests from northeastern Mexico to northern Argentina. There is almost universal acceptance of these genera, but the number of ways in which they have been assigned to subfamilies or otherwise arranged to show phylogenetic affinity is nearly equal to the number of authorities who have attempted such a task. Groves (1989) divided the Cebidae into five full families: the Cebidae, for *Cebus* and *Saimiri*; the Aotidae, for *Aotus*; the Callicebidae, for *Callicebus* and the extinct *Xenothrix*; the Pitheciidae, for *Pithecia*, *Chiropotes*, and *Cacajao*; and the Atelidae, with two subfamilies, the Atelinae for *Lagothrix*, *Ateles*, and *Brachyteles* and the Alouattinae for *Alouatta* (the extinct *Paralouatta* and *Antillothrix* had not yet been described). Subsequently Groves (*in* Wilson and Reeder 1993) reduced these families to subfamilial level. Hershkovitz (1977) recognized a seventh subfamily for *Saimiri*, the Saimirinae, whereas Martin (1990) accepted only five subfamilies, placing *Aotus* (see account thereof) in the Callicebinae. There has been a recent trend toward dividing the traditional cebids into two full families, the Atelidae and the Cebidae, and including the Callitrichidae in the latter, but there is further disagreement about the details of the division. Ford (1986*b*) considered the Atelidae to comprise the Alouattinae, the Atelinae, and the Pithecinae; Schneider et al. (1993) also included the Callicebinae; and Rosenberger, Setoguchi, and Shigehara (1990) added the Aotinae (thereby restricting the Cebidae to *Cebus*, *Saimiri*, and also the callitrichids).

Most of the authorities cited above, along with Kay (1990) and Thorington and Anderson (1984), agree that there is affinity between the genera *Lagothrix*, *Ateles*, *Brachyteles*, and *Alouatta* and between the genera *Pithecia*, *Chiropotes*, and *Cacajao*; a looser connection between both groups also seems to be recognized. There is a further partial consensus on some phylogenetic association of *Callicebus* and *Aotus*, and of those two genera and *Cebus*, *Saimiri*, and the callitrichids. These factors are taken into account in the following sequence of genera. The extinct *Xenothrix* is something of an enigma, having been assigned to various subfamilies and families over the years. MacPhee and Fleagle (1991) supported an early view that *Xenothrix* belongs in its own family, the Xenotrichidae, though they also recognized the Atelidae and the Cebidae as distinct families. Since such a division is not made here, *Xenothrix* is retained in the Cebidae and placed near *Callicebus* in keeping with the relationships suggested by Groves (1989) and Rosenberger, Setoguchi, and Shigehara (1990). The also extinct *Paralouatta* and *Antillothrix* are here placed adjacent to *Callicebus* in accordance with the finding by MacPhee et al. (1995) that they form a sister group to the latter (and also may have some phylogenetic association with *Xenothrix*). However, Rivero and Arredondo (1991) had concluded that *Paralouatta* is closely related to *Alouatta*.

Most cebids are much larger than any callitrichid. They range in size from *Aotus*, with a head and body length of 240–370 mm, to *Alouatta*, with a head and body length of up to 915 mm (*Brachyteles* may sometimes exceed *Alouatta* in weight). The tail is well haired in all members and long in all genera except *Cacajao*. The tail is fully prehensile in four genera: *Ateles*, *Brachyteles*, *Lagothrix*, and *Alouatta*. In these genera the underside of the tail is naked near the tip and soft and sensitive. The prehensile tail is used as though it were an additional hand—for grasping objects beyond the reach of the arms, in swinging and checking falls, and in steadying the animal by retaining a grasp with the tail when the hands and feet are being used in progression. The naked underside of the tip of the tail also serves as an organ of touch. In the squirrel monkeys *(Saimiri)* the tail is exceedingly mobile, but it is not prehensile. Monkeys of the genus *Cebus* carry the tail coiled and use it to brace or steady themselves, but it is not regularly used to grasp objects or to hang by.

Some zoologists use the general term Platyrrhine for the American monkeys, whose nostrils are widely separated and open to the sides, and the term Catarrhine for the Old World monkeys, in which the nostrils are set close together and open to the front. However, a number of species exhibit intermediate appearances.

In cebid monkeys the orbits are large and forwardly directed with large eyes and well-developed eyelids. The eyes of the nocturnal *Aotus* are very large. Except for this genus, the eye of cebid monkeys has the typical diurnal type of retina; that is, rods and cones are present and are arranged as in the human eye. The retina in *Aotus* contains rods only. None of the American monkeys possesses cheek pouches or rump callosities such as are present in many of the Old World monkeys. The digits, which are usually long and thin, bear flattened or curved nails. The thumb in cebid monkeys is not opposable, but the great toe is well developed and widely opposable to the other toes. A baculum is usually present but is lacking in *Aotus*, *Ateles*, and *Lagothrix*.

The dental formula is: (i 2/2, c 1/1, pm 3/3, m 3/3) × 2 = 36. The upper molars have four cusps, and the lower molars have four or sometimes five.

Most cebids are agile jumpers and runners, swinging and

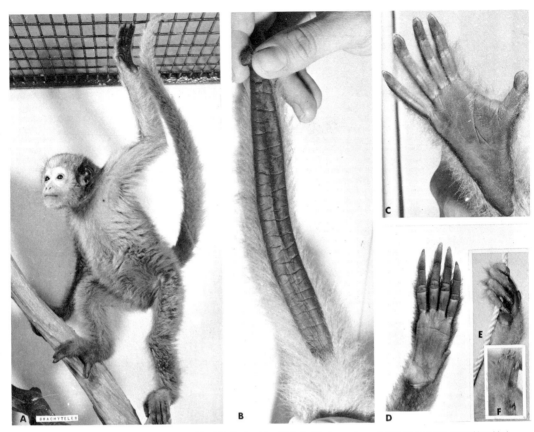

A. Woolly spider monkey *(Brachyteles arachnoides)*; B. Underside of tail; C. Plantar surface of right foot; photos from New York Zoological Society. D. Palmar surface of right hand, photo by G. E. Erikson, Harvard Medical School. E. Grasping hand; F. Dorsal view of hand; photos from New York Zoological Society.

leaping through trees, but some forms are rather deliberate in their movements. Their movements tend to be smoother and less jerky than those of marmosets and tamarins. The claim of early writers that monkeys made living chains and bridges to travel between trees too far apart for leaping has generally been discredited, but Dr. Carl Lovelace, of Waco, Texas, stated in a letter to William Mann that he had seen such a performance in 1905 in South America: "These monkeys made a living bridge . . . formed by one monkey swinging tail down from a limb and wrapping his tail around the head of another monkey, and so on until five successive monkeys made a pendant chain which then swung back and forth until the ultimate monkey attached himself to the tree across the interval. The remaining monkeys, including two females with little baby monkeys around their necks, then crossed this bridge. The last monkey to cross was . . . evidently the leader of the band. . . . The initial end of the bridge then turned loose and the whole band proceeded in the original direction in less time than it has taken me to describe it."

The known geological range of the Cebidae is upper Oligocene to Recent. Rosenberger, Setoguchi, and Shigehara (1990) listed 10 extinct genera but assigned all to living subfamilies. MacPhee and Iturralde-Vinent (1995a) reported the first specimen of a New World monkey of Tertiary age to be discovered outside of South America. It is an astragalus found at an early Miocene site in central Cuba.

PRIMATES; CEBIDAE; **Genus LAGOTHRIX**
E. Geoffroy St.-Hilaire, 1812

Woolly Monkeys

There are two species (Fooden 1963):

L. lagotricha, eastern slope of the Andes in Colombia to the Rio Tapajós and the Mato Grosso in central Brazil, eastern Ecuador and Peru, possibly Bolivia;
L. flavicauda, eastern slope of Cordillera Central in northern Peru.

The name of the first species sometimes is spelled *L. lagothricha.*

Among American monkeys this genus is exceeded in weight only by some *Alouatta* and *Brachyteles* and in length by some individuals of *Ateles,* with their very long tails. Head and body length is about 508–686 mm, tail length is about 600–720 mm, and weight is 5.5–10.8 kg. Captives may be much heavier. The hair is short, thick, and woolly, with a good growth of underfur. In some individuals the color of the head is distinctly darker than that of the back; in others it is almost the same color. The upper parts are hoary gray, blue gray, tawny, dark brown, or blackish brown, and the underparts are only slightly lighter. Some individuals are yellowish buff. The naked face is almost black. *L. flavicauda* differs from *L. lagotricha* in having ma-

Woolly monkey *(Lagothrix lagotricha)*. This photo, by Ernest P. Walker, illustrates the use of the prehensile tail.

hogany-colored pelage, a white or buffy circumbuccal patch on the muzzle, a long tuft of yellow scrotal hair, and a yellow band on the posteroventral surface of the tail (Mittermeier et al. 1977).

Woolly monkeys have a round and massive head, a relatively heavy body for tree monkeys, and a tail that is bare on the undersurface for several millimeters near the tip and prehensile. The thumbs and toes are well developed, and the fingers are short and thick with long pointed nails.

In Colombia *Lagothrix* lives in gallery forests, palm forests, flooded and nonflooded primary forest, and cloud forest as high as 3,000 meters but not in secondary forest. It is found in the crowns of the tallest trees as well as in the shrub layer (IUCN 1978). It seems to be less active in the wild than most American tree monkeys, and it definitely is less active in captivity. It is diurnal and arboreal but often comes to the ground, where it walks upright on its hind legs, using its arms to help maintain balance. When standing on its hind legs it sometimes uses the tail as a brace. Mittermeier and Coimbra-Filho (1977) reported the diet to consist mainly of fruits, supplemented by leaves, seeds, and some insects. Hernandez-Camacho and Cooper (1976), however, stated that leaves were not eaten.

In Peru, Neville et al. (1976) estimated a population density of 12–46/sq km. Studies in the upper Amazon region of Colombia showed that home range was at least 4 sq km for four groups of *L. lagotricha* and that one group of 42–43 individuals used a nomadic range of 11 sq km, moving about 1 km per day (Izawa 1976; Nishimura and Izawa 1975). Overall group size range in this region was 20–70 individuals, there usually being 30–40. Groups sometimes

included as many or more adult males than adult females. Nishimura (1990) added that group ranges overlapped extensively; two and occasionally three groups were observed to move in proximity for hours, usually without agonistic interaction. Individuals, especially subadult females, sometimes temporarily left their own group and spent a few hours or days with another. Males evidently maintained an intragroup dominance hierarchy through aggressive behavior. Hernandez-Camacho and Cooper (1976) observed groups of 4–6 *L. lagotricha* in Colombia, always including a rather old adult male. Available information on *L. flavicauda* indicates a group size of 4–35, with an average of 13 individuals (Mittermeier et al. 1977).

Female *L. lagotricha* carrying newborn young on their backs and bellies have been observed in the Amazon region of Brazil in November. In this species the estrous cycle is 12–49 days, estrus lasts 3–4 days, the gestation period is about 225 days, normal litter size is 1, weight at birth is 140 grams, lactation is estimated to continue for 9–12 months, and full sexual maturity is attained at 6–8 years by females and more than 5 years by males (Eisenberg 1977; Hayssen, Van Tienhoven, and Van Tienhoven 1993). A specimen of *L. lagotricha* lived for 24 years and 9 months in captivity (Jones 1982).

L. flavicauda, the yellow-tailed woolly monkey, is designated as critically endangered by the IUCN and as endangered by the USDI and is on appendix 1 of the CITES. This species, restricted to a small area of montane rainforest on the eastern slope of the Andes in northern Peru, where fewer than 250 individuals now survive, was discovered by Alexander von Humboldt in 1802. There was no new knowledge of this monkey until two specimens were collected in 1925 and three in 1926, and then nothing was heard again until the species was rediscovered by an expedition in 1974 (Mittermeier et al. 1977). Despite legal protection, this monkey is subject to intensive hunting by people for food. The area where it lives is being opened through construction of new roads, and this increased accessibility will lead to more hunting pressure, habitat destruction, and probably extinction (Butchart et al. 1995; Thornback and Jenkins 1982). The subspecies *L. lagotricha lugens* of Colombia also numbers fewer than 250 animals and is classified as critically endangered by the IUCN. Two other subspecies of *L. lagotricha*—*cana* and *poeppigii* of Ecuador, Peru, and Brazil—are designated as vulnerable by the IUCN. *L. lagotricha* also is on appendix 2 of the CITES, and Peres (1991*a*) argued that the entire species should be reclassified as endangered as it is being rapidly reduced in range and numbers by human encroachment. It is perhaps more susceptible to hunting than any other primate in the New World tropics. Its meat is esteemed as food, and its large size makes it a choice target. It is also avidly sought for local use as a pet, fetching up to U.S. $80 each. Generally the mother must be killed in order to capture her infant, and it is estimated that at least 10 females are sacrificed for every live young that reaches the market. In addition, this species seems unable to adapt to secondary forest and so is especially vulnerable to human habitat modification. It and *Ateles* tend to be the first primates to disappear following human encroachment (Thornback and Jenkins 1982).

PRIMATES; CEBIDAE; **Genus ATELES**
E. Geoffroy St.-Hilaire, 1806

Spider Monkeys

Based on the provisional conclusions of Groves (1989 and *in* Wilson and Reeder 1993) and on information from Cabrera (1957), Eisenberg (1989), and E. R. Hall (1981), the following six allopatric species are recognized to exist at present on the mainland of Middle and South America:

A. geoffroyi, Tamaulipas (northeastern Mexico) and Jalisco (west coast of Mexico) to Panama;
A. fusciceps, eastern Panama, Colombia and Ecuador west of the Andes;
A. belzebuth, eastern Colombia and Ecuador, Venezuela, northeastern Peru, northwestern Brazil;
A. paniscus, north of the Amazon River in the Guianas and northeastern and central Brazil;
A. marginatus, Brazil south of the lower Amazon River;
A. chamek, south of the Amazon River in eastern Peru, northern and central Bolivia, and western Brazil as far south as the Mato Grosso.

Several authors (e.g., Hernandez-Camacho and Cooper 1976 and Hershkovitz 1972) consider all of the above species to be intergrading subspecies of the single species *A. paniscus.* An additional named species, *A. anthropomorphus,* is represented by a series of teeth found in a cave in central Cuba, supposedly in direct association with pre-Columbian Amerindian remains. Although there is a possibility that the specimens are from an endemic Cuban species that became extinct before or shortly after the arrival of European explorers (Ford 1990), recent morphological analysis and radiometric dating indicate that they most likely represent a captive *A. fuscipes* that was brought

Spider monkey *(Ateles geoffroyi),* photo by Ernest P. Walker.

to Cuba by people within the last 300 years (MacPhee and Rivero 1996).

Head and body length is about 382–635 mm and tail length is about 508–890 mm. The weight is as much as 8 kg for captives, but it is probably nearer 6 kg for animals in the wild. Most forms have coarse, stringy hair, but some have a soft and fine pelage. Underfur is lacking. The coloration above is yellowish gray, darker grays, reddish brown, darker browns, or almost black; the sides may be golden yellow to rufous; and the underparts are lighter, usually whitish or yellowish. Most forms have a black face with white eye rings, but some forms have a flesh-colored face.

Spider monkeys have an exceptionally long tail and legs in relation to body length, a prehensile and extremely flexible tail, and thumbs that are poorly developed or lacking. The head is small, and the muzzle prominent. The clitoris of the female is greatly elongated, so that it is often mistaken for the penis of the male.

Spider monkeys are found in rain and montane forests and tend to occupy small branches of the high strata of the canopy. They are entirely diurnal, may feed intensively early in the morning, and rest for most of the remainder of the day (Hall and Dalquest 1963; Klein and Klein 1977). The diet consists largely of fruits, supplemented by nuts, seeds, buds, flowers, leaves, insects, arachnids, and eggs (Mittermeier and Coimbra-Filho 1977). In Surinam, Van

Roosmaien (1985) found groups of *A. paniscus* to move as much as 5,000 meters per day during the wet season (March–June), when fruit was abundant, and to have prolonged feeding periods at many feeding sites, apparently thereby building up energy reserves. During the subsequent dry season (July–September), when fruit was scarce, foraging covered as little as 500 meters per day, feeding times were short, and resting periods were long.

Probably only the gibbons exceed *Ateles* in agility in trees. Spider monkeys move swiftly through trees and use their tails as a fifth arm or leg. They pick up objects with the tail and can hang by a single hand or foot or by the tail alone. The normal movement through trees is along the upper surfaces of limbs, with the tail arched over the back. Only rarely do they descend to the ground. When approached in the wild, spider monkeys sometimes break off dead branches weighing nearly 5 kg and drop them, attempting to hit the observer. They also emit terrierlike barks when approached. The most frequently heard call resembles the whinnying of a horse and is made when the monkeys are separated.

Population densities of 1–35/sq km and home ranges of 1–4 sq km have been recorded (Eisenberg 1989; Wolfheim 1983). In studies of *A. paniscus* in Peru, Symington (1988) found group home ranges of 150–250 ha. Overlap with neighboring groups was only 10–15 percent. Females, particularly those with infants, tended to restrict their movements to a core area, 20–33 percent of the total group range, while males used the entire area. Apparently, the males were cooperatively defending this area as a joint territory. Although there generally was little contact between groups, at least some of the young females evidently emigrated to other groups. Males tended to stay with their natal unit. Van Roosmaien (1985) reported a similar pattern in Surinam and noted that older females apparently lead the foraging groups.

It is difficult to describe the social structure of these monkeys. Group size seems to vary considerably, partly with respect to habitat conditions, and there is a tendency for large bands regularly to divide into subgroups. Symington (1988) suggested that *Ateles* has a fission-fusion society such as has been described for the chimpanzee. Aggregations of up to 100 individuals have been reported, but the usual size range is 2–30 (Freese 1976; Hall and Dalquest 1963; Hernandez-Camacho and Cooper 1976; Izawa 1976; Leopold 1959; Mittermeier and Coimbra-Filho 1977; Napier and Napier 1967; Vessey, Mortenson, and Muckenhirn 1978). Probably the higher figures represent sightings when groups are united and the lower figures represent observations of subgroups. In a study in Colombia, Klein and Klein (1977) determined that in an 8-sq-km area there were three distinct and mutually exclusive social networks of 20 adults, almost always dispersed into subgroups averaging about 3.5 individuals each. In southeastern Peru, Durham (1971) found that average group size decreased sharply from 18.5 individuals at an elevation of 275 meters to 11 individuals at 576 meters and then less sharply to 7 at 889 meters and 4.5 at 1,424 meters. At higher altitudes there was only a single adult male in each group, but at 275 meters the number of adult males averaged 3.75.

There appears to be no regular breeding season. The estrous cycle is 24–27 days, estrus is 2 days, gestation is 200–232 days, the normal litter size is 1, birth weight is 300–500 grams, the interbirth interval averages about 34.5 months, lactation has been reported to last from 10 weeks to more than 2 years, and the age of sexual maturity is about 5 years in males and 4 years in females (Eisenberg 1977; Hayssen, Van Tienhoven, and Van Tienhoven 1993; Symington 1988). A captive *A. geoffroyi* lived for 48 years

and a specimen of *A. chamek* was still living at about 40 years (Marvin L. Jones, Zoological Society of San Diego, pers. comm., 1995).

In many parts of their range spider monkeys are avidly hunted for use as food. They are easily located because of their large size and noisy habits, and populations have been eliminated in some accessible areas (Hernandez-Camacho and Cooper 1976; Mittermeier and Coimbra-Filho 1977). Although they are relatively tolerant of a limited amount of logging and land clearing, they still require large tracts of tall forest and thus are vulnerable to human colonization (Leopold 1959). They formerly occurred throughout the tropical parts of Veracruz, Mexico, but now are found only in the south of that state (Hall and Dalquest 1963). Even in that area their range has become fragmented, and many populations have been exterminated (Estrada and Coates-Estrada 1984). The habitat of the two subspecies of *A. geoffroyi* in southern Mexico, *vellerosus* and *yucatanensis,* has now been reduced by at least 90 percent (Estrada and Coates-Estrada 1988). The subspecies *A. geoffroyi frontatus* and *A. geoffroyi panamensis,* of Central America, are listed as endangered by the USDI and are on appendix 1 of the CITES; all other species and subspecies are on appendix 2.

The IUCN classifies the species *A. marginatus* as endangered and the species *A. fuscipes* and *A. belzebuth* generally as vulnerable. However, the IUCN also classifies the following subspecies individually: *A. geoffroyi yucatanensis* (Mexico, Belize, Guatemala), vulnerable; *A. g. frontatus* (Nicaragua, Costa Rica), vulnerable; *A. g. ornatus* (Costa Rica), vulnerable; *A. g. panamensis* (Costa Rica, Panama), endangered, fewer than 2,500 individuals surviving; *A. g. azurensis* (Azuero Peninsula and adjacent parts of Panama), critically endangered, fewer than 250 individuals surviving; *A. g. grisescens* (eastern Panama and adjacent parts of Colombia), endangered, fewer than 2,500 individuals surviving; *A. fuscipes fuscipes* (Pacific slope forests of northern Ecuador), critically endangered, fewer than 250 individuals surviving, continuing to decline precipitously; *A. belzebuth brunneus* (Colombia), endangered; *A. b. hybridus* (northern Colombia and Venezuela), endangered. All of these species and subspecies are threatened by habitat destruction and hunting by people (Mittermeier et al. 1986; Thornback and Jenkins 1982). The same problems are leading to the rapid disappearance of *A. paniscus* in Surinam (Van Roosmaien 1985).

PRIMATES; CEBIDAE; Genus BRACHYTELES
Spix, 1823

Muriqui, or Woolly Spider Monkey

The single species, *B. arachnoides,* occurs in southeastern Brazil, formerly being found from the state of Bahia to the northern part of the state of Paraná (Cabrera 1957; Martuscelli, Petroni, and Olmos 1994). There are two distinct subspecies, *B. a. arachnoides* to the north and *B. b. hypoxanthus* to the south in the states of Sao Paulo and Paraná. Recent morphological assessments led Rylands, Mittermeier, and Luna (1995) to recognize *hypoxanthus* as a full species, though Strier (1995) continued to treat it as a subspecies.

Head and body length is 460–630 mm and tail length is 650–800 mm. Milton (1984) indicated that weight is 12–15 kg, which would make *Brachyteles* the largest New World monkey. Peres (1994) questioned this size range, noting that recent weights of live-captured animals have been

Woolly spider monkey *(Brachyteles arachnoides)*, photo by Andrew Young.

about 7–10 kg, but added that individuals of a Pleistocene population may have weighed up to 20 kg. Males and females of the living population are about equal in size (Strier 1992). The pelage is woolly, and the hairs of the head are short and directed backward. There may be a sexual difference in coloration, but this may be only individual and geographic variation. Most muriquis have a prevailing color of yellowish gray or ashy brown, but some are reddish. The head is blackish brown tinged with yellow, or dark gray washed with brown, or the forehead and back of the neck may be orange rufous and the top of the head chestnut. The flat, naked face is often brilliant red in color, particularly when the animal is excited.

The head is round, the body is heavy, the limbs are long and slender; the tail is longer than the body, naked beneath near the tip, and prehensile. *Brachyteles* resembles *Ateles* in such features as the long limbs. Its general appearance is even more like that of *Lagothrix*. The thumb in *Brachyteles* is vestigial or lacking, and the nails of the fingers and toes are sharp and laterally compressed. Lemos de Sá et al. (1990) found that *B. a. arachnoides* has a small but still obvious thumb, whereas *B. a. hypoxanthus* has no thumb at all.

The muriqui is found only in undisturbed high forest, including both lowland tropical and montane rainforest, and is predominantly arboreal (IUCN 1978). It long had been among the least known primates, but a series of recent studies have provided considerable new information. Strier (1992) found that during the hot summer groups begin to forage just after sunrise, rest in the shade during the middle of the day, and then move again in the late afternoon until sunset. During the winter they sun themselves for hours before beginning to forage and then are active

until about an hour before sunset. Daily movements cover about 1,000 meters in the winter and 1,400 meters in the summer. Most locomotion is through the trees, using the limbs and tail, but individuals occasionally come to the ground to drink or eat certain items. *Brachyteles* is strongly folivorous, spending about 51 percent of its annual foraging time feeding on leaves, 32 percent on fruits and seeds, and 11 percent on pollen and nectar. Fruit increases in importance at times, and individuals may remain for several days in areas with abundant fruit.

Strier et al. (1993) reported the results of a nine-year study of two groups of muriquis inhabiting an 800-ha. private reserve. At the start of the study the main study group had about 24 members, including 6 adult males and 8 adult females, and used a home range of 168 ha. The other group contained 18 individuals. The study group eventually grew in size to 42 members, mainly through the birth of 23 young, and its range nearly doubled in size. This group evidently formed an intact social unit for six years, but subsequently there was an increasing subdivision into two groups, which stayed apart for days at a time and increasingly interacted with the other original group on the reserve. The following additional observations of group and range sizes were listed by Torres de Assumpcao (1983): 7 or 8 individuals on 170 ha., 13 on 105 ha., 12 on 217 ha., and 30 on 580 ha. The group with 13 monkeys had 4 adult males, 4 adult females, and 5 young. Milton (1984, 1985) indicated that home ranges of females cover an estimated 70 ha., including a core area of 4 ha. where the animal often rests and sleeps, and that there is no territorial defense, though there is a dominance hierarchy among females. Milton's observations indicated somewhat smaller group size and looser social organization than found in Strier's study, but both agreed on a remarkable lack of aggression. Communication may be carried out through the depositing of urine on the hands and tail, and there are also eight known vocalizations, including a loud, piercing call for ascertaining the location of other individuals.

Strier (1990, 1992) reported that males and females are equal in rank and that there is little agonistic interaction between males. Receptive females attract numerous males and mate with several of them. In contrast to the behavior of most social primates, young male *Brachyteles* tend to remain in their natal group, but young females emigrate at about 5–6 years and join other groups. Females apparently are polyestrous and have an estrous cycle of about 4 weeks, an estrus of 2–3 days, and a gestation period of 7–8.5 months. Births are highly seasonal, being concentrated in the dry winter months (May–September). Females seem to give birth about once every 3 years. There usually is a single young, but twins have been observed on rare occasions. The infant initially is carried on its mother's belly or flanks but begins to ride on her back by 6 months. It is weaned at 18–30 months. The earliest known age of reproduction in a female is 7.5 years.

Although *Brachyteles* once presumably occupied all Atlantic coastal forests of eastern and southeastern Brazil, it is now only sparsely distributed in the states of Sao Paulo, Minas Gerais, Rio de Janeiro, and Paraná. Mittermeier (1987) estimated its total numbers at not over 400, compared with perhaps 400,000 in 1500. Martuscelli, Petroni, and Olmos (1994) reported recent surveys showing the current figure to be nearly twice as great and also cited still higher estimates but indicated that the species is continuing to decline because of hunting for food and clearing of forests for agriculture and settlement even within supposedly protected areas. More than 95 percent of the kind of forest on which it depends has been eliminated (Milton 1986). It is classified as endangered by the USDI and is on appendix 1 of the CITES. The IUCN recognizes *arachnoides* and *hypoxanthus* as separate species and designates both as endangered.

PRIMATES; CEBIDAE; Genus ALOUATTA
Lacépède, 1799

Howler Monkeys

There are nine species (Cabrera 1957; Froehlich and Froehlich 1987; E. R. Hall 1981; Hershkovitz 1972; Horwich 1983; Rylands, Mittermeier, and Luna 1995; Wolfheim 1983):

A. palliata, southern Mexico to northwest coast of South America, possibly as far south as Peru;
A. coibensis, Azuero Peninsula and Coiba Island of Panama;
A. pigra, Yucatan, Guatemala, Belize;
A. seniculus, Colombia, Venezuela, Guianas, eastern Ecuador, northeastern Peru, Amazonian Brazil, Bolivia;
A. arctoidea, northern Colombia and Venezuela;
A. sara, southeastern Bolivia;
A. belzebul, Amazonian Brazil and adjacent regions;
A. fusca, Bolivia, eastern Brazil, extreme northeastern Argentina;
A. caraya, eastern Bolivia, southern Brazil, Paraguay, northern Argentina.

Groves (*in* Wilson and Reeder 1993) included *A. arctoidea* in *A. seniculus*.

Howlers are among the largest of New World monkeys: head and body length is 559–915 mm, tail length is 585–915 mm, and adult weight generally ranges from 4 to 10 kg. Data compiled by Peres (1994) indicate that males are substantially larger than females and that those of *A. pigra* weigh up to 11.6 kg. The hair is coarse, and in some species it is long on the head and shoulders; the face is naked. Coloration is of three types—yellowish brown, deep reddish brown, and black—but there is considerable age and individual variation. Males are generally larger than females, and sometimes the sexes are dichromatic (Crockett and Eisenberg 1987). In *A. caraya* males average 6.7 kg and are all black, and females average 4.4 kg and are yellow-brown (Thorington, Ruiz, and Eisenberg 1984).

The angle of the lower jaw and the hyoid bone are both greatly enlarged, making it possible for the animal to produce such a remarkable voice. The lower jaw and neck are quite large, which, with other characteristics, such as the low facial angle, gives the animals a forbidding expression. The legs are shorter and stouter than those of the spider monkeys *(Ateles)*. The strongly prehensile tail is naked on the underside for the terminal third. The tail is so powerful that the animal may jump from a limb but check its flight by not releasing its tail hold. Also, if a monkey falls, it can keep from hitting the ground by grabbing onto a limb with its tail.

Howlers are arboreal, mainly diurnal forest dwellers. Mittermeier and Coimbra-Filho (1977) reported that *A. seniculus* inhabits cloud forest, rainforest, secondary forest, and mangroves. Hernandez-Camacho and Cooper (1976) stated that *A. seniculus* is an able swimmer. Freese (1976) observed that *A. palliata* prefers primary forest and usually avoids scrub forest. According to Carpenter (1965), the daily pattern of *A. palliata* on Barro Colorado Island in the Panama Canal Zone is as follows: a bout of roaring at dawn, rest, feeding in the immediate vicinity of the sleeping area, moving out to food trees in the midmorning, active feeding, rest till midafternoon, and then movement, with simultaneous vocalization, to the night lodging trees. C. C. Smith (1977) found that adults on Barro Colorado Island rested for 74 percent of the day and all of the night, fed for most of the remaining time, and used only 4 percent of their time for social activity. Glander (1978) reported a daily movement of 207–1,261 meters for *A. palliata*. Other reported day ranges for *Alouatta* average 123–706 meters (Crockett and Eisenberg 1987). The diet of *Alouatta* is varied but consists mainly of leaves, fruits, and other vegetable matter (Hernandez-Camacho and Cooper 1976; C. C. Smith 1977). According to Mittermeier and Coimbra-Filho (1977), this genus eats more leaves than does any other New World monkey. Crockett and Eisenberg (1987), however, suggested that *Alouatta* should be labeled a folivore-frugivore rather than simply a folivore.

In Venezuela, Sekulic (1982a) found that the abundance of fig trees was a principal determinant of group size for *A. seniculus*, with smaller groups being maintained when fewer figs were available. Four groups moved an average of 340–445 meters per day and slept in the tallest available trees at night. Population density in the area was more than 115/sq km, and yearly home range was 3.88–7.44 ha. Territoriality was not evident, with ranges overlapping by 32–63 percent. Groups contained an average of 8.9 (4–17) individuals, including 1–3 adult males and 2–4 adult females. Young of both sexes dispersed from their natal groups.

Klein and Klein (1976) reported population densities of about 12–30/sq km for *A. seniculus* and 12–15/sq km for *A. belzebul* in La Macarena National Park in Colombia. Thorington, Ruiz, and Eisenberg (1984) determined that a

Howler monkey *(Alouatta palliata)*, photo from New York Zoological Society.

population of *A. caraya* in northern Argentina had a density of 130/sq km and troop sizes of 3–19; groups included 1–3 adult males and 2–7 adult females. Baldwin and Baldwin (1976) reported a density of 1,040/sq km for *A. palliata* in southwestern Panama and noted that this figure was 12–21 times greater than densities found earlier on Barro Colorado Island. Crockett and Eisenberg (1987), however, pointed out that this figure was based on a relatively brief and unnatural situation and that other reported densities include 16–90/sq km for *A. palliata*, 8–22/sq km for *A. pigra*, and 15–118/sq km for *A. seniculus*. Group home ranges determined in Baldwin and Baldwin's study were 3.2–6.9 ha., and Crockett and Eisenberg cited other reported ranges of 10–60 ha. for *A. palliata*, 11–125 ha. for *A. pigra*, and 4–25 ha. for *A. seniculus*. In another investigation of *A. palliata*, Neville et al. (1976) found home ranges of 4.89–76.00 ha. Izawa (1976) reported that one group of *A. seniculus* used a nomadic range of 3–4 sq km. He also observed that group size averaged 5.4 and that there appeared to be 1 dominant male in the group. Various other reports indicate an overall range of 2–15 individuals in groups of *A. seniculus*, with 4–7 being most common (Hernandez-Camacho and Cooper 1976; Izawa 1976; Klein and Klein 1976; Mittermeier and Coimbra-Filho 1977; Neville et al. 1976; Vessey, Mortenson, and Muckenhirn 1978).

There have been a number of studies of social structure of *A. palliata* (Baldwin and Baldwin 1972, 1976; Carpenter 1965; Freese 1976; Heltne, Turner, and Scott 1976; Mittermeier 1973; Mittermeier and Coimbra-Filho 1977; C. C. Smith 1977). Group size in this species varies from 2 to 45, there generally being about 10–20 individuals. Each group contains about 2–4 adult males and 5–10 adult females. The males have a dominance hierarchy and appear to act in concert during travel and vocal battles with other groups. Fe-

males seem to maintain a peaceful relationship among themselves. A few males live alone but not in complete isolation. Groups generally are closed units, but sometimes they split up and new groups are formed. Occasionally several groups join in an aggregation of 38–65 animals. Most troops have little or no area of exclusive use, but they do defend the place where they happen to be at a given time. The dawn chorus of loud calls seems to function as a means of preventing direct contact between groups.

New studies (Crockett and Eisenberg 1987) indicate that altough troops of *A. palliata* have several adult males, other species of *Alouatta* usually have only 1–2 adult males and 2–3 adult females. Competition between males is severe, with invasions and infanticide by adult males reported for *A. palliata* and *A. seniculus*. Maturing individuals of both sexes usually emigrate from troops of these species, perhaps because of the limited opportunities for reproduction in an established hierarchy. As a result, the adults of both old and new groups have few or no close relatives therein.

All the species of this genus are particularly noted for their remarkably loud and persistent calls, described by some observers as deep howls and by others as deep, carrying growls comparable to the roars of lions. These calls have been heard by people 3 km away through the jungle and 5 km away across lakes. The cries are made by the males and may be emitted at any time the animals are active, though Horwich and Gebhard (1983) reported *A. pigra* to have intense periods of roaring at dawn and dusk during the dry season. Sekulic (1982b) suggested that the roars of *A. seniculus* elicit responses that allow it to assess neighboring groups, as an alternative to energetically expensive chases and fights. In addition to these calls, Carpenter (1965) identified a "deep cluck" used by a leader to coordinate group movement, "grunts" of adults when disturbed, and "wailing" by females to solicit help in retrieving separated young.

Alouatta appears to breed throughout the year, though Crockett and Rudran (1987*a*, 1987*b*) found *A. seniculus* in two habitats of Venezuela to have a consistently reduced birth frequency during the early wet season, in May–July; females there had an average interbirth interval of 17 months. Shoemaker (1982*a*) reported the interbirth interval in captive *A. caraya* to be 12–15 months for newly matured females and 7–10 months for older mothers. The estrous cycle in *A. palliata* is 13–24 days and gestation is 180–94 days (Eisenberg 1977). Figures for other species fall within the same range (Crockett and Eisenberg 1987). There normally is one young. That of *A. palliata* weighs 275–400 grams at birth (Hayssen, Van Tienhoven, and Van Tienhoven 1993). It clings to its mother's fur, and as it gets older it makes its permanent riding position on her back. This continues for about 1 year. Crockett and Rudran (1987*b*) found that 80 percent of the young of *A. seniculus* survived at least 1 year but that 44 percent of infant mortality was the result of infanticide by older males. Studies on Barro Colorado Island (Carpenter 1965; Froehlich, Thorington, and Otis 1981) indicate that the young of *A. palliata* are weaned at around 10 months and that sexual maturity is achieved at 3–4 years by females and 5 years by males but that several more years may pass before males attain social maturity and are allowed to mate. The average adult life span in this population is about 16 years, and maximum known longevity there is over 20 years. A captive *A. caraya* was still living at about 23 years (Marvin L. Jones, Zoological Society of San Diego, pers. comm., 1995).

The USDI lists *A. palliata* as endangered and *A. pigra* as threatened. Those two species also are on appendix 1 of the CITES, and all other species are on appendix 2. The IUCN classifies the species *A. fusca* and the subspecies *A. palliata mexicana* (southern Mexico, Guatemala) and *A. seniculus insulans* (Trinidad) as vulnerable, the subspecies *A. coibensis coibensis* (Coiba Island off Panama) as endangered, and the subspecies *A. coibensis trabeata* (Azuero Peninsula of Panama), *A. fusca fusca* (southeastern Brazil), and *A. belzebul ululata* (southeastern Brazil) as critically endangered. The last three of these subspecies have each been reduced to fewer than 250 individuals and are continuing to decline because of rapid human destruction of their limited habitat. Mittermeier et al. (1986) also considered the subspecies *A. fusca clamitans* of southeastern Brazil and adjacent Argentina to be endangered. Oliver and Santos (1991) reported that it still occupied an extensive range, but Di Bitetti et al. (1994) indicated that its range had been drastically reduced and fragmented. In addition, all species of *Alouatta* are hunted for food and are subject to commercial export. Mittermeier and Coimbra-Filho (1977) observed that *A. seniculus* was easy to locate because of its loud voice and had become rare in some areas but was still abundant in Amazonian Brazil.

Investigations by Estrada and Coates-Estrada (1984, 1988) and Horwich and Johnson (1986) indicate that habitable areas for *A. palliata* and *A. pigra* are rapidly being lost in Mexico and Central America and that troop sizes are falling as forests become fragmented. There was relatively little destruction of such habitat in southern Mexico until the 1940s, but in the last 40 years at least 90 percent of the original rainforests in the region have been converted to agriculture and pasture. The basic cause is rapid growth of the human population and a consequent demand for land to raise crops and graze livestock. Remaining monkey populations are small and scattered and could disappear entirely by the end of the century.

PRIMATES; CEBIDAE; **Genus PITHECIA**
Desmarest, 1804

Sakis

There are five species (Bodini and Pérez-Hernández 1987; Hershkovitz 1979, 1987*b*):

P. pithecia, southern and eastern Venezuela, Guianas, northeastern Brazil;

P. monachus, Amazon Basin of southern Colombia, Ecuador, Peru, and northwestern Brazil;

P. irrorata, south of the Amazon River in western Brazil, southeastern Peru, and northern Bolivia;

P. aequatorialis, known only from three areas in the upper Amazon Basin of eastern Ecuador and northeastern Peru;

P. albicans, south bank of the Amazon between the Jurua and Purús rivers in western Brazil.

According to Hershkovitz (1987*b*), the species now known as *P. monachus* was long called *P. hirsuta*, and the species now known as *P. aequatorialis* was previously identified as *P. monachus*.

Head and body length is 300–705 mm, tail length is 255–545 mm, and adult weight is generally 700–1,700 grams. *P. monachus*, *P. irrorata*, and *P. aequatorialis* are generally dark agouti in color with pale hands and feet. The underparts of *P. aequatorialis* vary from pale yellowish brown to rufescent; those of *P. monachus* and *P. irrorata* are predominantly blackish. The crown pelage of *P. aequatorialis* is short and stiff in males and long and lax in females, and the hood partly conceals the ears and face. *P. albicans* has a predominantly black back and tail, and the rest of the body is buffy to reddish. Male *P. pithecia* are uniformly blackish except that the crown, face, and throat are whitish to reddish; the hood does not cover the face. Female *P. pithecia* are predominantly blackish or brownish agouti and have a bright stripe extending from beneath each eye to the corner of the mouth or chin.

Sakis inhabit forests, usually at elevations of 210–750 meters. They are essentially diurnal and wholly arboreal but sometimes descend to the lower limbs of trees or even to bushes in search of food. Happel (1982) reported *P. monachus* to rest and feed mainly in the upper, emergent layers of the forest canopy, at heights of 15–24 meters, but to travel mostly at 10–19 meters. Sakis descend vertical supports tail first. They usually move about on all fours, but they are capable of making long leaps in trees and also have been observed running on horizontal branches in an erect posture with their arms held upward and their fingers extended. When alarmed, they can travel through trees with considerable speed. The pale-headed saki in its sleeping position coils up on a branch like a cat. Sakis, like many other monkeys, sometimes hang by their hind limbs when feeding. The diet consists of berries and other fruits, honey, leaves, flowers, small mammals such as mice and bats, and small birds. Sakis tear mammals and birds apart with their hands before eating them. In the Guianas, sakis have been observed going into tree hollows to collect bats, which they tore apart and skinned before eating.

Neville et al. (1976) reported a population density of 7.5–30.0/sq km in Peru. Oliveira et al. (1985) reported that *P. pithecia* occurred at a density of 40/sq km in Brazil and that groups used home ranges of 8–13.5 ha. These authorities and others (Hernandez-Camacho and Cooper 1976; Izawa 1976; Mittermeier and Coimbra-Filho 1977) indicate that sakis may occur alone or in small family groups com-

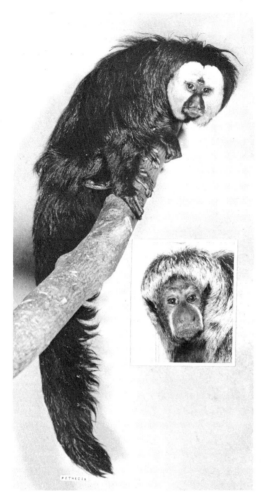

Saki *(Pithecia pithecia)*, male, photo by Ernest P. Walker. Inset: saki *(P. monachus)*, female, photo from Zoological Society of Philadelphia.

prising an adult pair and several young. In Guyana, Vessey, Mortenson, and Muckenhirn (1978) found group size of *P. pithecia* to average 3.3 and range from 1 to 5 individuals. In Peru, Happel (1982) observed four groups of *P. monachus* with a mean size of 4.5. Sakis are usually silent in captivity but emit a loud and penetrating call in the wild. Nervous and melancholy in appearance, they are essentially gentle in disposition and have made affectionate pets.

Well-developed young have been observed in Surinam at the beginning of the rainy season. Oliveira et al. (1985) indicated that a birth of *P. pithecia* evidently took place in late November or December. Johns (1986) recorded births of *P. albicans* in March, June, November, and December and noted that the young became totally independent after 6 months. Shoemaker (1982b) reported births of *P. pithecia* to occur throughout the year in captivity. He estimated the estrous cycle to last 18 days. A single young is produced. One infant was born to a male about 35 months old and a female about 26 months old. The gestation period of *P. pithecia* is 163–76 days (Eisenberg 1989). A specimen of *P. pithecia* living in the Pittsburgh Zoo at the end of 1994 was approximately 35 years old (Marvin L. Jones, Zoological Society of San Diego, pers. comm., 1995).

These monkeys are hunted for food and captured for use

as pets but do not appear to be endangered in Brazil (Mittermeier and Coimbra-Filho 1977). Even *P. albicans*, with its very limited range, is reported to be common and not threatened (Johns 1986), though Mittermeier et al. (1986) warned that it should be carefully watched. All species are on appendix 2 of the CITES, and the IUCN classifies the subspecies *P. monachus milleri*, of Colombia, as vulnerable.

PRIMATES; CEBIDAE; **Genus CHIROPOTES**
Lesson, 1840

Bearded Sakis

There are two species (Hershkovitz 1985):

C. albinasus, Brazil from the Amazon and Xingu rivers to the northern Mato Grosso;
C. satanas, southern Venezuela, Guianas, Brazil north of the Amazon, Brazil south of the Amazon and east of the Xingu.

Head and body length is 327–511 mm, tail length is 300–507 mm, and weight is 2,000–4,000 kg (Hershkovitz 1985). The species *C. albinasus* is shiny black in color. The nose and upper lip are covered with downward-growing, stiff, yellowish white hairs; the red skin in this area contrasts sharply with the blackness of the remainder of the body. The species *C. satanas* is colored as follows: the head and beard are black; the shoulders and back, including the flanks, vary from dark reddish chestnut to blackish brown; the limbs are black or chestnut; the hands and feet are blackish or reddish; and the tail is black. The beard of *Chi-*

Red-backed saki *(Chiropotes satanas)*, photo from New York Zoological Society.

ropotes is well developed, and the tail is so heavily furred that it is full and thick.

In Venezuela, Handley (1976) collected all 64 specimens of *C. satanas* in trees near streams within evergreen or deciduous forest. *C. albinasus* is reported to inhabit tropical rainforest in unflooded areas and sometimes flooded forest along riverbanks (IUCN 1978). The diet of the genus consists of fruits, nuts, flowers, leaves, apparently insects, and possibly small vertebrates (Ayres and Nessimian 1982; Mittermeier and Coimbra-Filho 1977; Mittermeier et al. 1983). Population densities of about 4–82/sq km have been recorded for *C. satanas* (Wolfheim 1983). In Guyana, Vessey, Mortenson, and Muckenhirn (1978) found group size in *C. satanas* to average 13.1 and range from 4 to 20 individuals. Eisenberg (1989) referred to troops of up to 25 and to one group with a home range of nearly 100 ha. Welker and Schäfer-Witt (*in* Grzimek 1990) wrote that the gestation period is about 5 months, there is a single young per birth, sexual maturity is attained at 4 years, and longevity is more than 18 years.

C. albinasus is classified as endangered by the USDI and is on appendix 1 of the CITES. It is declining because of the destruction of its forest habitat and excessive hunting by people (Thornback and Jenkins 1982). The subspecies *C. satanas satanas*, found in Brazil to the south of the Amazon and east of the Tocantins, is classified as endangered by the IUCN. Although reportedly intolerant of human disturbance, its range now is densely populated by people and subject to logging and other environmental disruption (Johns 1985). Mittermeier et al. (1986) regarded it as perhaps the most endangered Amazonian primate. The subspecies *C. satanas utahicki*, found from the Tocantins west to the Xingu, is classified as vulnerable by the IUCN.

PRIMATES; CEBIDAE; **Genus CACAJAO**
Lesson, 1840

Uakaris, or Uacaris

There are two species (Barnett and Da Cunha 1991; Boubli 1993; Cabrera 1957; Hernandez-Camacho and Cooper 1976; Hershkovitz 1972, 1987c):

C. melanocephalus, southeastern Colombia, southern Venezuela, northwestern Brazil;
C. calvus, upper Amazonian region of western Brazil, eastern Peru, and southern Colombia.

Head and body length is 300–570 mm, tail length is 125–210 mm, and weight is about 2.7–3.5 kg (Hershkovitz

Red uakaris *(Cacajao calvus rubicundus)*, photo from New York Zoological Society.

1987c). Uakaris are the only short-tailed American monkeys, the tail being less than half the length of the head and body, often much less. The face and cheeks are almost naked, but the jaws and throat are moderately haired, sometimes with a well-developed beard. In this genus the ears are large and may resemble those of *Homo*. In *C. calvus* the top of the head may appear nearly bald, having only a few scattered silky hairs, and the ears are exposed. In *C. melanocephalus* the top of the head is fully haired and the ears are concealed by fur.

There are several rather distinctive patterns of coloration (Hershkovitz 1987c). The subspecies *C. calvus calvus* of northwestern Brazil, known as the "white uakari," is whitish, yellowish, or buffy in color, except for the face, which is scarlet. The subspecies *C. calvus ucayali* of eastern Peru and immediately adjacent Brazil is entirely reddish orange or golden. *C. calvus rubicundus* of the Brazil-Colombia border region, sometimes called the "red uakari," is bright reddish brown to chestnut brown throughout, except for the strikingly paler nape of the neck, the hands, feet, and ears, which are brown, and the face and forehead, which are almost vermillion red. The subspecies *C. calvus novaesi*, found farther to the east, has the same general appearance, but the pale color of the nape extends down the middle of the back. *C. melanocephalus melanocephalus* of southern Venezuela and adjacent Brazil, the "black-headed uakari," is chestnut brown, except for the face, shoulders, arms, hands, feet, and lower surface of the tail, which are black. *C. melanocephalus ouakary* of northwestern Brazil and adjacent Colombia looks about the same, but the middle of the back is a contrasting pale orange, golden, or buffy.

Uakaris are more restricted in habitat than most South American primates; they seem to be found mostly along small rivers and lakes within forests and to avoid the margins of large rivers (Mittermeier and Coimbra-Filho 1977). They frequent the tops of large trees and rarely descend to the ground, at least during the part of the year when the forest floor is flooded. They are most active during the daytime and progress quite rapidly and nimbly on all fours but do not leap. Uakaris subsist chiefly on fruits but probably also eat leaves, insects, and other small animals.

In Brazil, Mittermeier and Coimbra-Filho (1977) observed groups of 10 to more than 30 individuals and heard reports that groups of 30–50 were not rare and that some troops contained as many as 100 animals. Fontaine and Du Mond (1977) reported that the members of a captive colony of *C. calvus rubicundus* in Florida established a linear dominance hierarchy through noisy fighting but without apparent injury. The breeding season in this and other captive groups in the Northern Hemisphere lasts from May to October. Females in the Florida colony were able to breed at 3 years, and one produced viable young until 11 years. Males became proven breeders at 6 years. Females gave birth to one young at a time, at intervals of about 2 years. One female appeared to be about 12 years old when brought into the colony and then lived for 11 more years. Hayssen, Van Tienhoven, and Van Tienhoven (1993) listed an estrous cycle of 14–48 days and a lactation period of about 21 months. According to Marvin L. Jones (Zoological Society of San Diego, pers. comm., 1995), a captive specimen of *C. calvus rubicundus* lived to at least 31 years.

Both species of *Cacajao* are listed as endangered by the USDI and are on appendix 1 of the CITES. The IUCN classifies the subspecies *C. calvus ucayali* as vulnerable and the subspecies *C. c. calvus, C. c. rubicundus*, and *C. c. novaesi* as endangered. All are declining because of human disruption of their limited habitat. *C. melanocephalus* is not now classified by the IUCN. *C. melanocephalus ouakary* of north-western Brazil and adjacent Colombia is the most widely distributed subspecies. Barnett and Da Cunha (1991) reported that it still is common, even in the vicinity of long-established villages, but that it is subject to heavy hunting pressure and is declining in those areas now being settled and disrupted. *C. calvus* has become rare in Peru because of excessive hunting and collection for use as a pet. The subspecies *C. calvus calvus* is naturally restricted to a very small area surrounded by rivers in northwestern Brazil. Although it is not often hunted at present, increased access to its range could lead to greater pressure by people and rapid extermination. Ayres and Johns (1987) reported the principal threat to *C. c. calvus* to be the growth of the timber industry.

PRIMATES; CEBIDAE; Genus **XENOTHRIX**
Williams and Koopman, 1952

Jamaican Monkey

The single species, *X. mcgregori*, is definitely known only by a subfossil partial mandible found in 1920 in Long Mile Cave in northwestern Jamaica. This specimen has been subject to much speculation; it sometimes was thought to be nothing more than the remains of a modern captive monkey, but Hershkovitz (1977) considered it to represent a distinctive New World family, the Xenotrichidae. The original describers (Williams and Koopman 1952) assigned it to the family Cebidae, and most subsequent authorities have agreed. Rosenberger (1977) and Rosenberger, Setoguchi, and Shigehara (1990) indicated affinity with *Aotus* and *Callicebus*. Ford (1986a, 1990), however, suggested that *Xenothrix* might be a callitrichid. Recently, MacPhee and Fleagle (1991) reported the existence of several postcranial bones apparently assignable to *Xenothrix*, found these specimens to show no evidence of affinity between *Xenothrix* and any other primate, and revived the position that the genus does indeed represent a separate family.

As in the living Callitrichidae (except *Callimico*), *Xenothrix* lacks a third lower molar, but the first and second molars are unlike those of marmosets, being bunodont, with the individual cusps greatly enlarged and the trigonid and talonid basins correspondingly reduced. The mandible is large and stout relative to that of the Callitrichidae, and the angle is expanded as in *Callicebus*. The mandible deepens posteriorly, also as in *Callicebus*, and the canine apparently was reduced, as in *Callicebus* and *Aotus*. The newly described postcranial elements indicate that the hind limbs were relatively short and robust.

The type specimen was discovered in detritus beneath an aboriginal kitchen midden in a rock shelter. The layer in which it was found has been dated by radiometric means at 2,145 years before the present (MacPhee 1984). The time of extinction and the natural history of the represented population are unknown. However, considering the morphology of the associated postcranial remains, MacPhee and Fleagle (1991) stated that *Xenothrix* was probably a heavy, slow-moving quadruped or climber. Those authorities also suggested that certain descriptions of living Jamaican animals by early European naturalists offer the possibility that *Xenothrix* externally resembled *Eulemur* or *Perodicticus* and that it may have survived at least until the eighteenth century.

PRIMATES; CEBIDAE; **Genus PARALOUATTA**
Rivero and Arredondo, 1991

Cuban Monkey

The single species, *P. varonai*, is known definitely only by a well-preserved and nearly complete skull found in 1987 and by a portion of a mandible of a different animal found in 1992 in a cave deposit in Cueva del Mono Fósil, Pinar del Río Province, western Cuba (Ford 1990; MacPhee 1993*a*; MacPhee et al. 1995; Rivero and Arredondo 1991; Salgado et al. 1992). Several additional teeth and postcranial elements found in the same vicinity are probably also referable to this species. Inclusion of *Paralouatta* here is provisional as the specimens have not yet been radiometrically dated, but associated fauna and conditions of discovery indicate a late Quaternary age.

The skull of *Paralouatta* is within the upper size range of those of living *Alouatta* and also superficially resembles those in the form of hafting of the neurocranium and face, the deep malar body, the marked lateral flaring of the maxillary root of the zygomatic process, and the low, posteriorly tapering cranial vault. The upper teeth of *Paralouatta* bear some resemblance to those of *Alouatta* and differ greatly from those of the possibly contemporary Cuban spider monkey, *Ateles anthropomorphus*. However, details of cusp morphology of the lower teeth appear to show less than expected correspondence with *Alouatta*. *Paralouatta* also differs from living *Alouatta* in having the foramen magnum more downwardly directed, the nuchal plane less vertically oriented, inconspicuous stylar shelves on the molars and premolars, three-rooted third and fourth upper premolars, and a completely incisiform lower canine. The cranial height is relatively great and the orbits relatively wide, being almost as large, relatively, as those of *Aotus*. Recent studies show that *Paralouatta* most closely resembles *Antillothrix* (see account thereof) in critical characters.

Mammalian remains found together with the new genus include almost exclusively those of living genera, such as *Solenodon* and *Capromys*, and very Recently extinct forms, such as *Nesophontes* and *Boromys*. It is therefore reasonable to think that, like those other mammals, *Paralouatta* survived until less than 5,000 years ago and became extinct in association with the human invasion and environmental disruption of the West Indies.

PRIMATES; CEBIDAE; **Genus ANTILLOTHRIX**
MacPhee, Horovitz, Arredondo, and Jiménez Vasquez, 1995

Hispaniolan Monkey

The single species, *A. bernensis*, is known only by skeletal remains found at Cueva de Berne and another site in eastern Dominican Republic and at and around Trou Woch Sa Wo in southwestern Haiti. This species, described in 1977, was then thought to represent the genus *Saimiri* and subsequently was associated with *Cebus* (see accounts of those two genera and of the family Cebidae), but detailed new morphological analyses by MacPhee et al. (1995) show that it represents a separate genus most closely related to the also extinct *Paralouatta* and the living *Callicebus*. These new studies also indicate that the living *Alouatta* is not a close relative of any of the other genera mentioned here.

The outstanding diagnostic feature of *Antillothrix* is a crest on the upper first and second molar teeth, not consis-

tently present in any other anthropoid. This "distal crest" runs directly distally from the protocone but is not distinct from it. *Antillothrix* also differs (1) from *Saimiri* and *Paralouatta* in lacking a continuous distal wall of the trigon running from protocone to metacone on the second upper molar; (2) from *Saimiri*, *Cebus*, and *Callicebus* in displaying a relatively lingual intersection of the protolophid and oblique cristid on the first lower molar; (3) from all of the other genera mentioned except *Paralouatta* in possessing a mesially projecting lobe of the lingual cingulum on the fourth upper premolar and an oblique alignment (relative to the midsagittal plane) of the protocone and hypocone of the first upper molar; and (4) from all of the other genera mentioned in the positioning of the infraorbital foramen above the interval between the third and fourth upper premolars.

The exclusive resemblances between *Antillothrix* and *Paralouatta* in the above critical dental characters are indicative of close phylogenetic relationship. In contrast, the resemblance between *Paralouatta* and *Alouatta* in cranial features is probably the result of convergence. The original association of the latter two genera was based largely on a skull with such worn teeth that no details of cusp morphology could be discerned.

Radiocarbon dates associated with the remains of *Antillothrix* range from about 3,850 to 9,850 years before the present. The genus still would have been present at least 2,000 years into the period of human occupation of Hispaniola. Details of its natural history and the exact time and cause of its extinction are not known.

PRIMATES; CEBIDAE; **Genus CALLICEBUS**
Thomas, 1903

Titi Monkeys

Hershkovitz (1988*a*, 1990*c*) recognized 13 species:

C. modestus, upper Beni River Basin in western Bolivia;

C. donacophilus, central and southern Bolivia, northern and western Paraguay;

C. olallae, upper Beni River Basin in western Bolivia;

C. oenanthe, upper Mayo River Valley in north-central Peru;

C. cinerascens, Madeira River Basin in central Brazil;

C. hoffmannsi, south of the Amazon and between the Canuma and Tapajós rivers in central Brazil;

C. moloch, south of the Amazon and between the Tapajós and Tocantins-Araguaia rivers in central Brazil;

C. brunneus, upper Madeira and upper Purús river basins in western Brazil and eastern Peru;

C. cupreus, upper Amazon and Orinoco basins in central and southern Colombia, eastern Ecuador, eastern Peru, and western Brazil;

C. caligatus, south of the Solimoes and between the Ucayali and Madeira rivers in western Brazil, eastern Peru, and extreme northern Bolivia;

C. dubius, south of the Amazon and between the lower Purús and Madeira rivers in central Brazil;

C. personatus, states of Bahia, Espirito Santo, Minas Gerais, Rio de Janeiro, and Sao Paulo in southeastern Brazil;

C. torquatus, upper Amazon and southern Orinoco basins in Venezuela, southern Colombia, eastern Ecuador, northeastern Peru, and northwestern Brazil.

In an earlier assessment of the genus, Hershkovitz (1963*c*) recognized only three species, *C. personatus* and *C. torqua-*

CALLICEBUS

tus, distributed largely as given above, and *C. moloch,* which essentially comprised the ranges of all the other species listed above.

Head and body length is 230–460 mm and tail length is 260–560 mm. Jones and Anderson (1978) stated that the weight range of *C. moloch* is 510–730 grams, and Hershkovitz (1990*c*) indicated that the genus can weigh up to 2,000 grams. The nonprehensile tail is well haired and sometimes tufted at the tip. The pelage consists of long, soft hairs. Coloration is variable; it ranges from reddish gray or yellowish through reddish brown to black on the upper parts and is usually paler on the underparts. A light-colored collar band is present on some species. Titi monkeys may have a black or white band on the forehead. These monkeys, with their harmonious colors and thick, soft fur, are attractive animals. They have a small, rounded head and a somewhat flattened, high face. Among the diagnostic characters pointed out by Hershkovitz (1990*c*) are: supraorbital ridges well defined, frontal bone above and between ridges depressed or flattened; interorbital septum broad, pneumatized, not perforated; dental (palatal) arcade more nearly V-shaped than U-shaped; mandibular angle greatly expanded; first upper incisor tooth with cutting edge bluntly pointed; second upper incisor staggered behind first; and molars heavy and brachyodont, their cusps high and sharp, buccal and secondary lingual cingula usually well developed. In general form and behavior titi monkeys bear some resemblance to *Aotus.*

Titis are arboreal and diurnal forest dwellers. In Venezuela, Handley (1976) found all specimens of *C. torquatus* in trees near streams or in other moist parts of evergreen forest. Jones and Anderson (1978) stated that *C. moloch* inhabits low and dense tropical forests, especially near riverbanks, and frequently occurs in understory vegetation. Hernandez-Camacho and Cooper (1976) reported *C. moloch* to occur mostly at heights of 3–8 meters, but it is said to descend to the ground occasionally. When progressing on all four limbs *Callicebus* hunches its body, and the hind limbs are rotated outward. In the characteristic resting position the body is hunched and all four limbs are brought together on the branch, with the tail hanging vertically down. From this position a titi monkey can jump by

Dusky titi monkey *(Callicebus moloch),* photo from New York Zoological Society.

rapid extension of the hind limbs while it stretches its arms outward to enable the hands to grasp another branch. In the usual sleeping position the body is extended, the side of the head resting on the folded hands. The hands are used to convey food to the mouth.

There is an intensive feeding period in the early morning, a major resting period at midday, considerable additional feeding after 1500 hours, and selection of a sleeping tree before dark (Jones and Anderson 1978). The diet consists mostly of fruits but also includes leaves, other vegetation, insects, other small invertebrates, birds' eggs, and small vertebrates (Kinzey 1977; Mittermeier and Coimbra-Filho 1977). A captive specimen of *C. personatus* exhibited considerable ingenuity; it was fond of cockroaches, and if these hid in a crevice, it used pieces of straw to pry them out. This individual also deliberately moved his sleeping box, presumably to obtain cockroaches that accumulated beneath.

Population densities of 2–31/sq km have been recorded (Wolfheim 1983). Mason (1968) reported a daily movement of 315–570 meters for *C. moloch*, and home ranges of three groups were 3,201, 5,093, and 4,762 sq meters. There was seldom more than 20 percent overlap between the ranges of different groups. Mason (1971) noted that each group of *C. moloch* occupied a small, well-defined, stable home range. There were regular territorial confrontations between groups where their ranges overlapped. These engagements involved displays, vocalizations, and vigorous chasing, but physical fighting was rare and never severe. Kinzey et al. (1977) reported that *C. torquatus*, like *C. moloch*, was territorial but that *C. torquatus* had a much larger group home range, about 20 ha.

Titi groups consist of two to seven individuals, including a strongly bonded pair of adults and their offspring (Jones and Anderson 1978; Kinzey 1977; Mittermeier and Coimbra-Filho 1977). The adult male searches for food, leads group movements, and also usually carries the infant when it is not being nursed by the female. There is a wide range of visual signals and vocalizations for communication. The intertwining of tails occurs frequently when two or more animals of the same group sit side by side. Litter size in this genus is one. Births in wild *C. moloch* occur from December to April. Birth weight is about 70 grams, and adult weight is attained at around 10 months. Hayssen, Van Tienhoven, and Van Tienhoven (1993) listed a gestation period of 5–6 months, a lactation period of 12–16 weeks, and an interbirth interval of 1 year. According to Marvin L. Jones (Zoological Society of San Diego, pers. comm., 1995), a specimen of *C. moloch* was still living in captivity at an age of at least 25 years.

All species of *Callicebus* are probably declining (Wolfheim 1983) and are on appendix 2 of the CITES. The IUCN gives an overall classification of vulnerable to the species *C. personatus* and singles out the subspecies *C. p. barbarabrownae* as critically endangered. *C. personatus* is restricted to the Atlantic coastal forests of eastern Brazil, at least 95 percent of which have been cut down for lumber and charcoal production and to make way for agriculture and pasture. The monkeys occur at low densities in widely scattered forest fragments that continue to decline as human settlement and activity proceeds (Mittermeier et al. 1986; Oliver and Santos 1991). The IUCN also now classifies *C. dubius, C. oenanthe, C. cupreus ornatus* of central Colombia, and *C. torquatus medemi* of southern Colombia as vulnerable. Each of these taxa occupies a relatively small area and is immediately jeopardized by human activity.

PRIMATES; CEBIDAE; Genus AOTUS
Humboldt, 1811

Douroucoulis, or Night Monkeys

Rylands, Mittermeier, and Luna (1995), based largely on the work of Hershkovitz (1983), recognized two species groups and ten species:

primitive gray-neck group

A. lemurinus, Panama, northern and western Colombia, western Venezuela;

A. brumbacki, eastern Colombia;

A. hershkovitzi, known only from the type locality on the east side of the Andes in central Colombia;

A. trivirgatus, southern Venezuela, north-central Brazil;

A. vociferans, southern Colombia, eastern Ecuador, northeastern Peru, northwestern Brazil;

derived red-neck group

A. miconax, north-central Peru;

A. nigriceps, eastern Peru, western Brazil;

A. nancymaae, northeastern Peru, northwestern Brazil;

A. infulatus, east-central Brazil;

A. azarai, Bolivia, Paraguay, northeastern Argentina.

Groves (*in* Wilson and Reeder 1993) accepted the above and used the spellings *A. nancymaae* and *A. azarai*, respectively, in place of the more commonly used *A. nancymae* and *A. azarae*. Based on analysis of cranial morphology, pelage characters, karyology, and blood protein variation, Ford (1994) generally accepted the above arrangement but concluded that *A. lemurinus* and *A. brumbacki* are part of *A. vociferans* and that the population of *A. azarai* in Bolivia actually is referable to *A. infulatus*; she also questioned whether *A. infulatus* is even specifically distinct from *A. azarai* and whether *A. nancymaae* is distinct from *A. miconax*. Pieczarka et al. (1993) reported chromosomal evidence that *A. azarai* and *A. infulatus* form a single intergrading species. Corbet and Hill (1991) considered the entire gray-neck group to compose only one species, *A. trivirgatus*, and the entire red-neck group to compose only one species, *A. azarai*. Other authorities, including Rathbun and Gache (1980) and Thorington and Vorek (1976), have recognized only the single species *A. trivirgatus*. E. R. Hall (1981) listed an additional species, *A. bipunctatus*, of the Azuero Peninsula of Panama, but stated that it was almost certainly no more than a subspecies of *A. trivirgatus*. Hall also noted that an old record of *Aotus* from Nicaragua probably involved a pet monkey that had been brought from elsewhere. On the basis of studies of chromosomes, Brumback (1974) suggested that there were three isolated species: *A. griseimembra* in Central America and northern South America, *A. trivirgatus* in the Amazon Valley, and *A. azarai* in the Parana-Paraguay River Valley in south-central South America.

In addition to this diversity of opinion regarding the intrageneric composition of *Aotus*, there is profound disagreement with respect to the proper phylogenetic position of the genus itself. Most authorities place it in the Cebidae, or sometimes in the Atelidae if the latter is recognized as a separate family, and affinity with *Callicebus* is often suggested. Rosenberger, Setoguchi, and Shigehara (1990) regarded *Aotus* as part of a tribe, together with *Callicebus* and *Xenothrix*, that in turn is part of the subfamily Pitheciinae. In contrast, Groves (*in* Wilson and Reeder 1993), Hershkovitz (1977), and Schneider et al. (1993) considered *Aotus* to represent a monotypic subfamily, the Aotinae. Earlier, Groves (1989) had treated this group as a full family, the Aotidae, and suggested that it might be the most primitive component of all the Platyrrhini (New World primates). Utilizing various morphological, immunological, and behavioral evidence, Tyler (1991a) suggested that *Aotus* actually diverged from both the Platyrrhini and Catarrhini (Old World monkeys, apes, and humans) before the latter two groups separated from one another and that it is thus the most primitive member of the entire Anthropoidea.

Head and body length is 240–370 mm and tail length is 316–400 mm. Weight is about 0.6–1.0 kg. The pelage is

Douroucoulis, or night monkeys (*Aotus* sp.), inset showing baby riding on mother's back, photos by Ernest P. Walker.

short, dense, semiwoolly, and soft. Coloration varies but is usually silvery gray to dark gray above and gray, buff, or brownish beneath. The markings on the face are usually three dark brown or black lines separated and bordered by grayish areas, but in some this pattern is only faintly indicated or is lacking. The head is relatively round and the eyes are very large. The body is so heavily covered with fur that it appears to be short and thick. The densely furred, slightly club-shaped tail is not prehensile. In most forms the ears are small and almost completely concealed in the fur. A sac under the chin can be inflated at will and gives resonance to the voice.

Douroucoulis live in primary and secondary forests from sea level to about 3,200 meters (Aquino and Encarnación 1994). In Colombia, Hernandez-Camacho and Cooper (1976) found them to be present at all levels in every major forest type except mangrove. Their day nests were located in tree hollows and/or woody climbing vines in accumulations of dry leaves and twigs. Aquino and Encarnación (1994) found *A. vociferans* only in tree holes but found *A. nancymaae* in a variety of sleeping sites, including tree hollows, thick vegetation on branches, and termite nests. *Aotus* is nocturnal; it has good vision at night and is adept at running on tree limbs, leaping, and performing remarkable acrobatics. Wright (1994) reported that the average nightly foraging path of a group was 708 meters, being about twice as long on moonlit nights as on nights with no moon. Mittermeier and Coimbra-Filho (1977) stated that the diet consists mainly of fruits, nuts, leaves, bark, flowers, gums, insects, and small vertebrates.

In Peru, Aquino and Encarnación (1986) reported an average population density of 8.75 groups or 25 individuals per sq km. Aquino and Encarnación (1994) listed other estimates of density, ranging from about 8 to 242 individuals and 2 to 70 groups per sq km, and indicated that the lower figures resulted from hunting pressure, deforestation and decreasing food resources, and cold temperatures at higher elevations. Thorington, Muckenhirn, and Montgomery (1976) determined that a radiotracked individual on Barro Colorado Island spent 72 percent of its time in an area of 800 sq meters. Wright (1994) indicated that groups maintain territories averaging 9.2 ha. in rainforests and 5 ha. in dry forest and that aggressive interaction sometimes is seen along boundaries.

Douroucoulis are usually seen in family groups of two to five individuals, including an adult pair and their young of up to three reproductive seasons (Aquino and Encarnación 1986, 1994; Izawa 1976). Subadults disperse from the natal group at about three years (Wright 1994). Any attempt to manage captives other than mated pairs invariably results in stress, conflict, and injury (Meritt 1980). The alarm or danger call is "wook," and individuals give a deep resonant call of several seconds' duration by inflating the throat sac. The night calls are referred to variously as squeaks, hisses, and barks and are sometimes likened to the calls of cats *(Felis)*. Ernest P. Walker learned from a pet that *Aotus* utters perhaps as many as 50 vocal sounds.

In northern Argentina, Rathbun and Gache (1980) found no evidence of a marked birth season. Most births in Peru occur from November to January (Aquino and Encarnación 1994). Observations of individuals in captivity (Dixson 1994; Gozalo and Montoya 1990; Meritt 1977) indicate that the estrous cycle lasts about 16 days, the gestation period is 133–41 days, females are capable of giving birth every eight months but average interbirth interval is about 12.7 months, there is almost always a single young, birth weight is about 96 grams, and sexual maturity is usually attained at about 3 years in both sexes. A female more than 13 years old and a male more than 11 years old were still reproductively active. Fathers play a major role in carrying and otherwise providing for the infant (Wright 1994). According to Marvin L. Jones (Zoological Society of San Diego, pers. comm., 1995), a captive *A. trivirgatus* was still living at about 27 years.

Although *Aotus* seems more tolerant of human activity than many other New World primates, its habitat is being rapidly reduced in some areas through the total elimination of forests. It also is killed for its meat and fur and has been collected extensively for the pet trade and biomedical research (Wolfheim 1983). *Aotus* has proved to be of particular value in studies of malaria and diseases of the eyes (Baer 1994). Importation to the United States declined substantially after federal restrictions were implemented in the 1970s, but it continues and most animals that are captured actually die before leaving the country of origin (Aquino and Encarnación 1994). All species are on appendix 2 of the CITES, but none are listed as endangered or threatened by the USDI. The IUCN now classifies the species *A. lemurinus, A. brumbacki,* and *A. miconax* as vulnerable and the subspecies *A. lemurinus griseimembra,* of Colombia and western Venezuela, as endangered. Aquino and Encarnación (1994) expressed concern for the restricted habitat of those taxa, as well as that of *A. nigriceps* and *A. trivirgatus.*

PRIMATES; CEBIDAE; Genus CEBUS
Erxleben, 1777

Capuchins, or Ring-tail Monkeys

There are six currently recognized species (Cabrera 1957; Cameron et al. 1989; Goodwin and Greenhall 1961; E. R. Hall 1981; Hershkovitz 1972; McCarthy 1982b; Ottocento et al. 1989; Queiroz 1992; Redford and Eisenberg 1992; Rylands and Luna 1993; Rylands, Mittermeier, and Luna 1995; Wolfheim 1983):

C. olivaceus, northern Colombia, Venezuela, Guianas, northern Brazil to north of the Amazon;
C. kaapori, small area of northeastern Brazil south of the lower Amazon along the Gurupi River;
C. capucinus, Belize and Honduras to western Colombia and Ecuador;
C. albifrons, the west coast of Colombia and Ecuador, parts of eastern Colombia, western and southern Venezuela, Trinidad, eastern Peru, northern Bolivia, much of Amazonian Brazil;
C. apella, eastern and southern Colombia, southern Venezuela, Margarita Island, the Guianas, eastern Ecuador and Peru, most of Brazil, Bolivia, Paraguay, northern Argentina;
C. xanthosternos, state of Bahia in southeastern Brazil.

Groves (*in* Wilson and Reeder 1993) included *C. xanthosternos* in *C. apella* and did not distinguish *C. kaapori* from *C. olivaceus.* Subfossil fragments, some of which have radiocarbon dates as recent as 3,860 years ago, suggested

the former presence of still another species of *Cebus* on the island of Hispaniola (Ford 1990; Hershkovitz 1988b; MacPhee and Woods 1982) but recently have been assigned to the new genus *Antillothrix* (see account thereof). Use of the name *C. olivaceus* in place of *C. nigrivittatus* is in accordance with Corbet and Hill (1991).

Head and body length is 305–565 mm, tail length is 300–560 mm, and weight is about 1,100–4,300 grams. Capuchins can be divided into two groups with respect to pelage. Those in the first group *(C. apella* and *C. xanthosternos)* have tufts, that is, horns of hair over the eyes or ridges of hair along the sides of the top of the head. These monkeys are usually without pronounced contrasting color pattern, the color normally ranging through grayish browns. The monkeys of the other group *(C. capucinus, C. olivaceus, C. kaapori,* and *C. albifrons)* do not have tufts on the head. The more characteristically marked are white on the face, throat, and chest and black elsewhere. From this pattern they vary toward pale yellowish grays and grayish browns, some clearly showing the basic pattern and others having but little evidence of the color pattern. In all the body and tail are well haired, some having silky or shiny hair. A well-defined thumb is present. The tail is slightly prehensile and often carried coiled at the tip, hence the name "ring-tail." The tail is strong enough to support the weight of the animal.

Like most other New World primates, capuchins are arboreal and diurnal. According to information summarized by Hernandez-Camacho and Cooper (1976) and Mittermeier and Coimbra-Filho (1977), all *Cebus* are flexible in choice of habitat. *C. apella,* which has the largest range of any New World monkey, is especially adaptable and occurs in nearly all kinds of humid forest, both seasonally flooded and nonflooded, up to an elevation of 2,700 meters. It is found in broken forest and secondary growth and crosses open ground when traveling from one forest segment to another. *C. albifrons* also can live in secondary forest and even has been collected in mangroves. *C. olivaceus* is found in dry forest on the llanos of Venezuela as well as in mature tropical forest in the Guianas. *C. capucinus* lives in a variety of forest types and in the western Andes occurs to an elevation of 2,100 meters. In Costa Rica it has been seen in every kind of forest and even in mangroves and sparsely forested areas (Freese 1976). The diet of *Cebus* is more varied than that of any other New World monkey. Foods include fruits, nuts, berries, seeds, flowers, buds, shoots, bark, gums, insects, arachnids, eggs, small vertebrates, and even certain kinds of marine life, such as oysters and crabs, found in coastal areas. Antinucci and Visalberghi (1986) reported that a captive *C. apella* used various objects as tools to crack nuts open. Visalberghi (1990) also cited cases of *Cebus* using sticks and other tools for various purposes but concluded that this genus does not actually have the ability, as found in *Pan,* to understand a problem and mentally devise a tool that will provide a solution.

Population densities reported vary from about 1/sq km to 111/sq km (Klein and Klein 1976; Neville et al. 1976; Wolfheim 1983). Izawa (1976) estimated that a nomadic group of *C. albifrons* in the upper Amazon used a range of 3 sq km. A group of 35 *C. albifrons* occupied a home range of 110–20 ha. in eastern Colombia and moved up to 5 km per day (Defler 1979a). Groups of this species are strongly intolerant of one another and defend largely exclusive territories. In contrast, home ranges of *C. apella* have been found to overlap by at least 40 percent, and groups may feed side by side with no apparent antagonism (Defler 1982). In Panama, Baldwin and Baldwin (1976) determined that a group of 27–30 *C. capucinus* had a home range of 24–40 ha. They also observed that in coastal areas where *C.*

A. White-fronted capuchin *(Cebus albifrons)*, photo from New York Zoological Society. B. White-faced sapajou *(C. capucinus)*, photo from New York Zoological Society. C. Weeping capuchins *(C. apella)*, photo from San Diego Zoological Garden.

capucinus was not hunted or harassed it traveled in groups of 20–30 individuals that would make bold displays at humans but that in disturbed inland forests the monkeys lived in small bands that fled silently from people. Overall group size ranges reported for the four species of *Cebus* are: *C. capucinus*, 6–50; *C. apella*, 1–20; *C. olivaceus*, 6–15; and *C. albifrons*, 10–35 (Defler 1979a, 1979b, 1982; Freese 1976; Hernandez-Camacho and Cooper 1976; Klein and Klein 1976; Neville et al. 1976; Oppenheimer 1969; Vessey, Mortenson, and Muckenhirn 1978). Several hundred individuals of *C. albifrons* were once observed to congregate to feed on fruits.

In general, groups appear to contain more adult females than males but to be dominated by a large older male. Defler (1982) reported that groups of *C. albifrons* were multimale and that one group had 7 adult and subadult males, but two groups of *C. apella* had only 1 adult male. Izawa (1980) found groups of the latter species to have a strong dominance hierarchy for both sexes and indicated that rank order was less rigid in other species. Capuchins have a variety of chatters, squeaks, shrieks, and other sounds for communication. Oppenheimer and Oppenheimer (1973) noted 11 different vocalizations for *C. olivaceus* in Venezuela. These calls were divided into four fundamental categories according to their purpose: (1) gaining or maintaining contact with the troop, (2) gaining or maintaining contact with an individual, (3) terminating a potential or actual attack, and (4) alerting the troop to potential danger from without.

Hernandez-Camacho and Cooper (1976) stated that *Cebus* evidently is not a seasonal breeder in Colombia. Oppenheimer (1969), however, reported most births of *C. capucinus* on Barro Colorado Island, Panama Canal Zone, to occur in the dry season or the early rainy season. In *C. apella* the estrous cycle is 16–20 days, estrus lasts 2–4 days, gestation is 180 days, weight at birth is about 200–250 grams, and lactation lasts 9 months (Eisenberg 1977; Hayssen, Van Tienhoven, and Van Tienhoven 1993). Female *Cebus* normally give birth to a single young; immediately thereafter the infant clings tightly to its mother's hair with both hands and feet. As it gains strength it moves about on the mother or leaves her for short periods to explore nearby, but it nurses for at least several months. When a helpless young loses its mother or becomes separated from her, other members of the troop often come in response to its distress cries and assist it. Oppenheimer (1969) observed that full size and sexual maturity were reached by female *C. capucinus* at 4 years and by males at 8 years. According to Marvin L. Jones (Zoological Society of San Diego, pers. comm., 1995), a captive *C. capucinus* lived about 55 years and a specimen of *C. apella* lived in captivity for more than 45 years.

Capuchins are such vivacious, intelligent little creatures that they have become the most numerous monkeys in captivity in both the United States and Europe. They were the monkeys most often used by itinerant organ grinders. More recently they have been trained to assist human quadripeligics. They must be kept fairly warm and free from drafts, receive a wide variety of food, and have an opportunity for plenty of exercise and sunlight. They are so active and mischievous that they not only are first-class entertainers but usually become serious nuisances if allowed freedom in a home. Importation for such purposes now is illegal.

Four of the species of *Cebus* are generally common, but all are on appendix 2 of the CITES and the survival of certain subspecies is in question. The IUCN classifies *C. kaapori* as vulnerable and *C. xanthosternos* as critically endangered; fewer than 250 individuals of the latter are thought to survive. *C. alibifrons trinitatis*, of Trinidad, and *C. apella margaritae*, of Margarita Island, also have populations of fewer than 250 and also are classed as critically endangered. *C. capucinus curtus*, of Colombia, and *C. apella robustus*, of southeastern Brazil, are designated as vulnerable. The forms *robustus* and *xanthosternos* are restricted to remnants of the Atlantic forest region, are jeopardized by widespread habitat destruction and hunting, and may soon be entirely eliminated outside of protected reserves (Mittermeier et al. 1986; Oliver and Santos 1991). The other forms classified by the IUCN all have restricted ranges in areas that are undergoing severe habitat disruption. *C. o. kaapori* may be in particular jeopardy as its range is small, isolated, and likely to be affected by a massive mining project that will be fueled through extensive local charcoal production (Ferrari and Queiroz 1994).

PRIMATES; CEBIDAE; Genus SAIMIRI
Voigt, 1831

Squirrel Monkeys

Based on the studies of Ayres (1985), Hershkovitz (1984b), and Thorington (1985), there are at most five living species:

S. boliviensis, upper Amazon Basin in western Brazil, eastern Peru, and Bolivia;
S. vanzolinii, known only from the vicinity of the confluence of the Amazon and Japura rivers in west-central Brazil;
S. sciureus, east of the Andes from Colombia and northern Peru to northeastern Brazil;
S. ustus, south of the Amazon in central Brazil;
S. oerstedii, a small area on the Pacific coast of Costa Rica and Panama.

Groves (*in* Wilson and Reeder 1993) accepted the content of *Sciureus* as listed above. However, based on analyses of pelage characters, dental morphology, biochemical data, and behavior, Costello et al. (1993) concluded that *S. oerstedii*, with a range as given above, is the only species distinct from *S. sciureus*. Thorington (1985) considered *S. boliviensis* and *S. oerstedii* to be subspecies of *S. sciureus* and used the name *S. madeirae* in place of *S. ustus*. Silva et al. (1992) reported natural hybridization between *S. sciureus* and *S. boliviensis* in the Amazon Basin of Peru. Ayres's (1985) description of *S. vanzolinii* was published subsequent to the work of Hershkovitz and Thorington. Another species, *S. bernensis*, originally was based on a few subfossil fragments with an estimated age of 3,860 years B.P. found on the island of Hispaniola. Subsequent studies suggested that this species may actually belong to the genus *Cebus* (Ford 1986a, 1986b; MacPhee and Woods 1982), and it recently was assigned to a new genus, *Antillothrix* (see account thereof).

Head and body length is usually 260–360 mm, tail length is 350–425 mm, and weight is 750–1,100 grams. The pelage is short, thick, soft, and brightly colored. The skin of the lips, including the area around the nostrils, is black and nearly devoid of hair. The most common coloration is: white around the eyes and ears and on the throat and sides of the neck; the top of the head black to grayish; the back, forearms, hands, and feet reddish or yellow; the shoulders and hind feet suffused with gray; the underparts whitish or light ochraceous; and the tail bicolored like the body except that the terminal part is black. The markings and coloration of the sexes and of animals of different ages are similar.

Squirrel monkey *(Saimiri sciureus)*, photo by Ernest P. Walker. Inset photo by John G. Vandenburgh.

The eyes are large and set close together, and the large ears are shaped about as in people. The thumb is short but well developed. The tail is covered with short hair, slightly tufted at the tip, and not prehensile. There is no laryngeal pouch. The facial part of the skull is small in proportion to the cranial part.

These monkeys are found in primary and secondary forests and in cultivated areas, usually along streams. They are diurnal and arboreal but occasionally descend to the ground. Thorington (1968) reported that a wild group was most active from early morning to midmorning and from midafternoon to late afternoon, resting for one or two hours at midday. The diet consists of fruits, berries, nuts, flowers, buds, seeds, leaves, gums, insects, arachnids, and small vertebrates (Mittermeier and Coimbra-Filho 1977).

Reported population densities are 130/sq km in Panama (Baldwin and Baldwin 1976), 20–30/sq km in Colombia (Klein and Klein 1976), and 151–528/sq km in Peru (Neville et al. 1976). In the upper Amazon, Izawa (1976) found that one group of 42 or 43 individuals had a nomadic range of about 3 sq km and made daily movements of 0.6–1.1 km. In Colombia, Mason (1971) observed that groups occupied an ill-defined home range and showed no evidence of territorial encounters with adjacent groups. In Panama, Baldwin and Baldwin (1976) determined that a group of 23 monkeys used a home range of 17.5 ha., of which 1.8 ha. was exclusive to that group. A group of 27 monkeys in a coastal marsh had a home range of 24–40 ha., but troops of 10–20 animals lived in isolated patches of forest as small as 0.8–2.0 ha.

Squirrel monkeys form larger groups than do any other New World monkeys. Based on a survey of areas of natural habitat, Baldwin and Baldwin (1971) reported troop size to range from 10 to 35 individuals in Panama and Colombia and from 120 to 300 or more in unaltered Amazonian rainforests. Mittermeier and Coimbra-Filho (1977) found group size to be 20–70 in the Amazon and mentioned a report of 550 individuals in one troop. Hernandez-Camacho and Cooper (1976), however, were unaware of any reliable accounts of more than 40–50 monkeys in a group.

A wild group observed on the llanos of Colombia was seen to split daily into subgroups composed of pregnant females, females with young, and adult males (Thorington 1968). A group in a 6-ha. enclosure in Florida appeared to be divided into socially independent subgroups (Du Mond 1968). The adult males generally had little to do with the other animals and had no relationship with the young. During the mating season they established a dominance hierarchy through fierce fighting, and the higher-ranking individuals then interacted with the females. The female subgroup began to form a separate identity, probably with its own dominance hierarchy, when pregnancy was evident. Following birth of the young, the adult males were excluded by the females. A third subgroup comprised young adult or preadult males. As these males matured they could move to the adult male subgroup, but they sometimes had to fight to win a place. In the course of this study squirrel monkeys were found to be among the most vocal of primates. The multiplicity of sounds included chirps and peeps for contact or alarm, squawks and purrs during the mating and birth seasons, barks of aggression, and screams of pain. Winter (1968) identified 26 different calls in these monkeys.

Observations of enclosed animals in Florida showed that there was a discrete, 2- to 4-month mating period during the dry season and a period of births from late June to early August (Baldwin 1970; Du Mond 1968). The estrous cycle in *Saimiri* has been reported to last 6–25 days, with estrus itself being 12–36 hours (Hayssen, Van Tienhoven, and Van Tienhoven 1993; Rosenblum 1968). Reported gestation periods have ranged from 152 to 172 days (Goss et al. 1968; Lorenz, Anderson, and Mason 1973). The newborn weigh about 100 grams, normally cling to the back of the mother for the first few weeks of life, and are independent at 1 year. Reproductive status is attained at about 3 years in females and about 5 years in males (Du Mond 1968). Several individuals have lived in captivity for about 30 years

(Marvin L. Jones, Zoological Society of San Diego, pers. comm., 1995).

The United States Department of the Interior (USDI 1976) reported that squirrel monkeys were being captured extensively for export. From 1968 to 1972 more than 173,000 were brought into the United States alone. In 1968 about half of those imported were for use in biomedical research, but in 1969 fewer than 20 percent were so used. The remainder went into the pet market or to zoos. Ultimately, the United States Department of Health, Education and Welfare (1975) issued regulations essentially ending the importation of all primates into the United States for use as pets and considerably restricting importation for all purposes. Despite the heavy commercial take and local use for food, bait, and pets, Mittermeier and Coimbra-Filho (1977) considered S. sciureus to be perhaps the most abundant and least threatened primate in Brazilian Amazonia. Long-term habitat alteration, however, could jeopardize this and other species in the region. Myers (1987b) suggested that expansion of colonization and beef production could deforest much of the Amazon Valley and Central America. The species S. oerstedii of Panama and Costa Rica has already declined drastically through clearing of the forests. It is listed as endangered by the IUCN and the USDI and is on appendix 1 of the CITES; all other species are on appendix 2. Boinski (1985) estimated the population in Costa Rica to number 3,000.

However, the IUCN now singles out the subspecies S. o. citrinellus, known only from the vicinity of the Pirris River in Costa Rica, as critically endangered and notes that both that subspecies and S. o. oerstedii, which occupies the remainder of the range of the species, have populations of fewer than 250 mature individuals. The IUCN also classifies S. vanzolinii as vulnerable, noting that it numbers fewer than 10,000 individuals and is continuing to decline because of habitat disturbance through logging. Ayres (1985) had suggested that it probably has the smallest geographic range of any South American primate and estimated the population at 50,000.

PRIMATES; Family CALLITRICHIDAE

Marmosets, Tamarins, and Goeldi's Monkey

This family of 5 Recent genera and 35 species is found in the tropical forests of Central and South America, mainly in the Amazon region. Although the family name sometimes is spelled Callithricidae, most authorities now are following Hershkovitz (1977, 1984a) in accepting Callitrichidae as the correct version. Much disagreement remains, however, relative to the phylogenetic position of this group. Traditionally the Callitrichidae have been considered a distinct family and the more primitive of the New World monkeys. That position basically was taken by Hershkovitz (1977), Groves (1989 and in Wilson and Reeder 1993), and Corbet and Hill (1991). However, there has been an increasing trend toward treating the Callitrichidae only as a subfamily of the family Cebidae (Rosenberger and Coimbra-Filho 1984; Rosenberger, Setoguchi, and Shigehara 1990; Schneider et al. 1993; Thorington and Anderson 1984). Martin (1990) continued to accept the Callitrichidae as a full family but presented substantial evidence that the family is a highly specialized group derived from cebidlike ancestral stock. Other recent authorities giving familial status to the Callitrichidae include Ford

(1986b), Rylands, Coimbra-Filho, and Mittermeier (1993), and Sussman and Kinzey (1984). An additional question involving the group is the affinities of the genus Callimico (Goeldi's monkey); it was placed in a separate family by Hershkovitz (1977), tentatively included in the Cebidae by Martin (1990), allocated to a subtribe of what was regarded as the subfamily Callitrichinae by Schneider et al. (1993), and considered to represent a separate subfamily of the family Callitrichidae by Groves (1989), Napier and Napier (1985) and Sussman and Kinzey (1984). Recently, Martin (1992) summarized extensive morphological, chromosomal, and biochemical evidence and concluded that Callimico and the four callitrichid genera form a monophyletic group but that Callimico branched away from the group prior to the divergence of the other genera. Based on that conclusion and a review of the various other authorities cited above, the Callitrichidae are treated here as a family, the components being arranged phylogenetically as follows: the subfamily Callimiconinae, with the genus Callimico; and the subfamily Callitrichinae, with Saguinus, Callithrix, Cebuella (included in Callithrix by some of the authorities), and Leontopithecus.

These are among the smallest of primates. Head and body length is 130–370 mm and tail length is 150–420 mm. Adults weigh from approximately 100 to 900 grams. The pelage is soft, dense, and in some forms silky. Often there are hair tufts and other adornments on the head. The face is naked or only sparsely haired. The coloration is quite variable in the family. In some species the tail is marked with alternate dark- and light-colored bands. The tail is not prehensile in any of the species.

The forelimbs are shorter than the hind limbs. The thumb is not opposable. The hand and the foot are elongate, and all the digits bear pointed, sickle-shaped nails, except the great toe, which has a flat nail. A baculum is present in the males, and the females have two mammae in the chest region.

The dental formula of the Callitrichinae is: (i 2/2, c 1/1, pm 3/3, m 2/2) × 2 = 32; the Callimiconinae have m 3/3. In Callithrix and Cebuella the lower canines barely extend beyond the adjacent incisors, the "short-tusked" condition, but in Saguinus and Leontopithecus the lower canines are longer than the incisors, the "long-tusked" condition. This shared characteristic is not necessarily indicative of phylogenetic affinity; Callithrix and Cebuella actually are considered more closely related to Leontopithecus than any of those genera is to Saguinus (Groves 1989; Rosenberger and Coimbra-Filho 1984).

Callitrichids sometimes are considered primitive, squirrel-like primates, but new evidence indicates that they are both morphologically and ecologically advanced (Sussman and Kinzey 1984; Thorington and Anderson 1984). Unlike squirrels, which range through the forest canopy mainly by ascending and descending large vertical trunks, callitrichids move by running quadrupedally along horizontal branches and leaping between thin terminal supports. Squirrels are adapted to feed on hard nuts, while callitrichids eat mainly insects, soft fruits, nectar, and plant exudates.

Callitrichids have acute sight, good hearing, and apparently a good sense of smell. Facial expression is indicated mainly by lip movements. Emotions are expressed by movements of the eyelids, ears, and the hairy adornments on the head if present. The voice has a variety of high-pitched trilling and staccato calls. These animals are diurnal, sheltering at night in tree holes and cavities. Most of their time is spent in trees or shrubs, though Callithrix argentata in the Matto Grosso sometimes goes into tall grass. They are active and agile, running, jumping, and occasion-

PRIMATES
CALLITRICIDAE

A mother pygmy marmoset *(Cebuella pygmaea)* with two babies clinging to her. The young of all the primates but *Homo* ride on either the mother or the father during their long period of dependence. Photo from San Diego Zoological Garden.

ally leaping among trees and shrubs. *Leontopithecus* is said to bound from tree to tree with remarkable rapidity. Movements are usually quick and jerky. At rest, these animals sometimes draw the fingers inward, so that the clawlike nails pierce the bark of the support, and rest with the belly in contact with the branch, all four limbs hanging on either side. They are fastidious in the care of their pelage and engage in both individual and mutual grooming, the former

being more common. The hands are used in securing food but not invariably for conveying it to the mouth.

Callitrichids live in small groups. Contrary to what is sometimes reported, they do not have a monogamous mating system. A female apparently mates with more than one male during the breeding season, and all raise her young cooperatively (Sussman and Kinzey 1984). Females give birth to one to three young after a gestation period of approximately 130–50 days. It has been reported that a female giving birth in captivity sometimes kills and eats its first litter or the third member of a set of triplets. Whether this occurs in the wild is not known. It is probable that this behavior is similar to that of many captive animals and is caused by anxiety. The males assist in birth and carry the offspring on their backs. The males transfer the young to the mother at feeding time and then accept them from the mother again after feeding. This takes place every two to three hours. Sexual maturity is attained at 12–18 months, and longevity in captivity has been as high as 28 years.

Callitrichids are usually docile and gentle in captivity but bite if handled against their will, especially in the presence of strangers. Sudden noises or movements sometimes cause them to panic. They do quite well in captivity if fed, housed properly, and exposed to ultraviolet radiation to replace the lack of sunshine. It has been reported that the members of this family do not seem to be as intelligent as monkeys of the family Cebidae; however, it is very difficult to determine relative intelligences of animals of different groups.

The name "marmoset" dates back to Middle English and is said to be adapted from the Old French word *marmouset,* meaning "a grotesque image" or "manikin." Members of the genera *Saguinus* and *Leontopithecus* are usually called "tamarins," a word of obscure origin.

There was no fossil record of this family until Setoguchi and Rosenberger (1985) reported specimens of a new genus, *Micodon,* from the middle Miocene of Colombia. Rosenberger, Setoguchi, and Shigehara (1990) indicated that a subsequently described genus from the same site, *Mohanamico,* may have affinity to *Callimico.* However, Kay (1990) suggested that *Mohanamico* is probably a primitive member of the subfamily Pitheciinae of the family Cebidae.

PRIMATES; CALLITRICHIDAE; **Genus CALLIMICO**
Miranda Ribeiro, 1911

Goeldi's Monkey

The single species, *C. goeldii,* occurs in the upper Amazonian rainforests of southern Colombia, eastern Ecuador, eastern Peru, western Brazil, and northern Bolivia. Although here referred to the Callitrichidae, Hershkovitz (1977) observed that it is not a callitrichid or a cebid, nor a link between the two, and thus should be placed in a separate family. A recent analysis of DNA (Montagnon, Crovella, and Rumpler 1993), however, indicates that *Callimico* is part of the group here designated the family Callitrichidae; most other authorities now seem to agree (see the family account, above).

Head and body length is 210–34 mm, tail length is 255–324 mm, and adult weight is 393–860 grams (Hershkovitz 1977). Adults are brownish black with buffy markings on the back of the neck and two or three light buff-colored rings on the basal part of the tail. Young animals lack the buffy tail rings and sometimes the buff on the back. The head or even the entire dorsal surface may be spotted with white. The hair is thick and soft. There are no ear tufts. A mane drapes from the neck and shoulders, and the elongated hairs of the rump extend skirtlike over the base of the tail.

Although the facial appearance, foot structure, and claw-like nails resemble those of other callitrichids, the dental formula and cranial configuration are like those of the Cebidae. *Callimico* differs from the Cebidae in the structure

of the molar teeth, for example, in the absence of external cingula on the lower molars (Hershkovitz 1977).

Information available to Hershkovitz (1977) indicated that *Callimico* prefers deep, mature forests and is relatively rare in areas generally accessible to people. Mittermeier and Coimbra-Filho (1977) observed that *Callimico* occurs in bamboo forest, secondary forest, and primary forest but is naturally rare and very sparsely distributed. Hernandez-Camacho and Cooper (1976) stated that this monkey is found most frequently in the understory and on the ground. The diet includes fruits, insects, and some vertebrates.

In a study in a seasonally dry rainforest in northwestern Bolivia, Pook and Pook (1981) located only a single group within an area of 4 sq km. Groups were largely isolated from one another, and no interaction was seen. The main study group had a home range of 30–60 ha. It traveled about 2 km per day, usually at a height of a few meters, but occasionally came to the ground or climbed to the tops of trees to feed. Leaps of up to 4 meters were made without loss of height. Average group size was 6 individuals, but the main study group had 8, including an adult male and more than 1 breeding female. Members usually remained well within 15 meters of one another and maintained contact by a shrill call. Studies by Masataka (1981*a,* 1981*b*) in the same area suggest a similar social structure and support other new evidence that callitrichids are not monogamous. The main study group contained about 6 individuals, including 1 adult male and 2 breeding females, and membership was not exclusive. Also in northern Bolivia, Christen and Geissmann (1994) observed groups of 2–5 individuals occupying adjacent home ranges, one of which was 80 ha. Old reports of troops of up to 40 individuals may have been exaggerated or mistaken. Masataka (1982) reported 40 different vocalizations in his wild study group. These sounds included a whistle for long-distance contact, trills for alarm and warning, and a "truuu" during agonistic behavior.

Births have taken place throughout the year in captivity (Beck et al. 1982), but in northwestern Bolivia they occur mainly in the early wet season, September–November (Masataka 1981*a;* Pook and Pook 1981). Females are polyestrous, the estrous cycle is about 22–24 days, estrus lasts about 7 days, and gestation averages about 155 days, but the reported range is 139–80 days. Twins occur rarely in captivity, but normally there is a single young weighing 30–60 grams at birth. It is handled solely by its mother for about the first 10–20 days of life, but subsequently there is increasing participation in its care by other members of the group. At 7 weeks the young is able to move about and forage on its own. Weaning is completed by 12 weeks and sexual maturity is attained at 18–24 months (Altmann, Warneke, and Ramer 1988; Hayssen, Van Tienhoven, and Van Tienhoven 1993; Heltne, Turner, and Scott 1976; Hershkovitz 1977; Mallinson 1977). Maximum recorded longevity is 17.9 years (Ross 1991).

Goeldi's monkey is classified as endangered by the USDI and as vulnerable by the IUCN and is on appendix 1 of the CITES. Its natural rarity and apparently specialized habitat make it vulnerable to such adverse factors as habitat destruction and hunting, and parts of its range may soon come under development (Thornback and Jenkins 1982).

CALLIMICO

Goeldi's marmoset *(Callimico goeldii)*, photo by Ernest P. Walker.

PRIMATES; CALLITRICHIDAE; **Genus SAGUINUS**
Hoffmannsegg, 1807

Long-tusked Tamarins

There are 12 species (Hershkovitz 1977; Moore and Cheverud 1992; Natori and Hanihara 1988; Rylands, Coimbra-Filho, and Mittermeier 1993; Skinner 1991; Thorington 1988; Wolfheim 1983):

S. nigricollis, eastern Ecuador, southern Colombia, northeastern Peru, a small part of northwestern Brazil;
S. fuscicollis, upper Amazonian region of eastern Ecuador, southern Colombia, eastern Peru, western Brazil, and northern Bolivia;

S. tripartitus, eastern Ecuador, northeastern Peru, possibly Colombia;
S. labiatus, middle Amazonian region of Brazil and adjacent fringes of southeastern Peru and northwestern Bolivia;
S. mystax, eastern Peru, western Brazil, northwestern Bolivia, possibly southern Colombia;
S. imperator, southeastern Peru, northwestern Bolivia, and adjacent parts of western Brazil;
S. midas, Guianas, northern Brazil;
S. inustus, southeastern Colombia and adjacent part of Brazil;
S. bicolor, an area north of the Amazon in north-central Brazil;
S. leucopus, north-central Colombia;
S. oedipus, northwestern Colombia;

S. geoffroyi, southeastern Costa Rica to extreme
northwestern Colombia.

Until recently this genus sometimes was known as
Tamarin Gray, 1870, or Leontocebus Wagner, 1840. Rylands, Coimbra-Filho, and Mittermeier (1993) suggested
that S. melanoleucus of the upper Juruá River is specifically distinct from S. fuscicollis and that S. niger to the south
of the Amazon River is specifically distinct from S. midas
to the north, but they did not actually proceed with such
designations. Morphological and genetic differentiation
between the various subspecies of S. fuscicollis was discussed by Cheverud, Jacobs, and Moore (1993), Cheverud
and Moore (1990), and Moore and Cheverud (1992) and
was found to be pronounced in some cases, but no divisions
at the specific level were indicated. Additional modifications of the above list as suggested by the work of Natori
(1988) and Natori and Hanihara (1988, 1992) are discussed
below.

Head and body length is 175–310 mm, tail length is
250–440 mm, and weight is usually 225–900 grams.
Hershkovitz (1977) divided the species into a number of
sections and groups depending largely on pelage. The following brief descriptions are based mainly on his detailed
analysis.

In the hairy-faced tamarin section the forehead, crown,
cheeks, and temples are covered with long hairs. In the S.
nigricollis group of this section there is no development of
a mustache, but an area around the mouth is covered with
short white hairs. The species of this group include S. nigricollis, which has uniformly dark-colored upper parts,
and S. fuscicollis, which, except for extreme albinistic races,
has upper parts with a trizonal pattern, there being a sharp
division between the colorations of the mantle, saddle, and
rump. According to Thorington (1988), S. tripartitus, which
Hershkovitz considered to be a subspecies of S. fuscicollis,
is distinguished from the latter by a golden, rather than
dark red, mantle and a prominent chevron between and
above the eyes. The S. mystax group of the hairy-faced
tamarin section has a mustache as well as a white area
around the mouth. The three species of this group are S.
labiatus, with a poorly developed mustache, a black back
marbled with silver, a golden or reddish crown, and mostly reddish underparts; S. mystax, with a well-developed but
not particularly elongate mustache and a black crown and
underparts; and S. imperator, the emperor tamarin, with a
long white mustache that extends to the shoulders when
laid back. The S. midas group of this section, with the one
species S. midas, lacks the white area around the mouth and
has a blackish face and orange or yellowish hands and feet.

The mottled-face tamarin section includes only the
species S. inustus. Its crown is densely haired, but the
cheeks and sides of the face are naked, or nearly so. The face
is mostly unpigmented and the pelage is melanistic.

The bare-faced tamarin section also has a densely haired
crown. The mostly naked face is black. The S. bicolor group
of this section contains two species: S. bicolor, with yellowish or white forequarters sharply defined from the grayish
brown hindquarters; and S. leucopus, with silvery hairs on
the head, buffy brown upper parts, and white limbs. The S.
oedipus group of this section, with the cotton-top marmosets S. oedipus and S. geoffroyi, has a general agouti
pelage but a prominent crest of white hairs extending back
over the head.

Somewhat different groupings than those set forth by
Hershkovitz are suggested by analyses of cranial and dental characters (Natori 1988; Natori and Hanihara 1988,
1992). These analyses place S. oedipus, S. geoffroyi, S. bicolor, and S. midas in one group of related species. Within

that group, S. oedipus and S. geoffroyi have affinity with
one another and S. bicolor and S. midas seem immediately
related. The other eight species are placed in another group,
with S. leucopus, S. imperator, S. nigricollis, S. fuscicollis,
and S. tripartitus forming one monophyletic subgroup and
S. inustus, S. labiatus, and S. mystax forming a second such
subgroup of related taxa.

Saguinus inhabits tropical forests, open woodlands, and
secondary growth. It is diurnal and arboreal, moving
through trees with rather quick and jerky movements. An
individual S. midas jumped 20 meters from the top of a tree
to the ground without injury. Neyman (1977) reported that
S. oedipus slept in broad tree forks, began moving about 80
minutes after dawn, and entered the sleeping tree at 1630
to 1830 hours. Data from studies of different species summarized by Buchanan-Smith (1991) and Garber (1993) indicate that groups move about 1,000–2,000 meters per day.
Garber also reported that the primary components of the
diet of Saguinus now are known to be insects (especially
large orthopterans), ripe fruits, plant exudates (sap, gums,
resin), and nectar. Other foods include some tender vegetation, spiders, small vertebrates, and probably birds' eggs.
With the exception of some of the smaller prey, animals are
killed by a bite on the head.

Freese, Freese, and Castro (1977) reported a population
density of 30/sq km for S. fuscicollis in eastern Peru. Density of S. nigricollis in the upper Amazon of Colombia was
10–13/sq km, and groups had extensively overlapping
home ranges of about 0.3–0.5 sq km (Izawa 1978). In a
study of S. geoffroyi in the Panama Canal Zone, Dawson
(1977) found a density of 20–30/sq km and home ranges of
26 ha. and 32 ha. In Colombia, Neyman (1977) estimated
densities of 30–180/sq km for S. oedipus and found well-
defined home ranges of 7.8, 7.8, and 10 ha. Neighboring
home ranges overlapped substantially, but contact between
groups was agonistic. Hershkovitz (1977) observed that
each troop of S. oedipus has a defended territory. Defense
of an area is associated with its importance as a feeding site;
overall reported group home range size for the various
species is 8–120 ha., with about 10–83 percent overlap between such ranges (Garber 1993).

The investigations of Neyman (1977) indicate that
groups of S. oedipus do not represent extended families.
Such groups may consist of a dominant mated pair, their
young of the year, and a transient complement of subordinate, probably young animals of both sexes. These subor-
dinates sometimes form small groups of their own within
the home range of the main groups. They enter and leave
the main groups and possibly sometimes remain and rise
to breeding position. The overall size range reported for
groups of S. oedipus is 1–19, with about 3–9 being most
common. Izawa (1978) reported much the same social
arrangement for S. nigricollis and observed that 2–3 family groups often joined to form assemblies of 10–20 individuals that usually remained together for less than a day.

Reported group sizes in some other species of Saguinus
are: S. nigricollis, 5–10; S. fuscicollis, 2–40; S. labiatus, 5–10;
S. imperator, 1–3; S. midas, 1–20; and S. leucopus, 3–12
(Buchanan-Smith 1991; Freese, Freese, and Castro 1977;
Hernandez-Camacho and Cooper 1976; Neyman 1977).
Castro and Soini (1977) stated that S. mystax lives in
parental groups of 2–6 individuals, the largest comprising
an adult pair, 2 subadults, and 2 dorsally carried young.
Garber, Moya, and Malaga (1984) and Garber (1993) reported a somewhat different social structure for this
species, with groups of 3–11 individuals, including a single
breeding female and 1 or 2 other adult females but up to 3
reproductively active males. The female may mate with
each of the males, and all of the latter assist in caring for

Long-tusked marmosets, or tamarins *(Saguinus):* A. Pied marmoset *(S. bicolor),* photo by Harald Schultz; B. White-lipped marmoset *(S. nigricollis),* photo by Ernest P. Walker; C. Geoffroy's marmoset *(S. geoffroyi),* photo from New York Zoological Society; D. Emperor marmoset *(S. imperator),* photo from New York Zoological Society; E. Yellow-handed marmoset *(S. midas),* photo by Ernest P. Walker; F. Cotton-top marmoset *(S. oedipus),* photo by Ernest P. Walker.

Long-tusked tamarin (Saguinus nigricollis), photo by Russell A. Mittermeier.

the young. There is frequent migration of adults and subadults between groups. Most troops of S. mystax have been observed to travel together with bands of S. fuscicollis in apparently stable associations. Garber and Teaford (1986) suggested that such mixed troops were able to utilize information obtained by both species relative to location and productivity of food sources and thus to increase foraging efficiency. Epple et al. (1993) reported that Saguinus possesses specialized scent glands in the gular-sternal region of the mid-chest and the area surrounding the genitalia. Secretions from these glands, together with urine and other discharges, are rubbed or deposited in various places to mark territories and convey information about identity, social status, sexual receptivity, and other factors.

A common characteristic of tamarin groups is the suppression of reproductive activity in all but the dominant female member. This effect results from a combination of inhibitory behavior by the dominant and loss of ovulatory capacity in the subordinate (Abbott, Barrett, and George 1993). Nonetheless, tamarins generally display minimal intragroup aggression, with a marked degree of cooperation and tolerance, even by sexually active males towards one another (Caine 1993).

Hernandez-Camacho and Cooper (1976) stated that there is a tendency to seasonal reproduction in the species S. leucopus and S. oedipus. Hershkovitz (1977) noted that in many years of observation of S. geoffroyi in Panama and Colombia pregnant females and suckling young were found only from January to June. Eisenberg (1977) cited studies showing that the estrous cycle in this species is about 15 days, gestation lasts about 140 days, litter size is usually 2, and sexual maturity is attained at about 18 months in females and 24 months in males. According to Gengozian, Batson, and Smith (1977), gestation in S. fuscicollis is estimated at 140–50 days; births in captivity have occurred in every month, with a peak from March to May; and 75 percent of births have been of twins, 20 percent single, and 5 percent triplets. Birth weight in Saguinus is 25–55 grams (Hayssen, Van Tienhoven, and Van Tienhoven 1993).

In Saguinus the father and sometimes other adult members of a group assist at birth, receiving and washing the young. The newborn have a coat of short hair and are helpless. They cling tightly with their hands and feet to the body of the mother or father. The father transfers the young to the mother at feeding time and then accepts them from the mother again after feeding. This behavior, which is common among callitrichids, takes place every two to three hours and lasts about half an hour. At about 21 days the young begin to explore nearby surroundings, but they continue to ride on the backs of the parents until they are about 6–7 weeks of age. At 4 weeks they begin to accept soft food in addition to their mother's milk. Several members of a group besides the mother and father may help carry and provision the young, such communal roles apparently being more important in Saguinus than in Callithrix (Tardif, Harrison, and Simek 1993). According to Marvin L. Jones (Zoological Society of San Diego, pers. comm., 1995), a captive specimen of S. imperator lived for more than 20 years, one of S. fuscicollis lived for about 24 years, and one of S. oedipus lived for nearly 25 years.

The USDI lists the species S. bicolor, S. oedipus, and S. geoffroyi as endangered and S. leucopus as threatened. Those four species also are on appendix 1 of the CITES, and all other species are on appendix 2. The IUCN classifies the species S. oedipus and the subspecies S. bicolor bicolor, of central Brazil, as endangered, noting that each has a population of fewer than 2,500 individuals and is continuing to decline. The IUCN also designates the species S. leucopus and the subspecies S. nigricollis hernandezi, of southern Colombia, and S. imperator imperator, found in a small area where Bolivia, Brazil, and Peru come together, as vulnerable; each has an estimated surviving population of fewer than 10,000. Clearing of forest habitat by people is the main problem for all these forms, and populations of S. oedipus also were depleted through taking for the animal trade. S. oedipus already has lost most of its habitat, and it is unlikely that many remaining tracts of forest are large enough to maintain populations (Thornback and Jenkins 1982). Rylands, Coimbra-Filho, and Mittermeier (1993) expressed particular concern for S. bicolor bicolor, which occupies a very small range near the large city of Manaus. They also considered the subspecies S. fuscicollis melanoleucus and S. f. acrensis, confronted with rapid development in the Brazilian state of Acre, as vulnerable.

PRIMATES; CALLITRICHIDAE; Genus CALLITHRIX
Erxleben, 1777

Marmosets

There are 17 named species (Coimbra-Filho and Mittermeier 1973b; Ferrari and Lopes 1992; Mittermeier, Schwarz, and Ayres 1992; Muskin 1984; Natori 1990, 1994; Rylands, Coimbra-Filho, and Mittermeier 1993; Rylands, Mittermeier, and Luna 1995):

C. jacchus, eastern Brazil;

C. penicillata, inland east-central Brazil;

C. kuhli, southern Bahia and northeastern Minas Gerais in coastal eastern Brazil;

C. geoffroyi, states of Espirito Santo and Minas Gerais in coastal eastern Brazil;

C. flaviceps, southern Espirito Santo and Minas Gerais in coastal southeastern Brazil;

C. aurita, states of Minas Gerais, Rio de Janeiro, and Sao Paulo in coastal southeastern Brazil;

C. argentata, south of the western part of the Amazon River in eastern Brazil;

C. leucippe, a small area between the Cuparí and Tapajós rivers in central Brazil;

C. melanura, eastern Bolivia, northeastern Paraguay, and adjacent parts of southwestern Brazil;

C. emiliae, states of Pará and Matto Grosso in central Brazil;

C. nigriceps, along the Madeira River in the states of Amazonas and Rondônia in west-central Brazil;

C. marcai, known only from near the confluence of the Aripuana and Roosevelt rivers in the state of Amazonas in central Brazil;

C. intermedia, Guariba River Basin in the state of Amazonas in west-central Brazil;

C. humeralifer, south of the Amazon and between the Tapajós and Canuma rivers in central Brazil;

C. chrysoleuca, between the Aripuana and Canuma rivers in the state of Amazonas in central Brazil;

C. mauesi, between the Urariá and Maués rivers in the state of Amazonas in central Brazil;

C. saterei, central Amazon region of Brazil.

Rylands, Coimbra-Filho, and Mittermeier (1993) pointed out that an additional species, unnamed but related to *C. melanura* and *C. emiliae,* occurs in much of the state of Rondônia in west-central Brazil. They noted also that a large part of the state of Bahia on the coast of east-central Brazil is apparently occupied by a hybrid zone involving *C. jacchus, C. penicillata,* and *C. kuhli.* Coimbra-Filho, Pissinatti, and Rylands (1993) reported that there also has been natural hybridization involving *C. geoffroyi, C. flaviceps,* and *C. aurita* and that experiments in captivity have produced fertile hybrids between a number of the eastern Brazilian species. Partly on the basis of such hybridization, Hershkovitz (1975a, 1977) recognized only three valid species: *C. jacchus,* which would include *C. penicillata, C. geoffroyi, C. flaviceps, C. aurita,* and *C. kuhli; C. argentata,* which would include *C. leucippe, C. melanura, C. emiliae,* and probably *C. nigriceps;* and *C. humeralifer,* which would include *C. intermedia, C. chrysoleuca,* and probably *C. mauesi.* Groves (*in* Wilson and Reeder 1993) included *emiliae, intermedia, leucippe,* and *melanura* in *C. argentata,* and *chrysoleuca* in *C. humeralifer.* Groves, like some other authorities, also included *Cebuella* (see account thereof) in *Callithrix.*

Head and body length is 180–300 mm, tail length is 172–405 mm, and weight is usually 230–453 grams. The general coloration of *C. jacchus* is agouti gray; its tail has alternating broad blackish and narrow pale bands. There are long tufts in front of, and often above and behind, the base of the ears, and these tufts, along with other areas of the head, vary among species as follows: *C. penicillata,* blackish circumauricular tufts, crown, and temples; *C. geoffroyi,* blackish circumauricular tufts, white or creamy crown and sides of face; *C. jacchus,* mostly white or grayish circumauricular tufts, blackish or brown forehead and temples; *C. aurita,* small white ear tufts, black throat, head, and cheeks; and *C. flaviceps,* small yellow to ochraceous ear tufts, other areas of head the same color. In *C. argentata,* the black-tailed or silvery marmoset, the pelage is fine and silky and silvery white, sometimes washed with gray or yellowish gray, especially on the back. The tail is black, contrasting sharply with the color of the body. The face and ears are devoid of hair and reddish in color. There is a complete absence of tufts, plumes, or other adornments. In *C. humeralifer* there are pronounced tassels on the ears. The entire pelage is predominantly whitish in the western part of the range of the species. In the east the tassels are silvery or buffy, the back is black flecked with white, the tail is black banded with silver, and the underparts are orange.

In this genus and in *Cebuella* the lower canines are incisiform, barely extending beyond the adjacent incisors. *Callithrix* differs from *Cebuella* in its larger size and in cranial and dental features.

These marmosets generally live in tropical or subtropical forests. They run and hop in trees and bushes and are capable of leaping. Their movements are usually quick and jerky. They are diurnal, sleeping at night in tree holes or other shelters. The diet includes insects, spiders, small vertebrates, birds' eggs, fruits, and tree exudates. Like *Cebuella,* they have specialized short lower canines for perforating tree bark and inducing the flow of gums and sap (Coimbra-Filho and Mittermeier 1977b; Mittermeier and Coimbra-Filho 1977). Reportedly more than 70 percent of the foraging time of *C. jacchus* involves the collection of such exudates (Bouchardet da Fonseca and Lacher 1984). *C. penicillata* also depends heavily on exudates; however, other species are less adapted for exploitation of such resources, and *C. humeralifer* and *C. argentata* are highly frugivorous (Rylands and de Faria 1993).

Information summarized by Sussman and Kinzey (1984) indicates that these marmosets live in groups of about 4–15 individuals that have overlapping home ranges of 0.5–28.0 ha. and do not defend territories. Studies in captivity suggest that groups are dominated by a monogamous pair, but the mating system in the wild may be comparable to that described above in the account of *Saguinus.* Groups may contain several adults of each sex; generally only the dominant female breeds, but groups of *C. jacchus* with two reproductively active females have been observed (Digby and Ferrari 1993). Observations of *C. jacchus* show that ovulation in subordinate females is inhibited through suppression of necessary hormone secretions, perhaps in response to pheromones from the dominant female (Abbott, Barrett, and George 1993). Groups of *C. jacchus* appear to be extended families, and emigration seems less significant than it is in *Saguinus* (Digby and Barreto 1993). Dominance is maintained through aggressive interaction. The vocal repertoire includes a soft "phee" for contact, a louder "phee" for territorial expression, a "tsee-tsee-tsee" as an aggressive threat, and a high-pitched whistle as a warning signal. *Callithrix* has scent glands and utilizes them in the same manner described above in the account of *Saguinus* (Epple et al. 1993).

Data cited by Hayssen, Van Tienhoven, and Van Tienhoven (1993) indicate that breeding is year-round in captivity but that births of *C. jacchus* in Brazil occur in late October–November and late March–April. Interbirth interval in that species is about 6 months, the estrous cycle is about 30 days, estrus lasts 2–3 days, the gestation period is around 130–50 days, birth weight is 20–35 grams, lactation lasts up to 100 days, and sexual maturity is attained at 11–15 months by males and 14–24 months by females. Litter size in *C. jacchus* is 1–4, with more than half of recorded births resulting in twins (Hershkovitz 1977). *Callithrix* carries and provisions its young for a shorter period than does *Saguinus,* and there is less involvement in this function by adults other than the mother (Tardif, Harrison, and Simek 1993). Life span in the genus seems to be about 10 years in the wild and up to 16 years in captivity.

The species *C. aurita* and *C. flaviceps,* found only along the rapidly developing southeastern coast of Brazil and each numbering fewer than 2,500 individuals, are classified as endangered by the IUCN and the USDI and are on appendix 1 of the CITES; all other species are on appendix 2. The IUCN also classifies *C. geoffroyi, C. nigriceps, C. leucippe,* and *C. chrysoleuca* as vulnerable. All of these species are jeopardized through destruction of their habitat by people. Oliver and Santos (1991) reported that *C. flaviceps* has

Short-tusked marmosets *(Callithrix):* A. Black ear-tufted marmoset *(C. penicillata),* photo by Ernest P. Walker; B. Silky marmoset *(C. chrysoleuca),* photo by Harold Schultz; C. Geoffroy's marmoset *(C. geoffroyi),* photo by Bernhard Grzimek; D. Black-tailed marmoset *(C. argentata);* E & F. White ear-tufted marmosets *(C. jacchus),* photos by Ernest P. Walker.

the most limited distribution of any marmoset but that *C. kuhli* and *C. geoffroyi* remain relatively widespread. The small ranges of *C. leucippe*, *C. nigriceps*, and *C. humeralifer* are bisected by the new Trans-Amazonian Highway in Brazil and are being subjected to much clear-cutting to make way for cattle ranches (Ferrari and Queiroz 1994; Rylands, Coimbra-Filho, and Mittermeier 1993; Thornback and Jenkins 1982). Recent studies of *C. flaviceps* indicate that although it has lost most of its original habitat, it is adaptable to secondary forest and can be expected to survive with proper protection (Ferrari and Mendes 1991).

PRIMATES; CALLITRICHIDAE; Genus CEBUELLA
Gray, 1865

Pygmy Marmoset

The single species, *C. pygmaea*, inhabits tropical forests of the upper Amazon region in western Brazil, southeastern Colombia, eastern Ecuador, eastern Peru, and northwestern Bolivia (Hershkovitz 1977; Rylands, Coimbra-Filho, and Mittermeier 1993). Groves (1989 and *in* Wilson and Reeder 1993) and Rosenberger and Coimbra-Filho (1984) included this species in the genus *Callithrix*, but *Cebuella* was maintained as a full genus by Schneider et al. (1993).

This is the smallest New World primate. Head and body length is 117–52 mm, tail length is 172–229 mm, and adult weight is 85–140 grams (Hershkovitz 1977; Soini 1993). The head and neck are dark brown and gray or dark brown and buff; the back is grayish, black mixed with buff, or brownish tawny, sometimes with a greenish cast; the hands and feet are yellowish or orangish; the tail is indistinctly ringed with black and tawny; and the underparts are often orangish but vary from white to tawny. From *Callithrix*, *Cebuella* differs in its smaller size, multibanded agouti pelage, penis shaft lacking spines, sessile scrotum, second upper incisor tooth and both lower incisors having a mesial as well as a distal capsule, and lower premolars being at least as long as they are wide (Groves 1989).

The pygmy marmoset prefers low second growth along streams where cover is thick, visibility is good, and insects are most abundant (Hershkovitz 1977). It is arboreal and completely diurnal, being most active on cool mornings and in late afternoons. Movement usually is by running along horizontal or diagonal branches with a galloping gait. It can make long, horizontal leaps of a meter or more and also clings and leaps vertically. Movements are sometimes exceedingly slow but more often occur in spurts; an

Pygmy marmosets *(Cebuella pygmaea)*, photo by Howard E. Uible. Inset photo by Ernest P. Walker.

alarmed individual continually turns its head in all directions (Moynihan 1976). The diet consists of fruits, buds, insects, and sap or exudates from trees, the latter item being an especially important food source. The short lower canine teeth are specialized for gouging holes in trees and inducing exudate flow (Mittermeier and Coimbra-Filho 1977). Each pygmy marmoset group has one or more trees in its range, which are riddled with small holes for feeding on the sap (Moynihan 1976). There may be hundreds of such holes per square meter of tree surface (Ramirez, Freese, and Revilla C. 1977).

In a floodplain forest of Amazonian Peru, Soini (1982) found a density of 51.5 independently locomoting individuals per sq km, but along river edges density increased to 274/sq km. Of these animals, 50 percent were adults; 83 percent were in stable troops, and the rest were incipient pairs and lone individuals. Forest troops lived in home ranges of 0.2–0.4 ha., while the ranges of riverside troops covered 70–90 meters of shoreline and extended 20–60 meters inland. Home ranges were contiguous but nonoverlapping. Stable troops contained a mated adult pair and their mature offspring of up to four generations. All members of a troop roosted together, usually huddled in a clump on a leafy branch 7–10 meters high.

In another Peruvian study, Castro and Soini (1977) calculated a population density of 5.6/ha. Of 4 troops of 5–10 animals each, 3 occupied contiguous, nonoverlapping home ranges along a river and 1 inhabited a single large "home tree" in which the marmosets foraged and slept. Also in Peru, Ramirez, Freese, and Revilla C. (1977) found that troops numbered 7–9 animals and utilized home ranges measuring about 75 × 40 meters. About a third of this area formed a core zone containing all of the troop's exudate trees, in which the group spent 80 percent of its time. The overall reported size range for pygmy marmoset groups is 2–15 (Mittermeier and Coimbra-Filho 1977). Although sometimes considered a relatively quiet primate, Cebuella has a variety of vocalizations, including a trill for communication over distance, a high, sharp warning whistle, and a clicking sound for threats (Pola and Snowdon 1975).

According to Soini (1993), some troops contain more than one adult of each sex, but in such cases only one female appears to be reproductively active and one male maintains exclusive mating access to her. Soini (1982, 1993) reported that births in the wild occur throughout the year but that in Amazonian Peru there are two peaks, in October–January and May–June. Hershkovitz (1977) wrote that no reproductive seasonality was evident in captivity, that gestation lasted about 20 weeks, and that sexual maturity was attained at 18–24 months. Of 21 recorded births, 14 were of twins, 6 were single, and 1 was of triplets. Hayssen, Van Tienhoven, and Van Tienhoven (1993) listed birth weights of 14–27 grams, a minimum interbirth interval of about 150 days, and a lactation period of 3 months. One captive specimen in Japan lived for 18 years (Marvin L. Jones, Zoological Society of San Diego, pers. comm., 1995).

Moynihan (1976) found that in the Putumayo region of Colombia the pygmy marmoset was subject to intensive collection for use as a pet. Mittermeier and Coimbra-Filho (1977), however, stated that the species was in no danger in Brazil as it was highly adaptable to change and could inhabit degraded as well as virgin forest. It is on appendix 2 of the CITES.

PRIMATES; CALLITRICHIDAE; **Genus**
LEONTOPITHECUS
Lesson, 1840

Lion Tamarins

There are four species, all in the lowland rainforests of southeastern Brazil (Coimbra-Filho and Mittermeier 1977a; Lorini and Persson 1990; Pinto and Tavares 1994; Rosenberger and Coimbra-Filho 1984; Rylands, Coimbra-Filho, and Mittermeier 1993):

L. chrysopygus (black lion tamarin), state of Sao Paulo;
L. caissara (black-headed lion tamarin), Superagui Island and nearby mainland of states of Paraná and Sao Paulo;
L. chrysomelas (golden-headed lion tamarin), state of Bahia and extreme northeastern Minas Gerais;
L. rosalia (golden lion tamarin), state of Rio de Janeiro.

The name *Leontideus* Cabrera, 1956, often has been used for this genus. Hershkovitz (1977) considered the species to be color grades of an otherwise morphologically uniform species, *L. rosalia*, but Rosenberger and Coimbra-Filho (1984) showed that they are distinct in cranial and dental characters as well as in color.

Head and body length is about 200–336 mm, tail length is 315–400 mm, and adult weight is 600–800 grams (Kleiman 1981). The pelage is long and silky; the common name "lion" refers to the mane on the shoulders. *L. chrysopygus* is mostly black with a gold rump and thighs; *L. caissara* is mostly gold with black head, hands, feet, and tail; *L. chrysomelas* also is mostly black but has a golden mane; and *L. rosalia* is entirely gold, reddish, or buffy to white (Hershkovitz 1977; Lorini and Persson 1990).

Lion tamarins prefer primary tropical forest but have been reported in secondary forest and areas under partial cultivation. They are usually found at a height of 3–10 meters in the trees, where interlacing branches, vines, and epiphytes provide optimum shelter and an abundance of insect and small vertebrate prey. They are diurnal, sleeping at night in tree holes or occasionally in vines or epiphytes (Coimbra-Filho and Mittermeier 1977a; Hershkovitz 1977). Groups forage along a path of about 1,300–2,600 meters each day (Rylands 1993a). The animals leap from branch to branch with great agility. They are primarily insectivorous and frugivorous but also eat spiders, snails, small lizards, birds' eggs, small birds, and plant exudates.

Leontopithecus occurs at population densities of about 1–17/sq km and groups occupy home ranges of up to 200 ha. (Rylands 1993a). A portion of the home range appears to be a territory that is defended from other groups (Peres 1989). Territory size has been found to average 75 ha. for *L. chrysomelas* and 42 ha. for *L. rosalia* (Dietz, De Sousa, and Da Silva 1994). Groups consist of 2–11, usually 3–7, related individuals, but temporary associations of 15–16 have been observed. The basic group is thought to consist of a mated pair plus their young of one or more years. Although adults of the same sex have been reported to be extremely aggressive toward one another, even fighting to the death if kept together, numerous observations now indicate that natural groups frequently contain more than one adult of each sex. The dominant adult male and female of a group form a permanent pair bond, are equal in rank, and share responsibility for raising the young. In the rate of dispersal of subordinate adults, groups of *Leontopithecus* probably are comparable to those of *Saguinus* rather than to the more stable groups of *Callithrix* (Coimbra-Filho and Mittermeier 1973a, 1977a; Rylands 1993a; Snyder 1974).

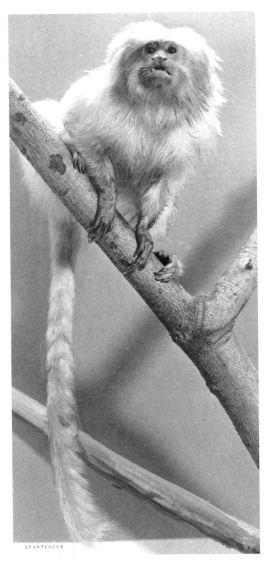

Golden lion tamarin *(Leontopithecus rosalia)*, photo from New York Zoological Society.

McLanahan and Green (1977) identified 17 types of vocalization, grouped into several classes with such associations as trills (for solo activity), clucks (for foraging), long calls (for vigilance), and whines (for contact).

The dominant female of a group may inhibit reproductive activity in other females, but unlike the process discussed in the account of *Callithrix*, this one seems to be strictly behavioral and does not involve physiological mechanisms. This results in a more flexible mating system, perhaps associated with resource availability, and in some groups there is mating by more than one female (Dietz and Baker 1993; Rylands 1993a). *Leontopithecus* apparently is a seasonal breeder. In Brazil births occur from September to March, the warmest and wettest period of the year (Coimbra-Filho and Mittermeier 1973a). In captivity in the Northern Hemisphere most litters are born from January to June. The estrous cycle averages about 2–3 weeks and gestation averages 128 days, ranging from 125 to 132 days. Of the recorded births in the National Zoo in Washington,

D.C., 116 were of twins, 41 were single, and 8 were of triplets (Kleiman 1977b). The young are born fully furred with their eyes open and weigh an average of 60.6 grams (Kleiman 1981). Within a few days after a birth the father begins to carry the infant at times, and he is the principal carrier after the third week. Juvenile members of the family also carry and care for the newborn (Kleiman 1981). Weaning is complete at 12 weeks (Hayssen, Van Tienhoven, and Van Tienhoven 1993). Sexual maturity is reached by males at about 24 months and by females at 18 months (Eisenberg 1977). Life span in captivity, when the animals are well cared for, can be 15 years or longer. One specimen of *L. rosalia* was still living in December 1995 at an age of 28 years and 2 months (Marvin L. Jones, Zoological Society of San Diego, pers. comm., 1995).

Lion tamarins are among the world's most critically endangered mammals. They are classified as endangered by the USDI and are on appendix 1 of the CITES. The IUCN now designates *L. chrysomelas* as endangered and the other three species as critically endangered. They are found in the part of Brazil that has the densest human population. They have declined largely because of destruction of their forest habitat for lumber, agriculture, pasture, and housing. Until the 1960s they also were subject to considerable exportation for use in zoos, laboratories, and the pet trade.

In the nineteenth century the species *L. rosalia* occurred across most of the present state of Rio de Janeiro. Attempts to establish reserves for *L. rosalia* failed in the 1960s, and two prime sites were deforested in 1971 when money could not be found for purchase (Coimbra-Filho and Mittermeier 1977a). *L. rosalia* now is restricted to the Sao Joao Basin in Rio de Janeiro, where the habitat it currently occupies totals about 105 sq km and is continuing to decline. A 5,500-ha. reserve for *L. rosalia* was established in the mid-1970s and is now the site of a reintroduction program. Intensive efforts to establish a captive breeding pool for *L. rosalia* were begun in the 1960s. By 1983 the captive population of that species contained about 370 animals and was increasing at a rate of 20–25 percent annually. Efforts to reintroduce some of these captives were initiated in Brazil in 1984 and eventually achieved a high degree of success, including substantial reproduction in the wild (Beck et al. 1991; Kleiman et al. 1986). The captive population has now leveled off at just over 500 and is being intensively managed for genetic viability (Rylands 1994). The total wild population of *L. rosalia* is now estimated at 560 individuals, including about 39 surviving reintroduced animals and 53 of their offspring (Mallinson 1994).

L. chrysomelas has always been restricted to a small coastal area of the state of Bahia and extreme northeastern Minas Gerais, but its range is now fragmented. A reserve of about 11,400 ha. was authorized for its protection, but only about 7,000 ha. has been purchased and much of that has been taken over by squatters (Mittermeier et al. 1986; Oliver and Santos 1991; Rylands 1994). The species also is able to use some of the cocoa plantations within its range since cocoa is grown in the shade and many large forest trees are left standing for this purpose. The great majority of wild individuals thus do not live in officially protected areas. Although the total remaining range is about 37,500 sq km, suitable habitat is fragmented and declining precipitously. During the 1980s a substantial number of *L. chrysomelas* were exported illegally from Brazil to various other countries. An international project to recover these animals was largely successful and contributed to the start of a captive breeding pool of *L. chrysomelas* (Mallinson 1987). Although once thought to number only about 200 individuals, intensive conservation efforts have resulted in a wild population of the species estimated at 6,000–14,000 and to at least 575 more in captivity (De Bois 1994; Mallinson 1994; Pinto and Tavares 1994; Rylands 1994).

L. chrysopygus, which once occupied a large forested part of the state of Sao Paulo, now has been reduced in range to a 37,157-ha. state park at the western edge of the state, two nearby private ranches, and two small areas in central Sao Paulo, one private and one state-owned. There were once estimated to be only 50–100 individuals left in the wild (Mittermeier et al. 1985), but conservation efforts seem to have led to a moderate improvement in status. There now are about 1,000 in the wild and another 80 in captivity (Mallinson 1994; Valladares-Padua, Padua, and Cullen 1994).

The newly discovered *L. caissara* probably is the rarest and most endangered member of the Callitrichidae. Its entire range may be less than 30,000 ha., though about a third of this area is within protected parks. There are only about 260 individuals in the wild, divided into three isolated groups, and none in captivity (Lorini and Persson 1994; Mallinson 1994; Rylands, Coimbra-Filho, and Mittermeier 1993).

PRIMATES; Family CERCOPITHECIDAE

Old World Monkeys

This family of 21 living genera and 96 species is found in Africa (and, possibly through introduction, in Gibraltar in extreme southern Spain), the southwestern part of the Arabian Peninsula, south-central and southeastern Asia, Japan, and the East Indies as far as Sulawesi and Timor. The sequence of genera presented here is an attempt to express phylogenetic relationships with consideration of the arrangements of both Hershkovitz (1977) and Groves (1989). Two subfamilies are recognized: the Cercopithecinae, with the genera *Erythrocebus, Chlorocebus, Cercopithecus, Miopithecus, Allenopithecus, Cercocebus, Lophocebus, Macaca, Papio, Mandrillus,* and *Theropithecus;* and the Colobinae, with *Nasalis, Simias, Pygathrix, Rhinopithecus, Presbytis, Semnopithecus, Trachypithecus, Colobus, Piliocolobus,* and *Procolobus.* Groves (1989) actually had divided the Cercopithecidae into two full families: the Cercopithecidae, with the subfamily Cercopithecinae for the genera *Erythrocebus, Chlorocebus,* and *Cercopithecus* and Papioninae for all other genera; and Colobidae, with the subfamily Nasalinae for *Nasalis* and *Simias* and Colobinae for all other genera. Subsequently, Groves (*in* Wilson and Reeder 1993) accepted only the single family Cercopithecidae with the two subfamilies Cercopithecinae and Colobinae. The same basic subfamilial divisions were given by Dandelot (*in* Meester and Setzer 1977), Delson (1994), and Thorington and Groves (1970). Hill (1970), however, placed the genera *Cercocebus, Lophocebus, Macaca, Papio, Mandrillus,* and *Theropithecus* in a separate subfamily, the Cynopithecinae.

Head and body length is 325–1,100 mm. The tail, when present, is 20–1,030 mm long; it is absent in the Barbary ape *(Macaca sylvanus).* The heaviest members of this family weigh up to 54 kg *(Mandrillus).* The tail is slightly prehensile in young guenons *(Cercopithecus).* Adult mangabeys *(Cercocebus)* sometimes coil their tails around a branch for support, but the tail is not prehensile. With these exceptions, the tail, when well developed, serves as a balancing organ. The pelage varies from short to long, and a cap or crest on the head, a mane, or a beard is present on some. The face is bare or nearly so and is often almost black, but adult male *Mandrillus sphinx* have brilliant red, blue,

or purple streaks on the face. The palms and soles are naked, and the underparts are usually sparsely haired. Naked buttock pads are often brightly colored. The muzzle is elongate to rounded, usually longer in male baboons than in females. The ears are rounded. The space between the nostrils is narrow, the nostrils being higher than wide and directed forward or downward. The members of the genera *Rhinopithecus* and *Simias* have a peculiar upturned nose, and in old males of *Nasalis* the nose is long and pendulous. Cheek pouches inside the lips are present in the macaques, mangabeys, guenons, and baboons but absent in the langurs, proboscis monkey, and colobus monkeys. The stomach in the latter group is large and has folds or pockets in the walls that are adaptations for handling bulky leafy food. The females of this family have one pair of mammae in the chest region.

The forelimbs are generally shorter than the hind limbs. The hands and feet are adapted for grasping and have five digits, but the thumb is absent, or nearly so, in colobus monkeys. The thumb, if present, is opposable, as is the large toe. The nails are flattened.

The dental formula is: (i 2/2, c 1/1, pm 2/2, m 3/3) × 2 = 32. The lower incisors are nearly upright and in contact with the canines, whereas the upper incisors are pressed close together and separated from the canines by a small space. The upper canines are elongate and tusklike, and the lower canines are curved inward and backward slightly. The premolars are smaller than the molars, and the molars are supplied with tubercles. The highly concave palate extends beyond the last molar, and the skull is rounded or flattened, with a large braincase.

The members of this family have good sight, hearing, and sense of smell, a moderate amount of facial expression, and a variety of calls. Nearly all Old World monkeys and baboons are diurnal. Baboons are mainly terrestrial; macaques are at home either in trees or on the ground; and the other members of the family are mainly arboreal. Cercopithecids can stand upright, but they generally progress on all four limbs. Langurs and colobus monkeys, perhaps the most arboreal members of the family, jump and leap through trees with amazing agility and speed. Sentinels may be posted to detect danger; the usual reaction when they are threatened is to flee, but male baboons often stay at the rear of a retreating group to fight an enemy that comes too close. Most forms are good swimmers.

Some members of this family raid crops or forage near villages. Some are hunted, usually illegally, for their meat and fur. The rhesus monkey *(Macaca mulatta)* and the crab-eating macaque *(M. fascicularis)* are used extensively in laboratory research. The constitution of most cercopithecids makes them better suited for a life in captivity than the New World monkeys. However, their manners are less gentle, and when they grow old they usually are not satisfactory pets. The capture of primates for use as pets is usually illegal in the countries where the animals occur, and importation into the United States for such purposes is now prohibited by federal law.

The geological range of this family is early Miocene to Recent in Africa and late Miocene to Recent in Europe and Asia (Thorington and Anderson 1984). There was an ancestral subfamily, the Victoriapithecinae, and the divergence of the modern subfamilies Cercopithecinae and Colobinae apparently occurred in the middle to late Miocene (Delson 1994).

Patas monkeys *(Erythrocebus patas):* A. Photo by Ernest P. Walker. B. Photo from U.S. National Zoological Park.

PRIMATES; CERCOPITHECIDAE; **Genus**
ERYTHROCEBUS
Trouessart, 1897

Patas Monkey, or Red Guenon

The single species, *E. patas,* inhabits open country from Senegal to Ethiopia and south to Tanzania. Dandelot (*in* Meester and Setzer 1977) and Lernould (1988), among others, treated *Erythrocebus* as a subgenus of *Cercopithecus,* but Groves (1989), Hershkovitz (1977), Napier and Napier (1985), and Thorington and Groves (1970) regarded it as a separate genus.

Head and body length is 600–875 mm and tail length is 500–750 mm. The adult males, weighing about 7–13 kg, are usually much larger than the females, which weigh about 4–7 kg. On the adult males the hairs of the nape and shoulders may be long and manelike, and there is a mustache of white hairs, along with whiskers that are usually white. The upper parts are usually reddish mixed with gray and the underparts are whitish, grayish, buff pink, or orange rufous. The thighs are often whitish, and the tail is dark above and light below. There may be a light forehead band and a dark dorsal stripe. The young are almost entirely red. The long limbs of patas monkeys are of about equal length. The digits are considerably reduced, but the more proximal bones of the feet are elongated (Kingdon 1988). According to Groves (1989), the limb specializations and unusual red coloration of *Erythrocebus* set it apart from *Cercopithecus.* The latter also is distinguished from *Erythrocebus,* as well as *Chlorocebus* (see account thereof), by various cranial characters.

Much of what is known about the natural history of this genus has resulted from Hall's studies (1967, 1968) in Uganda. The patas monkey occupies grass and woodland savannahs and avoids the more densely treed areas within its range. It is diurnal and largely terrestrial. It can climb trees when alarmed but usually relies on its speed on the ground to escape from danger. Perhaps the fastest of all pri-

mates, it has been timed running at about 55 km/hr. During a 10- to 12-hour day a group travels from as little as 500 meters, when food is abundant, to as much as 12,000 meters, when resources are scarce. During the 2–3 hottest hours of the day the animals rest in a large shade tree. At night each individual goes up a separate tree to sleep, except that females retain their infants at this time. A group may be spread over an area of 250,000 sq meters at night and thus be protected from severe loss to predators. The bulk of the diet consists of grasses, berries, fruits, beans, and seeds, occasionally supplemented by mushrooms, ants, grasshoppers, lizards, and birds' eggs. In the 311,200-ha. area studied by Hall there were 110 patas. Each group was found to occupy an extensive home range, that of one troop of 31 animals being 51.8 sq km. There is no core area, and almost any part of the range may be used at night for sleeping. Chism and Rowell (1988) reported that two groups used home ranges of 23.4 and 32.0 sq km.

Hall observed group size to average 15 and to range from 9 to 31 individuals. Each heterosexual group included only 1 adult male and 2–8 adult females. Isolated adult males and one group of 4 males were also seen. Presumably, the dominant male of a heterosexual group drives the other males out. In a study in Cameroon that in most respects confirmed Hall's findings, Struhsaker and Gartlan (1970) found mean group size to be 21 (7–34) patas. Although typically there was a single adult male per heterosexual group, one troop apparently had 2 adult males, and several all-male groups were also seen. In Kenya, Harding and Olson (1986) found as many as 10 males to enter a group of 74 individuals during the mating period and to mate with the females, thus suggesting that the patas monkey's social structure is more variable than previously thought. Perhaps the situation can be explained by the work of Chism and Rowell (1988), also in Kenya, who found the patas population to consist both of heterosexual groups, usually with 1 or a few adult males and numerous females, and of other males that lived alone or in all-male associations. These other males appeared to have large home ranges that overlapped those of several heterosexual groups. Males fre-

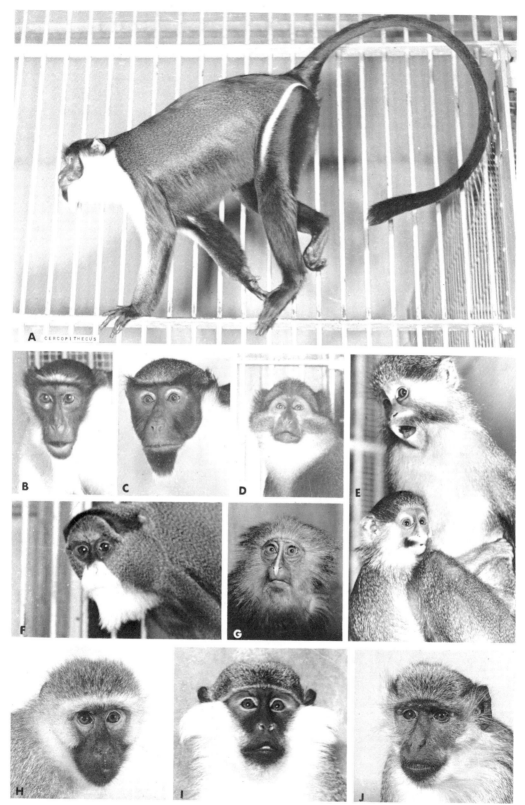

African guenons: A & C. Diana monkey *(Cercopithecus diana)*; B. Roloway monkey *(Cercopithecus diana roloway)*; D. Preuss's monkey *(Cercopithecus preussi)*; E. Moustached monkey and baby *(Cercopithecus cephus)*; F. De Brazza's monkey *(Cercopithecus neglectus)*, photos by Ernest P. Walker. G. Owl-faced monkey *(Cercopithecus hamlyni)*; H. Vervet monkey *(Chlorocebus pygerythrus)*, photos from San Diego Zoological Garden. I. Grivet monkey *(Chlorocebus aethiops)*, photo by Bob McIntyre through Cheyenne Mountain Zoo. J. Green monkey *(Chlorocebus sabaeus)*, photo from San Diego Zoological Garden.

quently moved between their own associations and the heterosexual groups, their presence in the latter coinciding with the mating season.

Whereas Hall found spacing mechanisms to be such that intergroup encounters were rare, Struhsaker and Gartlan reported much contact and agonistic behavior between groups concentrated around water holes at the peak of the dry season. Perhaps such concentrations are responsible for earlier reports of groups comprising as many as 200 individuals. Chism and Rowell also found aggressive interaction between heterosexual groups and observed females and juveniles to be the chief antagonists. According to Hall, however, the adult male is the guardian of the group. He is constantly watchful and often moves several hundred meters away from the group to survey a new feeding area or check for danger. Since he is often away, the adult females establish their own hierarchy to assure group order, and one or more of the females may lead routine movement. Chism and Rowell found that adult females usually initiate group movement, determine the daily route, and select sleeping sites; they maintain contact with a "moo" vocalization. Though usually silent, the male may bark upon encountering another patas group or create a noise to distract a predator. Laboratory investigation has shown that *Erythrocebus* has many vocalizations but that most are muted and can be heard by humans only within 100 meters.

The birth season apparently is restricted to November–January in Cameroon and to around February in Uganda. In Kenya the reproductive period also is limited, with mating occurring from June to September (Harding and Olson 1986) and most young being born in December and January (Chism and Rowell 1988). Studies of captives indicate that the estrous cycle averages about 33 days, estrus 13.5 days, gestation 167 days, and interbirth interval 384 days (Loy 1981). There is normally a single young. In a study of captives, Rowell (1977*b*) determined that females usually can first conceive at 2.5 years and that males reach sexual maturity 1–2 years later. Record longevity is 23 years and 11 months (Marvin L. Jones, Zoological Society of San Diego, pers. comm., 1995).

The subspecies *pyrrhonotus* of Nubia and Somaliland, known as the Nisnas, was probably the Cebus of the ancients, which is pictured so frequently on their monuments and which was described by Aelian as being of a bright flame color with white whiskers and underparts. The Teso natives formerly thought that the flesh of the patas monkey would cure leprosy.

Erythrocebus sometimes is destructive to crops. For that reason and for its meat, it often is hunted by people. Its habitat is being reduced in some areas by heavy cattle grazing and conversion of the savannahs to farmland (Wolfheim 1983). Although it is not dependent on large trees, it does require some woodland and will disappear from an area if such habitat is lost (Chism and Rowell 1988).

PRIMATES; CERCOPITHECIDAE; **Genus**
CHLOROCEBUS
Gray, 1870

Savannah Guenons

Dandelot (*in* Meester and Setzer 1977) and Lernold (1988) recognized four allopatric species:

C. sabaeus (green monkey), Senegal to the Volta River in Ghana;

C. tantalus, savannah zone from the Volta River in Ghana to Uganda;

C. aethiops (grivet), Sudan, Ethiopia;

C. pygerythrus (vervet), southern Ethiopia to Angola and South Africa.

Sineo, Stanyon, and Chiarelli (1986) treated *C. cynosurus* of Angola and some adjacent areas as a species distinct from *C. pygerythrus*. All of these species were considered only subspecies or synonyms of *C. aethiops* by Corbet and Hill (1991), Groves (1989 and *in* Wilson and Reeder 1993), and Thorington and Groves (1970). Of these various authorities only Groves actually recognized *Chlorocebus* as a genus separate from *Cercopithecus*.

Head and body length is 350–660 mm, tail length is 420–720 mm, and weight is 2.5–9.0 kg (Kingdon 1971; Van Hoof *in* Grzimek 1990). Average adult weight is 7.0 kg in males and 5.6 kg in females (Fedigan and Fedigan 1988). The fur is predominantly gray- to yellow-green and the face is black with a white fringe. Males have a light red penis and a blue scrotum. Groves (1989) pointed out that both *Chlorocebus* and *Erythrocebus* lack certain cranial specializations of *Cercopithecus*. In the latter genus the inferior suborbital region is regularly curved toward the dental arcade, the pyriform aperture is round to oval (not angled in the middle), the temporal lines anteriorly follow the posterior borders of the orbits, the nasal bones usually run straight across inferiorly instead of being pointed at the midline, and the second upper incisor tooth is small and pointed. *Chlorocebus* and *Erythrocebus* also share the following specializations: in side view the orbits do not slope forward inferiorly, but their lower borders are situated behind the upper margins; the auditory tube has a V-shaped lower margin; and the orbits themselves are angular instead of oval. The hand and foot of *Chlorocebus* exhibit some of the terrestrial adaptations of *Erythrocebus* but in less extreme form (Kingdon 1988).

According to Fedigan and Fedigan (1988), these monkeys are found in a variety of habitats, including savannah with minimal tree cover, open woodland, and gallery and rainforest edge. They prefer riverine woodland but seem limited only by the need for water and sleeping trees. They seem more adaptable than *Cercopithecus*, being able to utilize such marginal habitats as secondary growth, agricultural areas, and mangrove swamps. They are partly terrestrial, readily forage and run about on the ground, and are good swimmers. Like other guenons, however, they depend on trees as sleeping sites at night, and they take shelter in trees when alarmed. *C. aethiops* may even gather and hold well-leaved branches around itself. Savannah guenons eat a great variety of natural and cultivated fruits, other plant parts, insects, crustaceans, birds' eggs, and small vertebrates. In the mangroves of Senegal more than half of the feeding time of *C. sabaeus* involved hunting for fiddler crabs (Galat and Galat-Luong 1976).

According to Fedigan and Fedigan (1988), group home ranges have been calculated at 13–178 ha. and group size at 5–76, averaging 25, individuals; there usually are several adults of each sex in a group. Young males generally disperse from their natal groups, but young females remain and become part of an established social hierarchy. In some populations a number of groups seem to form a larger community, in which there is an interchange of males and other friendly interaction. The degree of territoriality displayed by groups seems to vary widely, depending on such factors as history of association with neighboring groups and availability and concentration of resources. In Kenya, Struhsaker (1967*b*) found groups of *C. pygerythrus* occupying home ranges of 18–96 ha., most of which was a defended territory. Galat and Galat-Luong (1976) found groups of *C. sabaeus* to defend their territories through loud barking and displays. *C. aethiops* has a creaking cry and a staccato bark that enable the members of a troop to keep in touch with one another. Other calls of this species express alarm, excitement, pain, rage, and melancholy. Struhsaker (1967a) found *C. pygerythrus* to make 36 sounds that conveyed at least 21–23 different messages.

In the Sahel region of Senegal, Galat and Galat-Luong (1977) found that over a three-year period a group of *C. sabaeus* numbered 33–41 animals, including 2–5 adult males and 10–15 adult females. Another troop in this region had 140–74 members (Galat and Galat-Luong 1978), but in a separate study of *C. sabaeus* in Senegal, Dunbar (1974) found group size to be only 4–16. In Uganda, Gartlan and Brain (1968) found group size in *C. pygerythrus* to range from 4 to 51 and to average 9–11. In contrast with what was observed in some other investigations, the adult males seemed peripheral to the group and may even have moved between groups. The nucleus of a group was formed by the adult females and their young. Rowell (1971), studying captive *C. aethiops*, observed that adult males were only equal to or lower than adult females in dominance. In an investigation of an introduced population of *C. sabaeus* on St. Kitts Island in the West Indies, however, McGuire (1974) found that all groups, each numbering 4–65 animals, had a single dominant adult male.

Dunbar (1974) observed that in Senegal there was an apparent birth peak in *C. sabaeus* from January to March. Fedigan and Fedigan (1988) reported that in any given population of *Chlorocebus* there is usually a synchronized peak of parturition, with approximately 80 percent of births occurring in a period of 2–3 months. This period seems usually to coincide with the beginning of the season during which resources are most plentiful. Females commonly produce young at intervals of 1 year. Weaning is usually completed when the young is 8.5 months old. Females are capable of giving birth at 2.5 months. Males are physiologically capable of reproduction at about 3 years but are not behaviorally mature until after 5 years. The estrous cycle lasts about 30 days (Hayssen, Van Tienhoven, and Van Tienhoven 1993). Van Hoof (*in* Grzimek 1990) noted that there normally is a single young, birth weight is 300–400 grams, gestation is 5.5 months, and the life span is more than 30 years.

All species of *Chlorocebus* are on appendix 2 of the CITES but are thought to be common and to be relatively adaptable to the presence of human activity (Fedigan and Fedigan 1988). Populations of *C. sabaeus* are established on St. Kitts, Nevis, and Barbados islands in the West Indies. They are probably descended from pets brought by African slave traders in the seventeenth century and have been the subject of a number of field studies (Horrocks 1986; McGuire 1974; Poirier 1972). Denham (1982) suggested that numbers initially increased rapidly on Barbados, then crashed in the eighteenth century because of loss of forest habitat and bounty hunting, and finally increased again in the 1950s after some areas had become reforested.

PRIMATES; CERCOPITHECIDAE; **Genus**
CERCOPITHECUS
Linnaeus, 1758

Guenons

There are 20 species (Colyn, Gautier-Hion, and Thys van den Audenaerde 1991; Dandelot *in* Meester and Setzer 1977; Dutrillaux et al. 1988; Harrison 1988; Kuroda, Kano, and Muhindo 1985; Lernould 1988; Oates 1988; Sineo, Stanyon, and Chiarelli 1986; Thorington and Groves 1970):

C. nictitans (white-nosed guenon), Liberia, Ivory Coast, Nigeria to Congo and Zaire north of Congo River, island of Bioko (Fernado Poo);

C. mitis (blue monkey), forested areas west of Rift Valley in southern Sudan, Ethiopia, Uganda, Kenya, Zaire, Angola, and Zambia;

C. albogularis (Sykes monkey), east of Rift Valley from Somalia south to eastern Cape Province in South Africa, islands of Phylax, Zanzibar, and Mafia;

C. ascanius (black-cheeked white-nosed monkey), Central African Republic to western Kenya and Angola;

C. petaurista (lesser white-nosed guenon), Guinea to Benin;

C. erythrogaster (red-bellied guenon), southern Nigeria to west of Niger River;

C. sclateri, southeastern Nigeria between Niger and Cross rivers;

C. erythrotis (russet-eared guenon), southeastern Nigeria east of Cross River, Cameroon, island of Bioko;

C. cephus (mustached monkey), southern Cameroon, Equatorial Guinea, Gabon, Congo, western Central African Republic;

C. campbelli, Gambia to Ghana;

C. mona (mona monkey), Ghana to southwestern Cameroon;

C. pogonias, southeastern Nigeria, Cameroon, Equatorial Guinea, Gabon, Congo, Central African Republic, northwestern Zaire;

C. wolfi, Zaire, western Uganda;

C. diana (Diana monkey), Sierra Leone to Ghana;

C. dryas, central Zaire;

C. neglectus (De Brazza's monkey), forested zones from southeastern Cameroon to southern Ethiopia and southern Zaire;

C. hamlyni (owl-faced monkey), eastern Zaire, Rwanda;

C. lhoesti (L'hoest's monkey), northeastern Zaire, Rwanda, Burundi, Uganda;

C. solatus, central Gabon;

C. preussi, southeastern Nigeria, Cameroon, island of Bioko.

De Brazza's monkeys *(Cercopithecus neglectus)*, photo from San Diego Zoological Society.

Chlorocebus (see account thereof) often is included within *Cercopithecus*. Various authorities also have included *Allenopithecus, Cercocebus, Erythrocebus, Miopithecus, Papio,* and *Theropithecus.* Thorington and Groves (1970) considered *C. erythrotis* part of *C. cephus* and thought *C. albogularis* might be a subspecies of *C. mitis.* Corbet and Hill (1991), Groves (*in* Wilson and Reeder 1993), and Lawes (1990) agreed with such allocation for *C. albogularis,* but Lernould (1988) retained it as a separate species. Groves, as well as Lernould, listed *C. sclateri* as a separate species, though Lernould indicated that it might be a hybrid between *C. erythrotis* and *C. erythrogaster.* Lernould also showed the presence of a hybrid zone between *C. erythrotis* and *C. cephus* along the Sanaga River in southwestern Cameroon, and he listed numerous other records of wild and captive hybrids between various species and subspecies of *Cercopithecus.*

Head and body length is about 325–700 mm and tail length is about 500–1,000 mm. Chiarelli (1972) listed the following weights: *C. ascanius,* 1.8–6.4 kg; *C. mona,* 3–6 kg; *C. neglectus,* 4.5–7.8 kg; and *C. mitis,* 6–12 kg. Some forms have long white whiskers associated with a blackish or dark red belly. There may be a white brow band and a tuft of hair on the chin; others have variously colored whiskers with or without a nasal spot and a patch of yellow or grayish yellow hair on the cheek defined by a black stripe above and below. The nasal spot may be white, red, or blue. Some forms have white or buff stripes on the thighs. Some guenons are known for the beauty of their pelage—the individual hairs ringed with different colors and combined with speckled black and gray. The upper parts are often greenish gray, but browns, grays, and intermediate shades are the common colors. The color of underparts also varies considerably. The skin of part of the face in some forms is blue or violet.

Cercopithecus is quite variable cranially but generally can be distinguished from *Chlorocebus* by the characters discussed in the account of the latter. Cheek pouches are present. Guenons may be characterized by their roundish head, slender body, long hind limbs, and long tail. The nostrils are close together, a beard is often present, and the side whiskers are well developed. The common name comes from the French word *guenon,* meaning "fright," and refers to the fact that these animals grimace and expose teeth when they are excited or angry.

Guenons inhabit forests, woodlands, and savannahs, usually near rivers and streams. *C. mona* has been found to live successfully in mangroves (Gartlan and Struhsaker 1972). Guenons are diurnal; they are most active in the early morning and again in the late afternoon or evening. Some forms spend nearly all of their time in trees, jumping from one tree to another. The tail in young guenons is prehensile, but adults normally use the tail as a balancing organ. Some guenons often run about on the ground, but all take shelter in trees when alarmed. They sleep in trees, probably in a sitting position holding onto branches or to each other. The diet is varied but consists mainly of fruits and the seeds thereof (Gautier-Hion 1988). A substantial amount of leaves also is eaten, some grain and roots, and, at least on occasion, young birds, birds' eggs, small reptiles, and insects.

Home range size varies greatly. In the coastal forests of the Ivory Coast a troop of *C. campbelli* stayed within 3 ha. from 1967 to 1970 and spent most of its time in an even smaller core area (Bourliere, Hunkeler, and Bertrand 1970). In a degraded forest of the Central African Republic a group of 17–23 *C. ascanius* used a 15-ha. home range with a 5-ha. core area (Galat-Luong 1975). The same species, however, has been reported to have a home range of up to 130 ha.

(Chalmers 1973). In Uganda a group of *C. mitis* used a 72.5-ha. home range but moved through only a small part of it on any given day (Rudran 1978). In Kenya troops of *C. neglectus* were found to use home ranges of 4.1–6.0 ha. and to move 330–1,001 meters per day (Wahome, Rowell, and Tsingalia 1993).

There appears to be considerable variation in the social structure of *Cercopithecus,* but data compiled by Cords (1988) suggest that permanent groups generally comprise a single dominant male, who may maintain his position for some years, and a number of adult females. Other males in the vicinity may sometimes move into the group and even mate with the females. Although one troop of about 200 *C. ascanius* has been reported, Struhsaker and Leland (1988) stated that this species typically is found in social groups of 30–35 individuals. There is a single adult male, who usually has a rather short tenure, while the females and their young form the core of the group and are the most active defenders of the territory; one large group was observed to gradually divide into two new permanent groups. Galat-Luong (1975) determined that a group of *C. ascanius* in the Central African Republic numbered only 17–23 individuals; 4 were adult males, 1 of whom was much larger than the others. The group often split into subgroups composed of either adult pairs and their young, females, or juveniles. Rudran (1978) reported groups of *C. mitis* in Uganda to average 20.8 (13–27) individuals and to include only 1 adult male but 4–12 adult females and 7–16 immatures. Apparently, most males dispersed from the groups upon reaching maturity and then lived alone. Tsingalia and Rowell (1984), however, observed 16 adult males in association with a troop of *C. mitis* during a six-month period. Although there was a dominant male, most of the others participated in mating with the group's females. Bourliere, Hunkeler, and Bertrand (1970) observed that a group of *C. campbelli* in the Ivory Coast had 9 members, including a single adult male. This male had a central role in the group and was the only member to emit loud calls, but it was not strongly dominant over the other animals. Younger males seemed to leave the group willingly at about 3–4 years and to take females of the same age with them. Wahome, Rowell, and Tsingalia (1993) found three troops of *C. neglectus* in Kenya to number 11, 13, and 16 individuals; each had a single adult male and at least 3 adult females.

Guenons apparently are territorial but generally avoid serious conflicts. DeVos and Omar (1971) reported *C. mitis* to have territories of 13.2–16.0 ha. Rudran (1978) stated that aggressive interaction in this species was rare as home ranges did not greatly overlap. Different species of guenons may travel together. In fact, guenons and monkeys of the genera *Cercocebus* and *Colobus* may associate. Baboons and chimpanzees, however, do not seem to mix with these smaller monkeys, though they may feed in the same area. Gautier (1988) identified 22 different vocalizations in *Cercopithecus,* including a variety of low- and high-pitched sounds for maintaining group cohesion, whistles and chirps for warning, and loud whoops and booms by adult males.

Data compiled by Butynski (1988) show that all guenon populations have peak mating and birth seasons but that breeding may occur year-round in areas of high rainfall. Most populations have synchronized mating seasons centered on July, August, or September and birth seasons centered on December, January, or February. Interbirth interval has been about a year in some of the populations studied but up to five years in others. A study of *C. mitis* in Kenya (Omar and DeVos 1971) indicated that conception occurred from July to November and that births were confined to a dry period from November to March. Most births took place at the end of this dry season, thereby allowing lactation to proceed when rainfall was high. Rudran (1978) also found seasonality for this species in Uganda, with births occurring from December to May. In the Ivory Coast *C. campbelli* was observed to give birth only from November to January, at the end of the rains or the beginning of the dry season, following a gestation period of about 6 months (Bourliere, Hunkeler, and Bertrand 1970). Gestation lasted approximately 5.6 months in most of the other species investigated but was only 4.7 months in *C. mitis* and up to 7.0 months in *C. badius* (Butynski 1988). Twins are born occasionally, but a single young is usual. Birth weight is around 400 grams (Hayssen, Van Tienhoven, and Van Tienhoven 1993). A young guenon clings to the fur of the underparts of the mother and entwines its tail with its mother's as they travel. Females generally produce their first young at about 4–5 years (Cords 1988). Several specimens of *Cercopithecus* are thought to have lived in captivity for more than 30 years, including one *C. campbelli* that may have reached an age of 38 years (Marvin L. Jones, Zoological Society of San Diego, pers. comm., 1995).

These guenons generally are less adaptable than *Chlorocebus* and have proved more vulnerable to human disruption of forests. The IUCN assigns the following specific and subspecific classifications: *C. mitis kandti* (Uganda, Rwanda, eastern Zaire), endangered; *C. erythrogaster*, vulnerable; *C. sclateri*, endangered; *C. erythrotis* (in general), vulnerable; *C. erythrotis erythrotis* (island of Bioko), endangered; *C. pogonias pogonias* (Cameroon, Equatorial Guinea), vulnerable; *C. diana diana* (Sierra Leone, eastern Guinea, Liberia), vulnerable; *C. diana roloway* (Ivory Coast, Ghana), endangered; *C. hamlyni*, near threatened; *C. lhoesti*, near threatened; *C. solatus*, vulnerable; and *C. preussi*, endangered. The species *C. diana*, *C. erythrotis*, *C. erythrogaster*, and *C. lhoesti* (including *C. preussi*) are listed as endangered by the USDI. *C. diana* is on appendix 1 of the CITES and all other species are on appendix 2. All of the species classified by the IUCN are thought to be declining because of destruction of forest habitat through agricultural expansion and logging and/or excessive hunting for meat by people (Lee, Thornback, and Bennett 1988). For the same reasons, Brennan (1985) regarded *C. neglectus* to be seriously endangered in Kenya, though Muriuki and Tsingalia (1990) reported that the species was somewhat more widely distributed in that country than had been previously thought. *C. solatus*, which was not described until 1988, and *C. sclateri*, a wild population of which was not found until that same year, both occupy highly restricted ranges that are undergoing severe disturbance (Gautier et al. 1992; Oates and Anadu 1989; Oates et al. 1992). *C. mona* was introduced on the island of Grenada in the West Indies, but Fedigan and Fedigan (1988) indicated that it did not adapt as well to the region as *Chlorocebus sabaeus* and may now have disappeared.

PRIMATES; CERCOPITHECIDAE; Genus **MIOPITHECUS**
I. Geoffroy St.-Hilaire, 1842

Talapoin

The single named species, *M. talapoin*, is found in southwestern Zaire, northwestern Angola, and possibly the Ruwenzori region farther east. *Miopithecus* also occurs in southern Cameroon, Equatorial Guinea, Gabon, Congo, and possibly the Central African Republic, but the populations in those countries are thought to represent a second,

unnamed species (Groves 1989; Lernould 1988; Wolfheim 1983). *Miopithecus* was listed as a subgenus of *Cercopithecus* by Dandelot (*in* Meester and Setzer 1977), Lernould (1988), and a number of other recent authorities, but Groves (1989) tentatively placed it in a subfamily entirely separate from that containing *Cercopithecus*.

This is the smallest Old World monkey. Head and body length is about 320–450 mm and tail length is 360–525 mm. Napier and Napier (1967) listed weights of 1,230 and 1,280 grams for two males and 745 and 820 grams for two females. The coloration above ranges from greenish gray or greenish black to greenish yellow or greenish buff. The underparts are white or grayish white. The face is naked except for a few blackish hairs on the lips (upper lip sometimes yellow) and nose. The hairs on the cheeks are golden yellow with black tips; a black streak extends from the corner of the eye halfway to the ear, which is also black. The skin around the eyes is sometimes yellowish or orangish. White hairs with black tips in front of the ear radiate from a point like a fan. The top of the head is ochraceous black. The color of the tail varies from grayish black to brownish black above; beneath, it is yellowish or yellowish gray at the base, blending into buff, yellowish black, or black at the tip. The outer side of the limbs is pale or chrome yellow sometimes tinged with red, and the hands and feet are chrome yellow with a buffy or reddish yellow tint. There is considerable color variation among individuals; some are nearly olive, while others have more of a true green tint.

The limbs and tail of the talapoin monkey are relatively long, but the hands and feet are short. There is some webbing between the fingers and toes. The facial part of the skull is small, and the cranium is relatively large.

Studies by Gautier-Hion (1971, 1973) in Gabon have added considerably to our knowledge of this monkey. All 25 troops observed were centered in inundated rainforest along rivers, and 17 of these troops were located near human settlements. At night the members of each group slept at a communal site in trees along the edge of the water. By day the animals made foraging trips of 1,500–2,950 meters, which sometimes took them into secondary growth or even plantations but never more than 450 meters from a river. There were two daily peaks of activity: from early morning to 1030 hours and from 1330 to 1830 hours. The diet was insectivorous and frugivorous.

On the average there was one troop along every 4,835 meters of river, and mean group home range was 1,216 sq meters. Gautier-Hion distinguished between "parasitic" troops, which lived near human settlements and depended in part on cultivated crops, and nonparasitic troops, which were found at a considerable distance from villages. Nonparasitic troops had on average about 66 members and lived at a population density of 30/sq km. Parasitic troops averaged 112 individuals and had a density of 92/sq km. Each group included a number of adult males that acted as leaders during daily movements and as sentinels at night. The adult males of a group were outnumbered by adult females, and the latter, along with their infants and year-old juveniles, formed a definite subgroup, at least during the night. No displays or interaction suggesting group territoriality were observed. Gautier-Hion noted that *Miopithecus* differed from *Cercopithecus* in its lower population density, larger troops, and lack of territoriality.

Based on a study of captives, Dixson, Scruton, and Herbert (1975) favored retaining the talapoin in a genus separate from *Cercopithecus*. Unlike *Cercopithecus*, the talapoin was found to be able to raise its eyebrows and move its ears during facial expression. Also, talapoins that groomed each other frequently sat or huddled with their tails entwined. Animals adjacent in rank groomed one an-

Talapoin monkey *(Miopithecus talapoin),* photo from San Diego Zoological Garden.

Allen's Monkey

The single species, *A. nigroviridis,* inhabits northeastern Congo and western Zaire. Dandelot (*in* Meester and Setzer 1977), Lernould (1988), and some other authorities regarded *Allenopithecus* as a subgenus of *Cercopithecus,* but Gautier (1985), Hershkovitz (1977), and Thorington and Groves (1970) treated it as distinct, and Groves (1989) placed it in a subfamily entirely separate from that containing *Cercopithecus.*

Head and body length is 400–510 mm and tail length is 350–520 mm (Van Hoof *in* Grzimek 1990). Gautier (1985) reported strong sexual dimorphism in size, with an average weight of about 6 kg for males and 3.5 kg for females. The hair is slightly longer on the nape and shoulders than on the back and scant on the underparts, hands, and feet. The skin of the face is dark grayish brown to black but lighter around the eyes. The hairs on the sides of the face extend outward to form a ruff from the ears to the mouth. The general coloration above is grayish to almost black, sometimes with a yellowish tinge, and the underparts are light gray or whitish, sometimes speckled with black and yellow. The tail is dark above and lighter below.

Groves (1989) observed that *Allenopithecus* differs from *Cercopithecus,* and shows some resemblance to *Papio,* as follows: the sexual area of the female's skin—vagina, perineum, anus—undergoes cyclic enlargement, being maximally swollen around the time of ovulation (there is no sexual swelling in *Cercopithecus*); the ischial callosities of adult males are fused across the midline (they are smaller and discrete in male *Cercopithecus*); there is some facial elongation; the molar teeth are large and have well-developed flare, a convexity of the buccal and lingual surfaces; and there is some facial elongation. Also unlike in *Cercopithecus,* the first and third lower molars of *Allenopithecus* have a small buccal cusplet between the anterior and posterior cusps. The body of *Allenopithecus* is rather heavy; in appearance it is similar to *Cercopithecus,* but anatomically it bears some resemblance to the baboons.

Allenopithecus apparently forages on the ground as the baboons do. Its diet is primarily frugivorous and insectivorous (Lee, Thornback, and Bennett 1988). It frequents swampy areas and probably goes into water freely. In Zaire, Gautier (1985) observed troops in swamp forests, with regular sleeping sites on riverbanks. These groups comprised more than 40 individuals, including several adult males. Captives appeared to enjoy wading in shallow water at the National Zoo in Washington, D.C. The animals there were friendly and comparable to *Cercopithecus* in behavior. The single young weighs about 200 grams at birth (Van Hoof *in* Grzimek 1990). Weaning occurs at 2.5 months (Hayssen, Van Tienhoven, and Van Tienhoven 1993). The record known longevity for *Allenopithecus* is 23 years (Jones 1982). The genus is classified as near threatened by the IUCN and is on appendix 2 of the CITES. Its numbers are thought to have declined in recent years, and remaining populations are threatened by hunting and habitat loss (Lee, Thornback, and Bennett 1988).

other far more than was the case in *Cercopithecus.* Earlier observations showed that the call of the talapoin is a short, explosive "k-sss!" like the splash of a rock in water, and is different from the calls of most guenons.

In Gabon, Gautier-Hion (1971, 1973) found that the birth season of any given troop lasted for about two months during the period November–April and that most adult females gave birth to a single young every year. In a study of captive talapoins, Rowell (1977*a*) determined the mean estrous cycle to be 35 days and gestation to be 158–66 days. The newborn were relatively enormous; one female weighing 1,100 grams gave birth to a 230-gram infant. Rowell (1977*b*) found that females usually were able to conceive at 4.5 years and that males reached sexual maturity 1–2 years after the females. One captive talapoin lived for 27 years and 8 months (Jones 1982).

Miopithecus is on appendix 2 of the CITES, but Lee, Thornback, and Bennett (1988) did not consider it to be threatened. Being so small, it is hardly worth hunting, and its northern population is centered in a region where human habitat disruption has been somewhat less intense than in most of Africa. However, the southern population is more restricted and susceptible to forest disturbance, and recognition of it as a separate species could cause increased conservation concern.

Allen's monkey *(Allenopithecus nigroviridis)*, photo by Ernest P. Walker.

Mangabeys

There are three species (Dandelot *in* Meester and Setzer 1977; Groves 1978*a*, 1989, and *in* Wilson and Reeder 1993; Homewood and Rodgers 1981; Struhsaker 1971):

C. agilis, Cameroon, Equatorial Guinea, Gabon, Congo, Central African Republic, Zaire, southern Tanzania;
C. galeritus, lower Tana River in eastern Kenya;
C. torquatus, Senegal to Congo.

Dandelot (*in* Meester and Setzer 1977) considered these species to form the superspecies *torquatus* and considered the two species here placed in *Lophocebus* (see account thereof) to form another superspecies of *Cercocebus.* Groves (1989) is followed here in recognizing *C. agilis* as a species separate from *C. galeritus,* though the two frequently are treated as conspecific. *C. atys* (sooty mangabey), found from Senegal to Benin, sometimes is regarded as a species distinct from *C. torquatus* (Lee, Thornback, and Bennett 1988).

Head and body length is about 382–888 mm, tail length is about 434–764 mm, and weight is about 3–20 kg. The height when seated is 35–40 cm. In *C. galeritus* the upper parts are golden brown speckled with black or dark grayish brown speckled with gold yellow; the underparts are ochraceous orange or whitish. There is neither a crown patch nor a dorsal line. This species has a whorl, or parting of the hair, on top of the head. In *C. agilis* this whorl is less developed and the pelage is strongly agouti-banded (Groves 1989). In *C. torquatus* the upper parts are slaty gray or grayish brown to dark brown and the underparts are generally grayish. There usually is no speckling in the pelage. The top of the head usually has a conspicuous patch of mixed olive and black, blackish brown, or chestnut red, and the individ-

ual hairs are tipped with black. There is no whorl on top of the head. The crown patch is surrounded by a black or white band that is continued behind into a neck patch. This gradually leads into a black line along the middle of the back that extends to the tail. The tip of the tail of *C. torquatus* is paler than the rest of the tail. A conspicuous cheek beard is present.

Mangabeys, though similar in most respects, are longer, taller, and of more slender proportions than are guenons *(Cercopithecus).* They also differ in that they have pale upper eyelids, the buttock pads are fused in the males, and the third lower molar has five tubercles (as in *Macaca*) rather than four (as in *Cercopithecus*). Mangabeys have an oval-shaped head with a fairly long muzzle. Large cheek pouches are present and constantly used. There are webs between the fingers; the second and third toes are united by a web for most of their length, and the fourth is united to the third and fifth as far as the middle joints.

According to Lee, Thornback, and Bennett (1988), *C. torquatus* is found mostly in high-canopy primary forest but also in mangrove, gallery, and inland forests, as well as in young secondary forests and around cultivated areas. In a study in Equatorial Guinea, Jones and Sabater Pi (1968) found *C. torquatus* occupying a broad range of habitats, including swamp and agricultural areas, and to be partly terrestrial. Groups had 14–23 individuals, of which several were large males. This species is mainly frugivorous but may eat significant amounts of animal matter.

In Gabon, Quris (1975) found *C. agilis* apparently to be restricted to periodically flooded areas and to prefer the lower forest strata. Fruit was the favorite food, and leaves, grass, and mushrooms were also eaten. Population density in the study area was 6.7–12.5/sq km, and one troop had a home range of 1.98 sq km. The number of individuals in a troop ranged from 8–9 to 17–18, but there was frequent division into subgroups. Each of three observed groups contained only a single adult male, who functioned as leader and defender. The loud calls of the adult males could be heard by people up to 1,000 meters away and perhaps played a role in regulation of group spacing. However,

Gray-cheeked mangabey and baby *(Lophocebus albigena)*, photo from New York Zoological Society. A. Red-crowned mangabey *(Cercocebus torquatus)*; B. Sooty mangabey *(Cercocebus torquatus atys)*; C. Black-crested mangabey *(Lophocebus aterrimus)*, photos by Ernest P. Walker.

groups sometimes met one another with no apparent antagonism and temporarily united. Exchange of members between troops sometimes occurred. The presumed season of birth in the area was December–February.

In a study of the Tana River mangabey, *C. galeritus,* in eastern Kenya, Homewood (1975) found population density to vary from about 3/ha., with a home range of as little as 15 ha., in optimal diverse evergreen forest to fewer than 1/2 ha., with a home range of 50–100 ha., on poor, low-diversity deciduous woodland. These mangabeys seemed equally at home 30 meters up in trees and down on the ground, where they traveled between adjacent forest patches and spent much of the day foraging in leaf litter. The diet consisted mainly of fruits, seeds, and insects. Group size was 13–36 individuals, usually including at least 2 adult males and about twice as many adult females. Some males lived alone. During the dry season, when food was limited,

the groups maintained discrete territories, with spacing maintained by the long-range vocalizations, displays, and occasional active combat between the dominant males. During the rains, however, when food was abundant, territories seemed to break down and groups became more tolerant of one another, sometimes joining in temporary associations of 40–50 mangabeys.

Apparently, a single young is born at a time. The gestation periods in three births of *C. torquatus* were 164, 173, and 175 days (Stevenson 1973). At the Yerkes Regional Primate Center, in Atlanta, Georgia, births of captive *C. torquatus atys* occurred in all months of the year but predominantly from March to August. Females first gave birth at an average age of 56.5 months and then had an average interbirth interval of 16.6 months. Females more than 22 years old were still reproductively active (Gust, Busse, and Gordon 1990).

All species of mangabeys are losing forest habitat to logging and other human activity (Wolfheim 1983). All are on appendix 2 of the CITES, except for the isolated Tana River mangabey *(C. galeritus)*, which is on appendix 1. *C. galeritus* (not including *C. agilis*) is designated as endangered by the IUCN and the USDI. It is restricted to a forested area of about 730 ha. along the Tana River in eastern Kenya. Suitable habitat has been reduced and severely modified through agriculture, logging, and construction of hydroelectric dams upstream. Recent population estimates of this species have ranged from about 600 to 1,200 individuals (Butynski and Mwangi 1995; Medley 1993). A subspecies of *C. agilis* discovered in the Uzungwa Mountains of Tanzania and not yet formally described also is classified as endangered by the IUCN (under the name *C. galeritus*) because of hunting and habitat loss (Lee, Thornback, and Bennett 1988). The IUCN designates the entire species *C. agilis* (under the name *C. galeritus*) as near threatened.

The species *C. torquatus* is listed as endangered by the USDI because large-scale logging apparently has greatly reduced its range in West Africa. Schlitter, Phillips, and Kemp (1973), however, stated that this species is adaptable and could survive in areas of cutover forest and agriculture. The IUCN recognizes *C. torquatus* and *C. atys* as separate species (see above) but designates both generally as near threatened. It singles out as vulnerable the subspecies *C. atys lunulatus*, of Ivory Coast and Ghana, which is threatened by habitat loss, hunting, and conflicts resulting from raids on fruit plantations (Lee, Thornback, and Bennett 1988).

PRIMATES; CERCOPITHECIDAE; **Genus**
LOPHOCEBUS
Palmer, 1908

Black Mangabeys

In designating *Lophocebus* as a genus distinct from *Cercocebus*, with consideration of morphological and immunological evidence, Groves (1978*a* and *in* Wilson and Reeder 1993) indicated that it comprised only the species *L. albigena* of southeastern Nigeria, Cameroon, Congo, Gabon, Equatorial Guinea, northeastern Angola, Central African Republic, Zaire, Burundi, Uganda, western Kenya, and western Tanzania. The populations to the south of the Congo River, in Zaire and Angola, were considered to represent a subspecies, *L. albigena aterrimus*. On the basis of sociological data, Horn (1987*b*) supported the subspecific designation of *aterrimus* (he did not comment on the generic status of *Lophocebus*). However, Groves (1989) indicated that despite evidence of some very slight gene flow across the Congo River, it would be appropriate to recognize *aterrimus* as a separate species. Dandelot (*in* Meester and Setzer 1977) treated *L. albigena* and *L. aterrimus* as full species that together formed the superspecies *albigena* of the genus *Cercocebus* (see account thereof). *Lophocebus* also was not accepted as a distinct genus by Corbet and Hill (1991) or Napier and Napier (1985). Such distinction, however, was supported by recent molecular and chromosomal analyses (Disotell 1994).

Head and body length is 400–720 mm, tail length is 550–1,000 mm, and weight is about 4–11 kg (Van Hoof *in* Grzimek 1990). In *L. albigena* the hair is soft and dull black. The cheeks have short hairs, a long brow fringe and crest are present, and there is a mantle on the shoulders. *L. aterrimus* has coarse and glossy black hairs, long grayish brown whiskers partly concealing the ears, and a high conical

crest, but the long brow fringe and mantle are not present. Groves (1989) wrote that *Lophocebus* differs from *Cercocebus* in the form of the nasal bones and interorbital region, cranial flexion, the form of the suborbital fossa, smaller cheek teeth and reduced molar flare, orbit shape, malar foramen size, and characters of the mandible. *Lophocebus* also differs in not having pale eyelids that contrast with an otherwise dark face.

Dandelot (*in* Meester and Setzer 1977) stated that *L. albigena* and *L. aterrimus* (which he considered to compose the superspecies *albigena* of the genus *Cercocebus*) are arboreal and have "supple" movements, very dark pelage, and a long, ruffled tail. In contrast, the species *Cercocebus torquatus* and *C. galeritus* (which he considered to compose the superspecies *torquatus*) are semiterrestrial and have a "stiff" gait, rather light pelage, and a medium-long, not very supple tail. *Lophocebus* has a throat sac that gives resonance to its calls. This sac is small in the female but large in the male. The call is a shrieking, howling sound resembling "hu, hu." According to Kingdon (1971), there are a variety of other vocalizations, including barks, chuckles, twitters, and grunts.

Kingdon noted also that black mangabeys are never found far from water or at elevations above 1,700 meters. In a study in Equatorial Guinea, Jones and Sabater Pi (1968) found *L. albigena* to inhabit mainly primary and secondary forest, whereas *Cercocebus torquatus* occupied a broad range of habitats. *L. albigena* was entirely arboreal and *C. torquatus* was partly terrestrial. Groups of *L. albigena* contained 9–11 animals, including only a single adult male, whereas groups of *C. torquatus* had 14–23 individuals, of which several were large males.

In Uganda, Chalmers (1968*a*, 1968*b*) observed *L. albigena* to spend most of its time in the high canopy of the forest but occasionally to come down to raid cultivated fields. Unlike most monkeys, these mangabeys had three, rather than two, daily feeding periods: 0800–0900, 1200–1300, and 1600–1700 hours. Food consisted mainly of fruits, buds, and shoots and was eaten mainly high in the trees. Population density in the study area was about 78/sq km and group home range was about 0.13–0.26 sq km. Groups often moved through their entire home range in a single day. They generally avoided one another, perhaps being spaced by the loud vocalizations of the males. Groups had 7–25 members, with an equal number of adult males and females. Males were aggressive but were solicitous of the young. Breeding seemed to continue throughout the year. Age of sexual maturity in males was about 5–7 years.

Additional studies of *L. albigena*, in the Kibale Forest Reserve of Uganda, showed population density to increase from 6–7/sq km in 1971 to 13–14/sq km in 1991, apparently in response to regeneration of forest cover. Groups moved about 1,000 meters per day and utilized home ranges of approximately 1 sq km (Olupot et al. 1994). In another investigation of *L. albigena* in Uganda, Waser (1977) observed group movement pattern apparently to center on large fruiting trees. A group would repeatedly visit a single such tree for days or weeks but also would spread out over an enormous area to search for other resources. One group had a home range of 4.1 sq km. Group size ranged from 6 to 28 animals. A particular group in this area studied by Waser and Floody (1974) was made up of 15–16 individuals, of which 6–7 were adult or subadult males.

In Zaire, Horn (1987*a*) found *L. aterrimus* to be completely arboreal and to use all levels of the forest canopy, but especially the middle layers. Foraging activity was concentrated in the early morning, sometimes even before daybreak. The diet consisted mainly of fruits and nuts. Two groups had home ranges of 48 ha. and 70 ha., which mostly overlapped. The groups contained 14–19 individuals, including several adults of both sexes. There was often division into subgroups, each with at least 1 adult male. Vocalizations were frequent and loud. Among them was the "whoopgobble," audible to humans more than 500 meters away and apparently given only by adult males for purposes of intergroup location and spacing.

Gevaerts (1992) found births in northern Zaire to occur mainly in July and August, at the start of the wet season. Kingdon (1971), however, indicated that in East Africa black mangabeys do not show any marked seasonal breeding and that in most troops there are usually babies and young of all sizes; the gestation period is about 174 days. A captive specimen of *L. albigena* lived for 32 years and 8 months (Jones 1982).

According to Lee, Thornback, and Bennett (1988), *L. albigena* is not threatened, but *L. aterrimus* is subject to intensive, uncontrolled hunting for use as food and also is vulnerable to loss of its restricted forest habitat. Both species are on appendix 2 of the CITES, and *L. aterrimus* is designated as near threatened by the IUCN.

PRIMATES; CERCOPITHECIDAE; Genus MACACA
Lacépède, 1799

Macaques

There are 20 species in 5 species groups (Corbet 1978; Corbet and Hill 1992; Cronin, Cann, and Sarich 1980; Delson 1980; Ellerman and Morrison-Scott 1966; Eudey 1987; Fa 1989; Fooden 1969, 1971, 1975, 1976b, 1988, 1989, 1990, 1991b; Fooden, Mahabal, and Saha 1981; Fooden et al. 1985, 1994; Groves 1980c; Lekagul and McNeely 1977; MacKinnon 1986; Tan 1985):

sylvanus group

M. sylvanus (Barbary ape), northern Africa, a possibly introduced population at Gibraltar in extreme southern Spain;

silenus group

M. silenus (liontail macaque), southwestern peninsular India;
M. nemestrina (pigtail macaque), Assam, Burma, southern Yunnan, Thailand, Indochina, Malay Peninsula, Sumatra, Bangka, Borneo and some small nearby islands;
M. pagensis, Mentawai Islands off western Sumatra;
M. tonkeana, central Sulawesi and nearby Togian Islands;
M. hecki, western part of northern peninsula of Sulawesi;
M. maura, southwestern peninsula of Sulawesi;
M. ochreata, southeastern peninsula of Sulawesi;
M. brunnescens, Muna and Butung islands off southeastern peninsula of Sulawesi;
M. nigra (Celebes ape), eastern part of northeastern peninsula of Sulawesi;
M. nigrescens, central part of northeastern peninsula of Sulawesi;

fascicularis group

M. fascicularis (crab-eating macaque), southeastern Burma, southern Thailand and Indochina, Malay Peninsula, Nicobar Islands, Sumatra, Java, Borneo,

throughout the Philippines, Lesser Sunda Islands as far east as Timor, but not Sulawesi;
M. mulatta (rhesus monkey), eastern Afghanistan and possibly formerly southeastern Pakistan, through much of India and Nepal, to northeastern China, Indochina, and Hainan;
M. cyclopis, Taiwan;
M. fuscata (Japanese macaque), Honshu, Shikoku, Kyushu, Yakushima;

sinica group

M. sinica (toque macaque), Sri Lanka;
M. radiata (bonnet macaque), peninsular India;
M. assamensis, mountainous areas from Nepal and Bangladesh to northern Indochina and western Thailand;
M. thibetana (Père David's stump-tailed macaque), Sichuan to Fujian and Guangxi provinces in southeastern China;

arctoides group

M. arctoides (stump-tailed macaque), Brahmaputra River of northeastern India to Guangxi (southeastern China), Viet Nam, and northern Malay Peninsula.

Fooden (1976b) combined the *sylvanus* and *silenus* groups, and Delson (1980) thought that *M. arctoides* belonged in the *sinica* group. Bernstein and Gordon (1980) and Fa (1989) reported that many species of *Macaca* are capable of interbreeding and producing offspring no less fertile than the parents. Groves (*in* Wilson and Reeder 1993) and Corbet and Hill (1992) included *M. pagensis* in *M. nemestrina*, though the latter authorities noted its distinctiveness and thought that it might warrant specific status. There is substantial disagreement regarding the systematic situation on Sulawesi, with some authorities recognizing a single intergrading species and others dividing various species into groups or subgenera (Fa 1989). Most recent workers have followed either Fooden (1976b), who recognized the seven Sulawesian species listed above, or Groves (1980c), who on the basis of evident interbreeding between certain populations considered *M. nigrescens* a subspecies of *M. nigra*, *M. hecki* a subspecies of *M. tonkeana*, and *M. brunnescens* a subspecies of *M. ochreata*. A recent series of more detailed investigations have suggested that although there is gene flow between the species of Sulawesi along most lines where their ranges meet, such interaction is more indicative of sporadic hybridization between normally separate species than of total intergradation between subspecies (Camperio Ciani et al. 1989; Watanabe, Lapasere, and Tantu 1991; Watanabe and Matsumura 1991; Watanabe et al. 1991).

The smallest species, *M. sinica*, has a head and body length of 367–528 mm and weighs about 2.5–6.1 kg (Fooden 1979). Other species range up to 764 mm in head and body length and 18 kg in weight. Males generally are about 50 percent heavier than females. In *M. sylvanus* there is no tail, and in the following species the tail length ranges from only about 10 mm to approximately 20 percent of the head and body length: *M. nigra, M. nigrescens, M. ochreata, M. brunnescens, M. maura, M. tonkeana, M. hecki, M. arctoides, M. thibetana,* and *M. fuscata*. In the remaining species the tail length is equal to at least 25 percent of the head and body length. In *M. sinica* the tail generally is longer than the head and body (Fooden 1976b, 1979). The coloration is usually yellowish brown above and lighter be-

A. Rhesus macaque *(Macaca mulatta);* B. Pigtail macaque *(M. nemestrina);* C. Barbary ape *(M. sylvanus);* D. Crab-eating macaque *(M. fascicularis)* with left cheek pouch filled; E. Liontail macaque *(M. silenus),* photos by Ernest P. Walker. F. Japanese macaques *(M. fuscata),* photo from U.S. National Zoological Park.

Barbary ape *(Macaca sylvanus)*, photo by Russell A. Mittermeier.

low, but some forms are olive and others almost black. Some species have a "cap" of hairs on the head. *M. silenus* has a ruff of long grayish hair on each side of the face, while *M. fuscata* has long whiskers, a beard, and long dense fur. *M. nigra*, of Sulawesi, has a conspicuous cone-shaped mass of long hairs on the crown of the head. Macaques are medium-sized monkeys with stout bodies and strong limbs. The snout is somewhat elongated.

M. mulatta has adapted to a wide range of habitats, from sea level to elevations of 2,500 meters, from snow to intense heat, and from near-desert situations to dense forests (Lindburg 1971). More than half of the rhesus monkeys in the northern Indian state of Uttar Pradesh actually live in cities and towns, where there is ideal habitat—shelter, large trees, abundant food, and water from wells (Southwick, Beg, and Siddiqi 1965). A group of 62 of these monkeys lived successfully in an area of less than 4 ha. within urban Calcutta, among large buildings, motor vehicles, crowds of people, and domestic animals (Southwick et al. 1974). In contrast, *M. silenus* is an obligate rainforest dweller, and *M. nemestrina* is restricted to dense, inland evergreen and deciduous forests. Both of these species seem unable to adapt to human encroachment and are declining in numbers (Green and Minkowski 1977; Roonwal and Mohnot 1977). Most species of macaques usually inhabit forests, sometimes at relatively high altitudes. *M. assamensis, M. arctoides,* and *M. sylvanus* are often present in upland areas or mountains at elevations of 2,000 meters or more (Lekagul and McNeely 1977; Taub 1977). *M. fascicularis* may also be found in such habitat and even in urban situations (Medway 1978), but it seems to prefer coastal mangroves; like *M. mulatta*, it is an expert swimmer (Lekagul and McNeely 1977; Roonwal and Mohnot 1977). *M. cyclopis*, of Taiwan, may once have been associated with the seacoast

but now has been restricted largely to inland hills because of human activity (Kuntz and Myers 1969).

All macaques are primarily diurnal, and all have an arboreal capability, but most species come down from the trees at least on occasion to forage or move over long distances. *M. radiata* has been observed to spend about one-third of its time on the ground (Simonds 1965), and *M. arctoides* is primarily terrestrial, sometimes making seasonal migrations from one mountain range to another (Lekagul and McNeely 1977). Groups of this species generally leave their sleeping trees at dawn, feed until about 1000 hours, rest till 1400 hours, and then feed again until 1800 hours (Roonwal and Mohnot 1977). *M. sinica*, of the tropical forests of Sri Lanka, spends about 75 percent of its time in the trees, always sleeping high in the canopy at night, and seems cautious and fearful when it does descend to the ground (Dittus 1977; Fooden 1979). *M. silenus* spends less than 1 percent of its time on the ground (Green and Minkowski 1977). Studies in Borneo indicated that one group of *M. fascicularis* slept in trees by night and came down to forage by day (Fittinghoff and Lindburg 1980) but that another group spent more than 97 percent of its time in the trees (Wheatley 1980). *M. mulatta* readily leaves the trees if attracted by food supplies and moves freely among and through human habitations if tolerated. The time and area of its activity may shift according to season. Lindburg (1971) found that in warmer periods it began feeding before dawn but took rest sessions during the day; in the cool season it began to feed later and lacked well-defined intervals of resting. Daily distance averaged 1,428 (350–2,820) meters. Like most other species of macaques, *M. mulatta* is largely vegetarian, feeding on wild and cultivated fruits, berries, grains, leaves, buds, seeds, flowers, and bark. All species, however, eat insects and other small invertebrates when they are available and occasionally take eggs and small vertebrates (Fooden 1979; Lekagul and McNeely 1977; Lindburg 1971; Roonwal and Mohnot 1977). *M. fascicularis* is known to feed on crabs, other crustaceans, shellfish, and other littoral animals exposed by the tide; and some of the macaques of Sulawesi may have a comparable diet (Fooden 1969; Lekagul and McNeely 1977). Captive *M. silenus* were found to manufacture and use tools for the extraction of syrup from a container experimentally provided to the animals (Westergaard 1988).

Fooden (1982) provided a summary of the distribution of the eight species of macaques found on the mainland of Asia, explaining that all are segregated either ecologically or geographically. The species *M. assamensis* and *M. thibetana* are found in subtropical broad-leaved evergreen forest, while *M. silenus* and *M. nemestrina* occur in tropical broad-leaved evergreen forest; none of these four species is sympatric. *M. arctoides* is present both in tropical and subtropical broad-leaved evergreen forest and is sympatric with *M. assamensis* and *M. nemestrina*. However, *M. arctoides* is largely terrestrial, *M. assamensis* is arboreal, and *M. nemestrina* is arboreal in those areas where its range overlaps that of *M. arctoides*. The other three mainland species—*M. mulatta, M. radiata,* and *M. fascicularis*—are rare or absent in broad-leaved evergreen forest but widely distributed in secondary, deciduous, coniferous, riverine, and mangrove forests and in disturbed habitats. They are not sympatric, *M. mulatta* being subtropical and the other two tropical.

MacKinnon (1986) listed the following population estimates and densities in Indonesia: *M. fascicularis*, 2,176,860 individuals occupying 73,371 sq km of primary habitat, at a density of 30/sq km, and 1,550,000 occupying 38,750 sq km of secondary habitat, at a density of 40/sq km; *M. nemestrina*, 895,700 individuals in 179,140 sq km, density

Macaque *(Macaca tonkeana)*, photo by Russell A. Mittermeier.

5/sq km; *M. pagensis,* 9,000 individuals in 4,500 sq km, density 2/sq km; *M. nigra* (including *M. nigrescens*), 144,000 individuals in 4,800 sq km, density 30/sq km; *M. tonkeana* (including *M. hecki*), 385,000 individuals in 38,500 sq km, density 10/sq km; *M. maura,* 56,000 individuals in 2,800 sq km, density 20/sq km; and *M. ochreata* (including *M. brunnescens*), 277,500 individuals in 18,500 sq km, density 15/sq km. Some other reported population densities of macaques are: *M. fascicularis,* 36–90/sq km (Kurland 1973); *M. mulatta,* 5–15/sq km in high forest, 57/sq km in low forest, and 753/sq km in towns (Roonwal and Mohnot 1977); *M. silenus,* 0.25/sq km (Green and Minkowski 1977); *M. nemestrina,* 5.5/sq km (Rodman 1978); *M. sylvanus,* 43–70/sq km (Deag 1977); and *M. sinica,* from 0.3/sq km in marginal habitats to 100/sq km in optimal habitats (Dittus 1977; Fooden 1979). It should be understood that these figures are speculative and based largely on assumptions regarding the overall extent and condition of the remaining habitat.

Estimated group home ranges are: *M. fascicularis,* generally 0.5–1.0 sq km but as small as 0.25 sq km in mangroves (Crockett and Wilson 1980); *M. mulatta,* up to 16 sq km in sub-Himalayan forests, 1–3 sq km in other forests, and 0.05 sq km in towns (Lindburg 1971; Roonwal and Mohnot 1977); *M. silenus,* 1–2 sq km during a period of 1–2 months and 5 sq km over a year (Green and Minkowski 1977); *M. nemestrina,* 1–3 sq km (Crockett and Wilson 1980); *M. sinica,* from 0.17 sq km for a group of 8 to 1.15 sq km for a group of 37 (Dittus 1977); and *M. radiata,* 0.4–5.2 sq km (Roonwal and Mohnot 1977). There is generally extensive and sometimes nearly complete overlap between the home ranges of groups, and there is little evi-

dence of territoriality. However, there may be defense of the area occupied at a given time, mutual avoidance by groups, and dominance of large groups over smaller ones. Lindburg (1971) found that most intergroup encounters of *M. mulatta* resulted in withdrawal of one group, but occasionally there was peaceful proximal feeding or agonistic displays and stalemate. Green and Minkowski (1977) indicated that a study group of *M. silenus* had a 300-ha. core area that was rarely entered by other groups. Roonwal and Mohnot (1977) stated that groups of *M. sinica* spend considerable time in the company of *Semnopithecus entellus* and that the two often play, feed, rest, and even groom together.

Overall heterosexual group size in *M. mulatta* is 8–180. Generally there are from two to four times as many adult females as adult males since most of the males leave to live alone or in small groups of their own (Roonwal and Mohnot 1977). The average composition of 399 groups in Uttar Pradesh was 17.6 animals, including 3.7 adult males, 7.7 adult females, 4.5 infants, and 1.7 juveniles (Southwick, Beg, and Siddiqi 1965). Another study in this state found 32 groups to number 8–98 individuals and to include 3 adult females for every adult male (Lindburg 1971). In *M. mulatta* there are dominance hierarchies in both sexes, but that of the females is less evident. Relationships among adult males range from peaceful to hostile, but the females generally live together in harmony. There is a tendency for mating between high-ranking adults. Although males appear to lead and defend the group, the females and infants form a central subgroup within which the young are raised. According to Sade (1967), the social status of the young is dependent on the rank of their mother. In *M. fuscata* of Japan, groups are known to be matrilineal. Young males emigrate, but females remain within the group of their birth and form close relationships among themselves (Kurland 1977). Groups of this species become very large; in one case a troop divided three times, in each instance a group of 70–100 animals separating after the mother troop had reached a population of 600–700 (Masui et al. 1975). Like most other macaques, *M. mulatta* has a wide variety of vocalizations. Lindburg (1971) identified a shrill bark for alarm, barking or screeching as a response to aggression, a scream when under attack, a growl of aggression, and a squawk for surprise.

The social structure of *M. fascicularis* appears close to that of *M. mulatta.* Reported group size is 6–100, and there are about 2.5 adult females for every adult male. Roonwal and Mohnot (1977) stated that in this species groups are divided into a central component, comprising a male leader and most of the females and young, and a peripheral component, made up of the younger males and perhaps a few females. Angst (1975) found dominance hierarchies based on age, kinship, and coalitions in each sex, with senile individuals dropping in rank. Dittus (1977) reported an average group size of 24.7 (8–43) in *M. sinica.* He observed that natural mortality was largely the result of lower-ranking animals' being prevented from foraging in areas of high quality. In *M. radiata,* the common performing monkey of southern India, group size is about 7–76, but usually there are more adult males than females (Roonwal and Mohnot 1977). Although there is a dominance hierarchy among the males, they seem more tolerant of one another and much less aggressive than males of *M. mulatta* (Simonds 1965). This situation may be attributable to the rich tropical habitat of *M. radiata,* which allows more large individuals (males) to live at close quarters without intense competition for resources (Caldecott 1986).

Some other reported information on group structure in macaques is: *M. nemestrina,* group size 3–15 in disturbed

Moor macaque *(Macaca maura)*, photo by Ernest P. Walker.

forest and 30–47 in undisturbed areas; *M. silenus*, group size 4–34, usually about 10–20, including 1–3 adult males; *M. assamensis*, group size 10–100; and *M. arctoides*, group size 20 to more than 100, usually, but not always, led by an adult male (Green and Minkowski 1977; Lekagul and Mc-Neely 1977; Roonwal and Mohnot 1977). According to Fooden (1969), the Sulawesi species form troops of 5–25 individuals, usually with an old male as leader; several groups may occasionally merge into a temporary aggregation of more than 100 animals. Ménard and Vallet (1993) observed a group of *M. sylvanus* to more than double in size over a four-year period, to 76 individuals, and to then undergo fission into three independent groups largely at the instigation of adult females. Maternally related individuals stayed together in the new groups. Most of the original males stayed in one group, immigrant males moved into another group, and the third new group had only one adult male. Wu and Lin (1992) also observed group fission in *M. cyclopis*, of Taiwan. Groups of that species are small, usually containing 10–20 individuals, and only one male, though solitary males sometimes move into the permanent units and mate with the females.

According to Roonwal and Mohnot (1977), there is mounting evidence that reproductive seasonality is the rule in all wild populations of *M. mulatta*. In south-central Asia

births occur mainly in February–May, with some also in September–October. The season is about the same in *M. radiata*, of southern India, but seems more variable in *M. sinica*, of Sri Lanka, perhaps being synchronized with the rainfall pattern (Fooden 1979). There apparently is a spring birth peak in *M. fascicularis* and *M. assamensis* in Thailand (Fooden 1971). In peninsular Malaysia *M. fascicularis* may give birth at any time of the year but is most likely to do so from May to July (Kavanagh and Laursen 1984). *M. cyclopis*, of Taiwan, mates mostly from November to January and gives birth mostly from April to June (Wu and Lin 1992). Female *Macaca* may enter estrus several times during the mating season, with receptivity being signaled by swelling and reddening of the genital region and vaginal discharge. Some reported estrous cycle lengths (in days) are: *M. nigra*, mean of 33.5 (Dixson 1977); *M. silenus*, mean of 31 days (Lindburg, Lyles, and Czekala 1989); *M. arctoides*, mean of 30.7; *M. fascicularis*, 24–52; *M. mulatta*, mean of 28; and *M. nemestrina*, mean of 42 (Roonwal and Mohnot 1977). Estrus commonly lasts about 9 days (Hayssen, Van Tienhoven, and Van Tienhoven 1993). In the laboratory a female *M. mulatta* has conceived as early as 45 days after giving birth, but in the wild usually only one birth occurs every two years. Interbirth interval in a captive colony of *M. arctoides* averaged about 619 days when

Celebes ape *(Macaca nigra)*, photo by Osman Hill.

weaning occurred naturally and outdoors but only 292 days following an abortion or stillbirth (Nieuwenhuijsen et al. 1985).

Roonwal and Mohnot (1977) listed the following gestation periods (in days): *M. nemestrina,* 162–86; *M. arctoides,* mean of 177.5; *M. fascicularis,* 160–70; *M. mulatta,* 135–94 (mean of 166); and *M. radiata,* 150–70. Ross (1992) added these means: *M. fuscata,* 173; *M. silenus,* 180; and *M. sylvanus,* 165. Based on 700 pregnancies in captivity, Silk et al. (1993) reported a range in gestation of 133–200 days (mean 166.5) for *M. mulatta,* with older females having significantly longer periods and heavier infants than younger mothers. Normally there is a single young, but twins have been reported on rare occasion for several species. A newborn weighs about 400–500 grams (Ross 1992) and nurses for about 1 year. It initially clings to the belly of the female and later may ride ventrally or dorsally. Sexual maturity is usually attained at around 2.5–4 years in females and 2–3 years later in males (Roonwal and Mohnot 1977; Simonds 1965). In *M. silenus,* however, females do not have their first offspring until they are 5 years old and males do not reach sexual maturity until 8 years (Green and Minkowski 1977). Most reproductive activity of female *M. fuscata* occurs between 6 and 18 years (Koyama, Norikoshi, and Mano 1975). Perhaps females of that species must attain maximum size before they can successfully bear young in the harsh climate of Japan (Ross 1992). Female *M. mulatta* reach menopause at about 25 years (Roonwal and Mohnot 1977). Although the average life span for wild *M. sinica* is less than 5 years, a captive individual lived for 33–35 years (Fooden 1979). Several other species have survived for more than 25 years in captivi-

Japanese macaques *(Macaca fuscata),* photo by Carl B. Koford.

ty, and record longevity for the genus appears to be held by a specimen of *M. fascicularis* that died at 37 years and 1 month (Jones 1982).

Several species of macaques are closely associated with people. *M. mulatta,* the rhesus monkey, is sacred to the Hindu religion and is commonly found in the vicinity of temples and urban areas in northern India and Nepal. This species has been used extensively in the West for experiments involving biology, behavior, medicine, and even space flight. It served in the studies that first demonstrated the Rh factor in blood. It also does well and is widely exhibited in zoos and circuses. The demand for *M. mulatta* was such that nearly 200,000 were exported each year to the United States alone in the late 1950s (Mohnot 1978). Subsequent restrictions by the governments of both India and the United States have greatly reduced this commercial traffic, but major problems continue. *M. mulatta* seems to have been adversely affected both by the loss of juveniles for export and by changing environmental and cultural conditions. After a lengthy study of population trends, Southwick and Siddiqi (1977) cautioned that food shortages in India were leading to a shift in villagers' attitude from traditional reverence and protection of the rhesus to the view that the species is a threat to crops and a nuisance to be eliminated. Southwick and Siddiqi stated that unless the rhesus received protection from local people, it would be extirpated from most agricultural areas of India within 25 years. Further investigations confirmed that habitat changes, intensification of agriculture, and loss of traditional protection were resulting in drastic overall declines and the disappearance of most populations along roads and canals, in villages, and even around temples (Southwick et al. 1982; Southwick, Siddiqi, and Oppenheimer 1983). The number of *M. mulatta* in India was estimated to be as high as 20 million in the 1940s, less than 2 million in 1960, and 180,000 by 1980 (Southwick and Lindburg 1986). Remarkably, a 1978 ban on exportation, a renewed interest in protection for both religious and bioconservation reasons, and a striking increase in agricultural production, apparently in association with greater tolerance of the monkeys, evidently led to recovery in some areas. In 1985 the total

rhesus population of the country was estimated at 410,000–460,000, and more recent surveys indicate that the number continues to grow (Southwick and Siddiqi 1988, 1994*a*). This information is tempered, however, by the finding that more than 85 percent of the rhesus in India are now living in a commensal or semicommensal status with people (Southwick and Siddiqi 1994*b*).

The population of *M. mulatta* in China, estimated at 150,000, also is declining through habitat loss and commercial trade (Wang and Quan 1986). A colony of about 50 individuals northeast of Beijing, once thought to be the result of an introduction, now is known to have been the last relict of a population that occupied much of northeastern China during the historical period (Tan 1985; Zhang et al. 1991). Unfortunately, this group, the northernmost nonhuman primates on the mainland of Eurasia, evidently was extirpated in the late 1980s through hunting and habitat destruction by people (Zhang et al. 1989). Another population, with more than 2,000 monkeys, survives about 500 km south of Beijing, in the Taihang Mountains on the Shanxi-Henan border (Qu et al. 1993). That population is nonetheless considered seriously endangered and has been found to represent a genetically distinctive subspecies, *M. m. tcheliensis* (Zhang and Shi 1993). *M. fascicularis,* which replaces *M. mulatta* to the southeast, has the reputation of being a pest in some areas because of its raids on fields and gardens (Medway 1978). This species, which is frequently referred to in medical terminology as the cynomolgus monkey, was used extensively in studies that led to the development of vaccine for the control of polio. The IUCN now designates both *M. fascicularis* and *M. mulatta* as near threatened.

A number of other species of macaques are of concern to the conservation community. Zhang et al. (1991) considered all six species that occur in China to be endangered or vulnerable. Eudey (1987) listed all the species and subspecies of Sulawesi as either highly vulnerable or endangered through human usurpation of habitat; the IUCN now classifies *M. maura* and *M. nigra* as endangered, *M. brunnescens* as vulnerable, *M. nigrescens* as conservation dependent, *M. tonkeana* and *M. hecki* as near threatened, and

M. thibetana as conservation dependent. Supriatna et al. (1992) found the habitat of *M. maura* to be restricted to small and fragmented nature reserves, though *M. tonkeana* still occupied large and protected areas of forest. Hamada, Oi, and Watanabe (1994) indicated that *M. nigra* is critically endangered on Sulawesi, though a population apparently introduced long ago on Bacan Island, in the Northern Moluccas, seems to be in good condition. *M. nemestrina*, a species often domesticated and trained to harvest coconuts, is reportedly declining rapidly throughout its range and is classified as vulnerable by the IUCN. Its flesh is sought by people for food and medicinal purposes. Because of intensive shooting, it has become rare in parts of Borneo (Medway 1970, 1977; Roonwal and Mohnot 1977). Khan (1978) estimated that the population in peninsular Malaysia had declined from 80,000 in 1958 to 45,000 in 1975. *M. sinica*, of Sri Lanka, is also declining because of the logging of its forest habitat and is listed as threatened by the USDI and as near threatened by the IUCN. Dittus (1977) estimated that the most threatened subspecies, *M. sinica opisthomelas*, numbered only 1,469 animals.

Also listed as threatened by the USDI are *M. arctoides*, *M. cyclopis*, and *M. fuscata*. The IUCN classifies *M. cyclopis*, *M. assamensis*, and *M. arctoides* as vulnerable, *M. fuscata* as endangered, and *M. pagensis* of the Mentawai Islands as critically endangered. Eudey (1987) already had regarded this last species, which sometimes is treated only as a subspecies of *M. nemestrina* (see above), as highly endangered. *M. pagensis* is estimated to have declined by at least 80 percent over about the past two decades and is forecast to decline by at least another 80 percent in the next two decades. The habitat of the species, which is restricted to a few small islands off southwestern Sumatra, is rapidly being destroyed by logging. In addition, the monkeys are being hunted by the native people for food and use as pets. *M. arctoides* is thought to have declined because of deforestation and environmental disruption resulting from military activity. Roonwal and Mohnot (1977) referred to *M. arctoides* as rare throughout its range. *M. cyclopis*, now restricted to the remote highlands of Taiwan by an expanding human population, is jeopardized by taking for use as food, pets, medicinal preparations, and research. Poirier (1982) estimated that at least 1,000–2,000 *M. cyclopis* are lost annually for such purposes but stated that the main threat is loss of forest habitat; every locality that was found to support the species in a 1940 survey is now occupied by a village or city. More recent surveys have found that these problems are continuing but that *M. cyclopis* is still widespread on Taiwan (Lee and Lin 1990). *M. fuscata*, while traditionally protected in Japan, is declining through cutting of the forests on which it depends and through persecution by people who consider it a threat to crops (Eudey 1987). Iwano (1975) warned that there had been a rapid population decrease throughout Japan since 1923 and that free-living monkeys could disappear within the next few decades. Maruhashi, Sprague, and Matsubayashi (1992) reported that about 5,000 *M. fuscata* were being removed annually in response to claims of agricultural damage. Hill and Sprague (1991) indicated that more than 500 of those taken each year represented the subspecies *M. f. yakui*, of Yakushima Island, which had begun to raid orange groves after their natural habitat had been reduced.

One of the most seriously jeopardized macaques is *M. silenus* of the Western Ghats Mountains in southern India. This species seems to have a low reproductive rate and to be unable to adapt to human encroachment. Suitable habitat has greatly diminished through clearing of the forests for agriculture. The largest remaining population contains about 300 animals and occupies about 160 sq km of forest, just about the minimum area needed to maintain viability. The total estimated number in the wild in 1975 was 405, and there were another 275–300 in captivity (Green and Minkowski 1977). A subsequent investigation resulted in an overall estimate of about 915 individuals, but it also revealed numerous new threats, including the construction of dams and roads in the habitat of the largest populations (Ali 1985). More recent reports indicate that the area occupied by the largest population has been incorporated into a biosphere reserve, that the total number in the wild has increased to more than 1,700, and that there is a rapidly growing captive population (Kurup 1988; Lindburg and Gledhill 1992). *M. silenus* is classified as endangered by the IUCN and the USDI and is on appendix 1 of the CITES; all other species of *Macaca* are on appendix 2.

The only macaque to occur naturally outside of Asia is *M. sylvanus* of northern Africa. It may have survived in Libya and Egypt in the early nineteenth century and as far east as Tunisia around 1900, but currently it is restricted to high cedar and oak forests in the Atlas and nearby ranges of Morocco and northern Algeria. It is continuing to decline because of the commercial logging of the forests. In 1975 the total population was estimated to number about 14,000–23,000 individuals, 77 percent of which were in Morocco (Deag 1977; Fa et al. 1984; Taub 1977, 1984). In 1992 the total estimate was 10,000–16,000 (Lilly and Mehlman 1993). The IUCN classifies *M. sylvanus* as vulnerable.

A small number of *M. sylvanus* also occupy the Rock of Gibraltar, a British possession at the tip of southern Spain. This population is fed and protected by the British army and long was artificially maintained, through capture and export to zoos, at about 33 individuals. Recently numbers have increased to approximately 105 and there are concerns that the health of the population is being adversely affected through excessive feeding by people and exposure to human diseases. The monkeys are a popular tourist attraction, and there is a superstition that if they should disappear from the Rock, British rule would end. Aside from that, the most interesting question about these animals is whether they are descended from stock originally introduced from Africa through human agency or are the natural remnant of a formerly widespread European monkey population. It is known that macaques were present at Gibraltar when the British arrived in 1704, but numerous introductions have occurred since then. Fossil remains indicate that *M. sylvanus* occurred in much of Europe during the late Pleistocene, and some animals may possibly have survived in southern Spain, outside of Gibraltar, as late as the 1890s. More recently, some *M. sylvanus* escaped from captivity in Spain and began to live and reproduce in the wild (Deag 1977; O'Leary 1993; Taub 1977, 1984).

Successful introductions of macaques are not unusual. In 1763 a troop of M. sylvanus was set loose in Germany, where it thrived until deliberately exterminated in 1784 (Grzimek 1975). There is currently a large, free-ranging colony of *M. sylvanus* in France (R. W. Thorington, Jr., U.S. National Museum of Natural History, pers. comm., 1981). *M. fascicularis* apparently was brought to Mauritius in the Indian Ocean by Dutch or Portuguese sailors in the early sixteenth century, and the monkey population on the 1,865-sq-km island now numbers 25,000–35,000 (Sussman and Tattersall 1986). *M. mulatta* has been established in Puerto Rico and on an island near Rio de Janeiro, Brazil (Hausfater 1974; Roonwal and Mohnot 1977). A feral population of *M. mulatta* in central Florida, probably established originally as a tourist attraction, has existed since the 1930s (Wolfe and Peters 1987). M. cyclopis has been introduced on the Izu Islands, south of Tokyo (Corbet 1978), *M. nemestrina* on Singapore and the Natuna Islands (Corbet and Hill 1992), and *M. nigra* on Bacan Island in the Moluccas (Flannery 1995).

PRIMATES; CERCOPITHECIDAE; **Genus PAPIO**
Erxleben, 1777

Baboons

There are five species (Corbet 1978; Dandelot *in* Meester and Setzer 1977; Harrison and Bates 1991; Yalden, Largen, and Kock 1977):

P. hamadryas (hamadryas baboon), upper Egypt, northeastern Sudan, eastern Ethiopia, northern Somalia, western Arabian Peninsula north to about 24° N;

P. anubis (olive baboon), savannah zone from Mali to Ethiopia and northern Tanzania, several mountainous areas in the Sahara Desert;

P. papio (western baboon), Senegal, Gambia, Guinea;

P. cynocephalus (yellow baboon), Angola, southern Zaire, Tanzania, Zambia, Malawi, northern Mozambique, Kenya, Somalia;

P. ursinus (chacma baboon), southwestern Angola, southern Zambia, southern Mozambique, Namibia, Botswana, Zimbabwe, South Africa, Swaziland, Lesotho.

The authority for the name *Papio* sometimes is given as Müller, 1773. *Mandrillus* (see account thereof) sometimes is regarded as a subgenus of *Papio*. The ranges of the above five species are largely, if not entirely, allopatric, and it may be that all intergrade and are no more than subspecies of a single widespread species (De Vore and Hall 1965; Jolly 1993; Maples 1972; Thorington and Groves 1970). Groves (1989) indicated that the five taxa are strongly differentiated and that at least some might be specifically distinct, but subsequently (*in* Wilson and Reeder 1993) he recognized only the species *P. hamadryas*, with the other four included therein. Corbet and Hill (1991), Meester et al. (1986), and Napier and Napier (1985) treated all five as separate species. *P. anubis* and *P. hamadryas* have formed an apparently stable hybrid zone where their ranges meet in Ethiopia (Gabow 1975; Kummer, Goetz, and Angst 1970; Nagel 1971). Limited hybridization also evidently occurs between *P. anubis* and *P. cynocephalus* and between *P. cynocephalus* and *P. ursinus*, but a recent odontomorphometric analysis by Hayes, Freedman, and Oxnard (1990) indicates that all are valid species. In contrast, Jolly (1993) concluded that the degree of gene flow between the various populations implies that they should be called a single species. Jolly also questioned whether *P. hamadryas* ever occurred naturally in Egypt and suggested that the population in Arabia is the result of introduction by human agency within the last 4,000 years.

In *P. hamadryas* head and body length is 610–762 mm and tail length is 382–610 mm. One individual weighed 18 kg. Young animals are brown, becoming ashy gray with age; females retain the brown coat longer than the males. The heavy mane around the neck and shoulders of the older males is lacking in the younger males and in the females. In the other four species head and body length is 508–1,143 mm, tail length is 456–711 mm, and weight is 14–41 kg. As with other baboons, the males are usually considerably larger than the females. The hair is rather coarse. The face and neck are thinly haired, but long hairs develop around the neck and shoulders in males. General coloration is olive brown in *P. anubis*, olive rufous in *P. papio*, yellowish in *P. cynocephalus*, and dark olive brown in *P. ursinus*. These four species lack the heavy mane of the hamadryas baboon and are not as brilliantly colored as *Mandrillus*.

Baboons inhabit open woodland, savannahs, grassland, and rocky hill country. All species are primarily diurnal and terrestrial but capable of climbing. If available, trees may be used for sleeping, but populations in Ethiopia and South Africa generally spend the night on the faces of cliffs (Davidge 1978; De Vore and Hall 1965; Kummer 1968). The mean daily distance traveled by *P. cynocephalus*, *P. ursinus*, and *P. anubis* is about 3–4 km (Altmann and Altmann 1970). They walk with an awkward swaggering gait and run in a "rocking horse" gallop. They are said to be good swimmers and to have excellent vision. All five species are omnivorous, taking a variety of vegetable matter but seeming to concentrate on whatever is easily available at a given time (De Vore and Hall 1965). During the dry season in Kenya, grass may make up 90 percent of the diet. On the coast of South Africa, mollusks, crabs, and other marine creatures are regularly eaten. Altmann and Altmann (1970) observed *P. cynocephalus* chasing and capturing hares *(Lepus capensis)* and digging out small fossorial mammals. Hamilton, Buskirk, and Buskirk (1978) found that when there were massive outbreaks of grasshoppers or scale insects *P. ursinus* ignored other food sources.

Wolfheim (1983) listed the following recorded population densities: *P. anubis*, 1.12–63.00/sq km; *P. papio*, 2–15/sq km; *P. cynocephalus*, 9.65–60.00/sq km; and *P. ursinus*, 2.31–43.2/sq km. These species reportedly have definite home ranges from which it is difficult to drive them. Group home range size is 2–40 sq km, though there may be smaller core areas where most time is spent. There is extensive or total overlap between the ranges of different groups and little evidence of territorial defense (Altmann and Altmann 1970; Chalmers 1973; Davidge 1978; De Vore and Hall 1965).

P. hamadryas generally occurs at low densities, recorded at 1.8–3.4/sq km, and groups move across a considerable distance, 6.5–19.6 km, each day (Wolfheim 1983). The detailed studies of Kummer (1968) in Ethiopia demonstrate that *P. hamadryas* has developed a trilevel social organization enabling the species to adapt to an environment with (1) sparse food resources, thereby favoring dispersed, small groups; and (2) scarce sleeping sites, thereby favoring group concentration. The largest social unit is the "troop," which comprises all the animals sleeping at a particular cliff. Troop size generally is 100 or more individuals, and one sleeping party of 750 animals was observed. In the morning the troops assemble above the cliffs and then move out in separate foraging "bands" of about 30–90 animals. The largest moving group seen had 494 baboons. Each band is composed of several single-male units, containing an adult male and 1–9 (usually 2–5) adult females with their young. Often the male leaders of two units will cooperate as a team, the younger individual initiating movement for both units and the older male bringing up the rear and accepting or rejecting the indicated direction. All females live in single-male units, and young males or adult males with no unit to lead may either maintain a loose association with a unit or live alone. Of the various social groupings, the single-male unit has the most stable membership. In six such units studied from one to six months there was no change in the male leader. He consistently defends the same set of females from the advances of other males. If a female strays from a unit, the adult male will drive her back, though if she is in estrus, the male may force the entire unit to follow her.

Of all the intragroup social activities, grooming takes up

Chacma baboon *(Papio ursinus)*, photo by Bernhard Grzimek.

most of the adults' time. If two females simultaneously try to groom the adult male, they will soon begin to scream and fight until one withdraws. Aggressive displays by females seem always to be associated with winning the attention of the unit leader and are never directed against females of other units. Adult males sometimes fight one another for possession of females. Such fighting involves rapid fencing with open jaws and hitting out with the hands, but there is almost never actual physical contact. Although the adult males of a given band seem to tolerate one another, there may be general conflicts between the males of different bands. When danger threatens, female *P. hamadryas* gather up the young and move along at a fast pace; the males usually lag behind, often turning around in threatening poses and uttering grunts and deep-throated sounds.

Sigg et al. (1982) reported a significant fourth level of organization, the "clan," between the band and the single-male unit. A clan is an association of interacting single-male units. Although young males may leave their natal unit, they always remain within their original clan. Females also usually transfer to a unit within the clan, though sometimes they switch clans and even bands. Age of departure from the natal unit is about 2 years in males and 3.5 years in females.

Studies of *P. cynocephalus*, *P. anubis*, and *P. ursinus* in eastern and southern Africa indicate a somewhat different social structure (Altmann and Altmann 1970; De Vore and Hall 1965; Hall and De Vore 1965). Group size is about 8–198, usually 30–60. These groups often contain several adults of each sex, though usually more females than males. There is no division into single-male units; rather, all the males compete for estrous females. The males form a dominance hierarchy, with the position of each animal depending on how well he fights and on his ability to form alliances with other males. Females have a less evident hierarchy, but several may sometimes join in aggression against others. Group movement is directed by the adult males, though females may walk in front. A young or subadult male moves well ahead of the column and sounds an alarm bark in case of danger, whereupon an adult male will move forward to investigate. Numerous other vocalizations have been identified in these baboons.

The differences in the social life of *P. anubis* and *P. hamadryas* have stimulated special interest in a zone of hybrids formed by these two species where their ranges meet along the Awash River in Ethiopia (Gabow 1975; Kummer, Goetz, and Angst 1970; Nagel 1971, 1973; Phillips-Conroy, Jolly, and Brett 1991). Since male *P. anubis* lack the ability to herd females, they are unable to control and mate with female *P. hamadryas*. Male *P. hamadryas*, however, can bring female *P. anubis* under their domination, and this process apparently was responsible for establishment of the

Hamadryas baboons *(Papio hamadryas)*, photo by Erik Parbst through Zoologisk Have, Copenhagen.

hybrid zone. Male *P. hamadryas* have been observed to enter groups of *P. anubis,* establish "harems" therein, and sire young, though there is no evidence that they actually abduct females. The hybrid young are viable, but the zone has not expanded beyond 20 km, evidently because male offspring do not inherit the herding ability of their fathers. Mori and Belay (1990) reported another zone of extensive hybridization between *P. anubis* and *P. hamadryas* along the Wabi-Shebeli River, about 100 km south of the Awash.

The following reproductive data on *P. hamadryas* were summarized by Kummer (1968): average estrous cycle, 30 days; gestation, 170–73 days; average lactation period, 239 days; and normally a single young. Breeding occurs throughout the year, but there are birth peaks in Ethiopia in May–June and November–December. The age of sexual maturity is 5 years in females and 7 years in males. Hayssen, Van Tienhoven, and Van Tienhoven (1993) listed birth weights of 600–900 grams and lactation periods of

Chacma baboons *(Papio ursinus)*, photos by Ernest P. Walker.

6–15 months. According to Marvin L. Jones (Zoological Society of San Diego, pers. comm., 1995), a captive specimen of *P. hamadryas* lived 37 years and 6 months.

In the other species of *Papio* there seems to be seasonal reproduction in some areas but not in others. In Kenya, births may occur at any time, but most take place from October to December. The interbirth interval is at least 15 months, the gestation period is 6 months, and lactation lasts 6–8 months. There is usually a single young, rarely two. At first the young is carried clinging to the mother's breast, but it soon begins to ride on her back. Females have their first estrus at 3.5–4.0 years and reach full size and bear their first offspring at 5 years. Males are sexually mature at 5 years but are not fully developed and able to mate successfully until 7–10 years (De Vore and Hall 1965; Hall and De Vore 1965). A specimen of *P. ursinus* lived 45 years in captivity.

The hamadryas was the sacred baboon of the ancient Egyptians and was often pictured on temples and monoliths as the attendant or representative of Thoth, the god of letters and scribe of the gods. It was also mummified, entombed, and associated with sun worship. Today *P. hamadryas* is generally tolerated and even actively fed by people in Saudi Arabia, and a number of populations have become commensal in the vicinity of cities and garbage dumps (Biquand et al. 1992*a*, 1992*b*). The species reportedly has recently extended its range northward on the Arabian Peninsula and increased in numbers, sometimes reaching pest proportions, with troops of up to 1,000 baboons moving freely through settled areas and showing little fear of people (Harrison and Bates 1991). *P. hamadryas*, however, has been exterminated in Egypt and reduced in numbers in some other areas. Irrigation projects in Ethiopia, home to 90 percent of the hamadryas population, have brought parts of its range under cultivation, led to conflicts with people, and often resulted in the baboons' being killed or harassed (Lee, Thornback, and Bennett 1988). The other four species reportedly are more adaptable to human environmental disturbance; nonetheless, they are losing habitat to agricultural expansion in many areas. They become subject to organized extermination efforts when they turn to raiding crops, and they also are hunted for food and sport (Wolfheim 1983). All species are on appendix 2 of the CITES. *P. hamadryas* and *P. papio* are designated as near threatened by the IUCN.

PRIMATES; CERCOPITHECIDAE; **Genus**
MANDRILLUS
Ritgen, 1824

Drill and Mandrill

There are two species (Dandelot *in* Meester and Setzer 1977; Grubb 1973; Wolfheim 1983):

M. sphinx (mandrill), south of Sanaga River in Cameroon, Equatorial Guinea, Gabon, Congo;
M. leucophaeus (drill), from Cross River in southeastern Nigeria to Sanaga River in Cameroon, island of Bioko (Fernando Poo).

Mandrillus was regarded as a subgenus of *Papio* by Dandelot (*in* Meester and Setzer 1977) and Wolfheim (1983) but as a distinct genus by Corbet and Hill (1991), Groves (1989), Grubb (1973), Hill (1970), and Napier and Napier (1985). Recent molecular and chromosomal analyses suggest that *Mandrillus* is less closely related to *Papio* than is *Theropithecus* (Disotell 1994).

Head and body length is about 610–764 mm and tail length is about 52–76 mm. Weight is usually about 25 kg in males and 11.5 kg in females (Napier and Napier 1985). *M. sphinx*, the heavier of the two species, is the largest of all monkeys, having a shoulder height of about 508 mm and a weight of up to 54 kg. Both species have prominent ridges on each side of the nasal bones; these are outgrowths of ridged and grooved bone. There are usually six grooves in each main ridge on the mandrill's face. In the adult male mandrill the skin is purple in the grooves and blue on the small ridges. The median tract between the outgrowths is brilliant scarlet. The scarlet extends forward to the muzzle and on the area around the nostrils. Female and young male mandrills have less pronounced ridges and lack the purple color in the grooves, and their noses are black, not scarlet. In the drill the face is black, lacking bright coloration, and each outgrowth has only two grooves. In both species the buttock pads and the skin around them have a lilac tinge, which becomes reddish purple at the edges. The red color of the skin is the result of the distribution and richness of blood vessels. The bright color of the skin of these animals becomes more pronounced when they are excited. Jouventin (1975*b*) suggested that the rear coloration serves as a signal facilitating group movement through thick vegetation.

Both species have a beard, a crest, and a mane. The hair on the upper parts of the mandrill is tawny greenish, and that on the underparts is yellowish. The pelage of the drill is brownish.

The drill and mandrill usually are found in thick rainforest. In a study of *M. sphinx* in Gabon, Jouventin (1975*a*) generally encountered the large adults on the ground and smaller females and juveniles at midlevel in the trees. These animals were observed to begin activity at dawn, rest at midday, and enter sleeping trees at 1700–1800 hours. On the average they foraged over about 8 km per day. In Cameroon, Hoshino (1985) determined that *M. sphinx* moved 2.5–4.5 km per day. Both *M. sphinx* and *M. leucophaeus* feed mainly on the ground, searching for fruits, nuts, other plant material, mushrooms, invertebrates, and occasionally small vertebrates. They turn over stones and debris when seeking food. Kudo and Mitani (1985) observed a pair of mandrills attacking and killing a duiker *(Cephalophus)*.

In a study in Gabon, Jouventin (1975*a*) found *M. sphinx* to use very large home ranges, two such areas being estimated at 40 and 50 sq km. The basic social unit comprises 1 large adult male, 5–10 adult females with or without infants, and approximately 10 juveniles. During the dry season, however, 6–7 such harems occasionally associate to form troops of 200 or more individuals. Excess males live alone. When a group is quietly foraging, the adult male keeps to the rear, but in the event of danger he moves out to the front. Three vocalizations were identified: a contact call by the females and young, an alarm call by females and subadults, and the rallying call of the male leader. Studies of the same species in Cameroon (Hoshino et al. 1984) indicated a generally similar social structure, with an average of 13.9 individuals per adult male, sometimes larger aggregations with several adult males and some solitary males. Observations of a group of *M. sphinx* in a large enclosure in Gabon (Dixson, Bossi, and Wickings 1993) revealed a multimale organization with a strict dominance hierarchy. The highest-ranking males had the most prominent facial coloration and the heaviest rumps, and they associated almost exclusively with the females and sired young. Other males tended to be peripheral and solitary. Gartlan (1970) observed troops of 14–170 *M. leucophaeus* in western Cameroon but suggested that the larger groups were tem-

A. Mandrills *(Mandrillus sphinx)*, female with baby riding in usual position, photo from Cheyenne Mountain Zoo. B. Mandrill *(M. sphinx)*, male, photo by Bernhard Grzimek. C. Drill *(M. leucophaeus)*, photo from Zoological Society of London.

porary aggregations and that social structure in this species is much like that of *M. sphinx.*

Jouventin (1975*a*) reported apparent seasonal reproduction for *M. sphinx* in Gabon, with most births occurring from January to April. Sabater Pi (1972) stated that young less than 3 months old had been taken in Equatorial Guinea in December, February, March, and June. Carman (1979) reported gestation periods of 168–76 days for captive *M. sphinx.* Based on observations of a captive group of *M. sphinx* in an enclosure of 5.3 ha., Feistner (1992) calculated an average gestation period of 176 days, an interbirth interval of 13–14 months for multiparous females, and age of sexual maturity in females as about 3.5 years. Hadidian and Bernstein (1979) listed the median estrous cycle as 33.5 days in *M. sphinx* and 35 days in *M. leucophaeus.* According to Hill (1970), mating in *M. leucophaeus* takes place throughout the year and the gestation period is 7.5 months. He also reported that record longevity for *Mandrillus* is 46 years.

Both Hill (1970) and Sabater Pi (1972) indicated that despite their formidable appearance and reputation for ferocity, specimens of *Mandrillus* are gentle and adapt readily to captivity. In the wild both *M. sphinx* and *M. leucophaeus* occupy relatively small ranges and are declining because of habitat loss and excessive hunting by people for food. Recent surveys indicate that hunting is mainly commercial, the meat being resold in urban centers (Blom et al. 1992). Numbers of *M. leucophaeus* are estimated to be fewer than 10,000 (Cox and Boer *in* Bowdoin et al. 1994). Considering both degree of threat and taxonomic distinctiveness, Oates (1996) ranked *M. leucophaeus* the African primate species most in need of conservation action. Both species are listed as endangered by the USDI and are on appendix 1 of the CITES. The IUCN classifies *M. leucophaeus*, the more nar-

rowly distributed and seriously jeopardized of the two, as endangered and *M. sphinx* as near threatened. Blom et al. (1992) indicated that the range of *M. sphinx* is now more restricted in Gabon than previously reported but that there had been recent unconfirmed reports of *M. leucophaeus* in the country.

PRIMATES; CERCOPITHECIDAE; **Genus THEROPITHECUS**
I. Geoffroy St.-Hilaire, 1843

Gelada

The single living species, *T. gelada,* is restricted to the mountains of central Ethiopia (Yalden, Largen, and Kock 1977). A previously unknown population about 300 km south of the others was discovered in 1989 (Mori and Belay 1990). During the Pliocene and Pleistocene various extinct species occurred across much of Africa and southwestern Asia (Pickford 1993). *Theropithecus* is sometimes considered a subgenus of *Papio.*

Head and body length is about 500–725 mm and tail length, including the tuft, is about 700–800 mm. Napier and Napier (1967) recorded weights of 20.5 kg for a male and 13.6 kg for a female. The whiskers are long, the head is crested, a mane is present on the shoulders, and the tail is long and tufted. Coloration is brown and black, but there is a bare red area on the chest that is conspicuous when movements of the animal bring it into full view. The underparts are lighter than the upper parts, usually being grayish or grayish white. The legs are grayish, or pale yellowish from the knees to the ankles. The buttock pads are black, not red as in *Papio hamadryas*. The gelada also may be distinguished from all other baboons by its nostrils, which open on the side of the nose, not terminally as in the other species. The body form is massive.

Jablonski (1993*a*) pointed out that *Theropithecus* is distinguished by a large number of skeletal characters associated mostly with adaptation to a terrestrial, grazing lifestyle and the stresses of mastication. These features include a skull with large infratemporal fossa, marked postorbital constriction, deep posterior maxilla, and deep mandibular body and symphysis; cheek teeth high-crowned with steep sides and increased relief and showing a distinctive pattern of wear with the exposure of completely curved infoldings of enamel and dentinal lakes as cusps become worn; and in the postcranial skeleton relative elongation of the forearm, greater flexibility of the elbow and wrist joints, relative elongation of the thumb, and relative shortening of the index finger.

The gelada is strictly associated with rocky gorges and precipices at altitudes between about 1,700 meters and the montane moorland zone at 4,400 meters (Iwamoto 1993; Yalden, Largen, and Kock 1977). It is dependent on the gorge systems for sleeping sites and refuge from predators and is never found more than 2 km from the rim of a gorge. About an hour after dawn groups climb out of the gorges to the plateaus above. They spend about two hours in social activity, and then around 0900–1000 hours they move out to feed. By 1600 hours they are back at the sleeping cliffs. *Theropithecus* is the only true grazing primate. One study showed that 90–95 percent of its diet consists of the leaves, roots, and seeds of grasses; the remainder includes

other vegetable matter and occasionally insects. Specialization in eating grass means that the gelada does not have to disperse very far to find food. Small groups may average as little as 600 meters a day, and even large herds of 300–400 animals rarely travel more than 2 km per day. This situation has resulted in population densities as great as 75–80/sq km in undisturbed areas (Dunbar 1977*a*, 1977*b*).

The complex social structure of *Theropithecus* has been the subject of several recent major studies, the results of which do not entirely coincide (Dunbar 1983*a*, 1983*b*, 1993*b*; Dunbar and Dunbar 1975; Kawai 1979; Ohsawa and Kawai 1975). Some of the differences were resolved by Kawai et al. (1983). The overall gelada population seems to be divided into "bands" of about 30–350 animals. Each band utilizes a home range of about 1.5–2.0 sq km, which overlaps to some extent with the ranges of neighboring bands. When grazing conditions are good, several bands may join in temporary aggregations of as many as 670 individuals. Bands are relatively stable, socially cohesive entities, but they are divided into several "reproductive units" and "all-male units" that may act independently.

All-male units, averaging about eight individuals each, contain mainly young animals, though an older male may also be present. Most reproductive units have only a single adult male, but about 25 percent have two or three. These units usually have about three to six adult females and their young. There seems to be general agreement that the main factor in keeping a reproductive unit together is the strength of bonds between the females rather than male dominance and herding, the case in *Papio hamadryas*. Through this female relationship and its descent from

Gelada baboons *(Theropithecus gelada)*, females, photo from Los Angeles Zoo.

Gelada baboon *(Theropithecus gelada)*, male, photo by P. R. McCann.

mother to daughter the reproductive unit persists generation after generation even if males are not present at times. Since there also is movement of females between groups, some or most of these animals may not be related. There may be little or no social interaction, particularly grooming, between unrelated individuals. The females take at least partial responsibility for group defense, and they may unite in aggression against the male. If the unit has more than one adult male, the females sometimes fight amongst themselves regarding which of the males is to be allowed to mate.

There is some disagreement regarding the role of the males, but it seems likely that at least certain males are capable of controlling group movements. This may be especially true if the females are young and have not yet developed strong bonds. In established groups, however, the male may have only a peripheral role. He does not necessarily interact with all of the females but has a single predominant social partner who, in turn, is likely to be an individual without female relatives in the unit. Males of reproductive units fight younger males attempting to move in from all-male units. Sometimes, however, a second or third male attaches itself to a reproductive unit. Such "follower" males may be tolerated and may even cooperate with the first male, but they appear to be merely waiting for the opportunity to obtain the females for themselves. A follower male may eventually try to take over the entire reproductive unit, which usually involves fierce combat with the first male, or he may gradually develop a relationship with one or two of the younger females, ultimately departing with them to form a new unit. If combat occurs, the actual issue to be decided is which male will have interaction with the females, and it is the latter who may choose the winner. If the first male is defeated, he sometimes remains in the unit as a follower. *Theropithecus* has a large number of visual signals and calls for communication; about 25 different vocalizations have been recognized.

Dunbar and Dunbar (1975) found birth peaks in January and June in their study area but noted that such peaks differed in other areas. They also stated that there was an evident interbirth interval of 24 months. Hayssen, Van Tienhoven, and Van Tienhoven (1993) listed estrous cycle lengths of 30–39 days, estrus lengths of 7–10 days, and lactation periods of 12–18 months. Mallinson (1974) reported gestation periods of 147, 171, and 192 days. Males are considered to be adults at 8 years and females, at 3.5 years. An individual in the San Antonio Zoo in December 1994 was probably well over 30 years old (Marvin L. Jones, Zoological Society of San Diego, pers. comm., 1995).

Dunbar (1977*b*) estimated the total gelada population at about 600,000 animals but warned that suitable natural habitat was being lost to agriculture. Population densities had been reduced to only about 15–20/sq km in areas disturbed by cultivation and topsoil erosion. In addition, in the southern part of the range of the species natives intensively hunt adult male geladas to obtain their manes for use in headdresses and capes. Such hunting was thought to reduce average life expectancy of males and to upset the social structure of reproductive units, though Dunbar (1993*a*) suggested that it has little or no impact on overall numbers. Dunbar and Dunbar (1974*a*) stated that human activity had resulted in a breakdown of the natural ecological separation of *Theropithecus* and *Papio anubis* and that hybridization between the two species might have occurred. Lee, Thornback, and Bennett (1988) stated that numbers may have fallen during the droughts of the 1980s, though Dunbar (1993*a*) considered any marked reduction to be unlikely. The gelada is listed as near threatened by the IUCN and as threatened by the USDI and is on appendix 2 of the CITES.

PRIMATES; CERCOPITHECIDAE; **Genus NASALIS**
E. Geoffroy St.-Hilaire, 1812

Proboscis Monkey

The single species, *N. larvatus,* occurs on Borneo (Groves 1970*a*). *Simias* (see account thereof) sometimes is considered part of *Nasalis.*

Males have a head and body length of 660–762 mm and weigh 16.0–22.5 kg; females have a head and body length of 533–609 mm and weigh 7–11 kg. Tail length is 559–762 mm. The upper parts vary from chestnut brown to cinnamon rufous, and the lighter underparts are usually creamy, creamy buff, or grayish. The skin of the face is entirely rufous.

The outstanding feature of the genus is the protruding nose, which becomes long and pendulous in old males; it is less developed in females. In young animals the nose is turned upward as in *Simias.* The large nostrils open downward. Bennett (1987) suggested that the nose may be involved with attracting females or radiating excess body heat. She stated also that the feet are partially webbed, thereby facilitating swimming and perhaps walking on mangrove mud.

The proboscis monkey is usually found near fresh water in lowland rainforest or mangrove swamp. Kawabe and Mano (1972) noted that it seemed to depend on mangrove trees for resting and sleeping. Salter et al. (1985) found it to sleep primarily along the edge of a river by night and to move an average of 1,312 meters during the day. Kern (1964) reported it to be active throughout the day, with a peak extending from late afternoon to dark. He observed it swimming both on the surface and underwater and considered it the best swimmer among the primates. A dive of an entire group from a height of about 16 meters was witnessed. Kern found that 95 percent of the diet consisted of leaves, pedada leaves being favored. Fruits and flowers were also eaten. In contrast, Yeager (1989) recorded leaves being eaten in 52 percent of observations and fruits in 40 percent, and seed-eating was a specialization.

Kern (1964) reported a population density of about 9/sq km. Three troops each had a home range of about 2 sq km, and they moved through these areas randomly, with the animals spread out over about 0.5 km. At night each individual slept alone in a tree. Group size averaged 20 (12–27) animals. Kawabe and Mano (1972) reported troops to contain 11–32 monkeys, of which several were adult males and considerably more were adult females and young. The adult males would confront intruders while the others departed. Macdonald (1982) observed group sizes of 2–63 animals and suggested that the larger parties were temporary foraging aggregations. The call of the male is a drawn-out, resonant "honk" or "kee-honk"; that of the female "has a milder, petulant, rather resounding cry faintly suggestive of a goose."

In forests along the Samunsam River in Sarawak, Bennett and Sebastian (1988) found a population density of 5.93/sq km. Most of the animals were organized into groups that generally stayed within 600 meters of the river and had a daily average movement of 483 meters. The basic social unit was a relatively stable harem that used an average length of 6.4 km of river and had an average size of 9 (6–16) individuals. Each harem had a single adult male and 2–7 adult females. Different groups frequently joined together by day to forage and often spent the night in close proximity. The harems were never observed to split into subgroups. Young females commonly transferred from one group to another, either before or after reaching reproductive age. Juvenile males regularly left their natal groups and moved about alone for a while or joined to form bachelor bands.

Based on a year of observations in the Indonesian portion of Borneo, Yeager (1989, 1990, 1991, 1992) reported an average population density of 63/sq km, an average group home range of 130.3 ha., and an average group size of 12 (3–23) animals. Social organization was found to closely resemble that of *Theropithecus*, there being "bands" comprising a number of associated groups that moved and rested in proximity and both "all-male" and "one-male" groups. The latter units, aside from the single male, contained 1–9 adult females and their young. Female bonding was the most important factor in holding the groups together, with young females remaining in their mother's unit and young males leaving to join an all-male group. Although agonistic displays were common, serious fighting was rare and bands tended to avoid one another. The females of a one-male unit seemed to initiate movement, and the male sometimes acted to stop agonistic behavior between other group members.

There appears to be no restricted breeding season; a single young is the rule, after a gestation period of about 166 days (Schultz 1942). Weaning is complete by 7 months (Hayssen, Van Tienhoven, and Van Tienhoven 1993). For many years the proboscis monkey was considered difficult or impossible to maintain in captivity, but according to Olney, Ellis, and Fisken (1994), zoos held 16 specimens, of which 14 had been born in captivity. One captive individ-

Proboscis monkeys *(Nasalis larvatus)*, photo by Hilmi Oesman.

Proboscis monkey *(Nasalis larvatus)*, photo by Bernhard Grzimek.

Like *Nasalis, Simias* is a large, stoutly built monkey. Both genera have arms nearly as long as the legs. The hair of the cheeks is brushed back toward the midline of the neck and does not have the "untidy" appearance found in *Pygathrix.* In males of both *Simias* and *Nasalis* the ischial callosities are united across the midline (Groves 1970a).

An investigation by Tilson (1977) on Siberut Island added considerably to our knowledge of this rare primate. *Simias* is found mainly in hilly primary forests of the interior, not in swamp or mangrove. It is totally arboreal, descending only when disturbed. The diet consists predominantly of leaves, but fruits and berries are also taken. The group foraging area is about 25–30 ha. The social unit is an adult pair with one to three immature offspring. Noticeable vocalizations are rare. Newborn young were observed only in June and July.

S. concolor is listed as endangered by the USDI and the IUCN and is on appendix 1 of the CITES. It is thought to have declined by at least 50 percent over the last decade and is expected to decline by at least another 50 percent during the next decade. It is intensively hunted for food, and its conspicuous size makes it especially vulnerable in this regard. More significantly, it is restricted to a few islands and is dependent on primary forests, which are now being logged. No specimens are known to be in captivity. MacKinnon (1986) estimated that the remaining 4,500 sq km of suitable habitat is occupied by 31,500 individuals, though

ual lived for about 23 years (Marvin L. Jones, Zoological Society of San Diego, pers. comm., 1995).

N. larvatus is classified as vulnerable by the IUCN and as endangered by the USDI and is on appendix 1 of the CITES. This species was long considered safe from severe human disturbance because of its prime habitat in inaccessible mangrove swamps. Now, however, improved technology is allowing the economical clear-cutting of mangroves, and the proboscis monkey is declining. MacKinnon (1986) estimated that suitable habitat has fallen from 19,622 to 10,438 sq km and that the latter area is occupied by 260,950 individuals.

PRIMATES; CERCOPITHECIDAE; **Genus SIMIAS**
Miller, 1903

Pig-tailed Langur

The single species, *S. concolor,* occurs on Siberut, Sipura, North Pagai, and South Pagai in the Mentawai Islands off western Sumatra. Groves (1970a) included *Simias* within *Nasalis* based on the close resemblance of the two in cranial features, limb proportions, and hair pattern. The differences in the nose and tail were regarded as superficial. Groves (*in* Wilson and Reeder 1993), as well as Brandon-Jones (1984) and Corbet and Hill (1992), continued to follow that arrangement, but Corbet and Hill (1991), Eudey (1987), Napier and Napier (1985), and Oates, Davies, and Delson (1994) all regarded *Simias* as a distinct genus.

Head and body length is usually 450–525 mm and tail length is only about 130–80 mm. A female specimen weighed 6.9 kg (Groves 1970a). The general coloration throughout is dusky brown; this may be lightened slightly on the nape, shoulders, back, and upper arms by buffy rings on the hairs. The hands and feet, as well as the naked palms, soles, and large buttock pads, are black. Creamy buff individuals have been noted. The short tail is bare except for an inconspicuous tuft of brownish hair at the tip. Unlike in *Nasalis,* the nose shows no trace of tubular elongation. The openings of the nostrils are on the surface of the wide upper lip, and only the upper nasal margin is lengthened to give a snub-nosed appearance.

Pig-tailed langur *(Simias concolor)*, mounted specimen, photo by H. Unte.

Eudey (1987) indicated that the total population numbered fewer than 10,000.

PRIMATES; CERCOPITHECIDAE; **Genus PYGATHRIX**
E. Geoffroy St.-Hilaire, 1812

Douc Langur

The single species, *P. nemaeus,* is found to the east of the Mekong River in central and southern Viet Nam, central Laos, and eastern Cambodia (Groves 1989). *P. nigripes* of southern Viet Nam and adjacent parts of Cambodia was treated as a separate species by Brandon-Jones (1984), but was considered a subspecies of *P. nemaeus* by Corbet and Hill (1991, 1992), Groves (*in* Wilson and Reeder 1993), Jablonski (1995), Jablonski and Peng (1993), and Oates, Davies, and Delson (1994). *P. nemaeus* also has been reported to occur on Hainan, but Groves (1970*a*) stated that

there is no evidence for its existence there. *Rhinopithecus* (see account thereof) sometimes is treated as a subgenus of *Pygathrix.*

Head and body length is usually 610–762 mm and tail length is 558–762 mm. *Pygathrix* is among the most strikingly colored of all mammals. It is marked with sharply contrasting patches and areas of color. The head is brown with a bright chestnut band below the ears. The body is mottled grayish, darker above than below, and the rump, tail, and forearms are white. The upper parts of the arms and legs, and the hands and feet, are black. In the subspecies *P. n. nemaeus* the face is bright yellow, the whiskers are white, and the lower legs are reddish chestnut. In *P. n. nigripes* the face is dark and the lower legs are black. One or two other distinctively colored subspecies may exist (Wirth, Adler, and Thang 1991), though Jablonski (1995) recognized only *nemaeus* and *nigripes.*

The douc langur inhabits tropical rainforests from sea level to 2,000 meters (IUCN 1972). Most of the available information about this monkey was obtained by Lippold

Douc langur *(Pygathrix nemaeus),* photo by Bernhard Grzimek.

(1977), who studied *P. nemaeus* both in captivity in the San Diego Zoo and in the wild on Mount Son Tra, near Da Lat, Viet Nam, from June to August 1974. The species is diurnal and arboreal. It eats a wide variety of leaves and also certain abundant fruits. Early reports suggested a group size of 30–50 individuals, but recent observations indicate a usual range of 4–15. Groups contain 1 or more adult males and generally about twice as many adult females. Each sex apparently has its own dominance hierarchy, and males are dominant over females. In the wild a peak in births probably occurs between February and June, being correlated with maximum seasonal availability of fruit. The estrous cycle is 28–30 days long, and estimates of gestation have ranged from 165 to 190 days. An individual born at the San Diego Zoo was still living in December 1995 at an age of 25 years and 9 months (Marvin L. Jones, Zoological Society of San Diego, pers. comm., 1995).

The douc langur is classified as endangered by the IUCN and the USDI and is on appendix 1 of the CITES. There has been recent confirmation that both the subspecies *nemaeus* and *nigripes* survive in small numbers, but they still are being actively collected for the pet market and hunted for use as food (Eames and Robson 1993; Wirth, Adler, and Thang 1991). They also are thought to have declined because of environmental disruption brought on by military activity during the Viet Nam War. The population studied by Lippold has now disappeared (Eudey 1987).

PRIMATES; CERCOPITHECIDAE; **Genus**
RHINOPITHECUS
Milne-Edwards, 1872

Snub-nosed Langurs

There are two subgenera and four species (Brandon-Jones 1984; Eudey 1987; Groves 1970*a*; Jablonski and Peng 1993):

subgenus *Presbytiscus* Pocock, 1924

R. avunculus, mountains of northern Viet Nam;

subgenus *Rhinopithecus* Milne-Edwards, 1872

R. roxellana, mountainous region on southeastern slopes of Tibetan Plateau in Chinese provinces of Hubei, Shaanxi, Gansu, Sichuan, and Yunnan;
R. bieti, Yun-ling Range of eastern Tibet and Yunnan;
R. brelichi, Fan-jin Range south of Middle Yangtze in Guizhou Province of China.

Groves (1970*a*, 1989, and *in* Wilson and Reeder 1993) considered *Rhinopithecus* not more than subgenerically distinct from *Pygathrix* on the basis of similarities in the skull and in limb proportions and in this regard was followed by Corbet and Hill (1991, 1992), Oates, Davies, and Delson (1994), and various other authorities. However, exhaustive analyses of characters of the skeleton, internal anatomy, and pelage of many specimens by Jablonski (1995) and Jablonski and Peng (1993) supported the validity of the above arrangement. A study of cranial morphology by Peng, Pan, and Jablonski (1993) even suggested that *Presbytiscus* might best be treated as a full genus with closest affinity to *Pygathrix*.

Known measurements for head and body length and tail length, respectively, are as follow: *R. roxellana*, 570–760 mm and 510–720 mm; *R. bieti*, 740–830 mm and 510–720 mm; *R. brelichi*, 730 mm and 970 mm; and *R. avunculus*,

Snub-nosed langurs *(Rhinopithecus roxellana)*, photos by Wang Sung.

510–620 mm and 660–920 mm. The males of *R. roxellana* are grayish black on the top of the head, nape, shoulders, upper parts of the arms, back, and tail (the back being more or less overlaid with long silvery hairs). The forehead, sides of the head, sides of the neck, and underparts are golden. The coloration of the females is much the same except that the head and upper parts are brownish black. *R. bieti* has mostly blackish gray upper parts, long yellow-gray hairs scattered on the shoulders, and white underparts. *R. brelichi* has a slaty gray back with a patch of white in the midline between the shoulders. The crown is yellowish, the forelimbs are yellowish and whitish, the hind limbs are grayish, and the belly is gray. In *R. avunculus* the forehead and cheeks are creamy, the sides of the neck are orange buff, and the back and limbs are black. There is a buffy white patch on the rump on either side of the tail. The hairs are long, measuring 150–80 mm in some forms. A crest of hair may be present on the crown.

The nose is flat in *Rhinopithecus*, whereas it is turned upward in *Pygathrix*. According to Napier and Napier (1985), who also regarded the two as generically distinct, the nose of *Rhinopithecus* is strongly retroussé and set well back from the large rounded muzzle. The nostrils are widely open and face forward, and two flaps of skin on the upper rim of the nostrils form two little peaks that almost touch the forehead. In *Pygathrix* these flaps lie flat.

Very little is known about snub-nosed langurs. *R. avunculus* lives at relatively low elevations in tropical monsoon forests. The other three species inhabit high mountain forests up to about 4,500 meters but may descend to lower elevations in winter. Part of their range is covered by snow for more than half the year. Recent studies in China (Chen et al. 1983; Happel and Cheek 1986; Kirkpatrick 1995; Li et al. 1982; Long et al. 1994; Mu and Yang 1982; Tan 1985) have added to our knowledge. Most activity takes place in trees, but some feeding and social interaction occurs on the ground. If frightened, the animals flee through the upper levels of the canopy at great speed. Normal daily foraging movement is about 1,000–1,500 meters. The diet seems to consist largely of leaves, grass, lichens, bamboo shoots, buds, and fruits. Troops of more than 100 and even up to 600 individuals have been reported, though such groups may be divided into many units, each containing one dominant male, 3–5 females, and young. The male secures his position by fighting other males. The social pattern is thought to be comparable to the fission and fusion system of *Pan troglodytes*. Groups have nonoverlapping home ranges of about 4–50 sq km, through which they move about once every month. If two groups meet, they quickly go off in different directions. The most common vocalization is "ga-ga," uttered loudly upon the discovery of a rich food source. Births take place during the spring and summer, following a reported gestation period of about 200 days. The number of young is usually 1, occasionally 2. Males reach sexual maturity at 7 years, females at 4–5 years.

Formerly the pelage of *Rhinopithecus* was believed to ward off rheumatism and could be worn only by Manchu officials. There still is intensive hunting to obtain pelts and other parts for supposed medicinal purposes. In some cases entire communes have joined in mass roundups of these monkeys in China (Tan 1985). Environmental disruption also is a problem. The IUCN classifies *R. roxellana* as vulnerable, *R. bieti* and *R. brelichi* as endangered, and *R. avunculus* as critically endangered. All species now are listed as endangered by the USDI and are on appendix 1 of the CITES. *R. avunculus* evidently has declined because of military activity and hunting. It occupies a restricted habitat, and fewer than 300 individuals are thought to survive (Nisbett and Ciochon 1993). Population sizes of the other species are estimated at 10,000–15,000 individuals for *R. roxellana*, 2,000 for *R. bieti*, and 800–1,200 for *R. brelichi* (Bleisch 1991; Eudey 1987; Long et al. 1994; Wang and Quan 1986). Eudey (1987) rated *R. avunculus*, *R. bieti*, and *R. brelichi* among the four primate species in Asia with the highest priority for conservation action (the fourth is *Simias concolor*).

PRIMATES; CERCOPITHECIDAE; Genus PRESBYTIS
Eschscholtz, 1821

Langurs, or Leaf Monkeys

There are eight species (Aimi and Bakar 1992; Chasen 1940; Corbet and Hill 1992; Eudey 1987; Fooden 1976a; Lekagul and McNeely 1977; Medway 1977; Weitzel and Groves 1985):

P. femoralis, Malay Peninsula (including southern Thailand), Riau Archipelago, east-central Sumatra, western Borneo, North Natuna Islands;

P. melalophos, southern and west-central Sumatra;

P. thomasi, northern Sumatra;

P. potenziani, Siberut, Sipura, North Pagai, and South Pagai in the Mentawai Islands off western Sumatra;

P. comata, western Java;

P. hosei, northern and eastern Borneo;

P. frontata, Borneo;

P. rubicunda, Borneo and nearby Karimata Island.

Several species formerly assigned to *Presbytis* now are placed in the genera *Semnopithecus* and *Trachypithecus* (see accounts thereof). The species listed above are the same as those accepted by Corbet and Hill (1992) and Groves (*in* Wilson and Reeder 1993). Oates, Davies, and Delson (1994) included *P. femoralis* in *P. melalophos*. Brandon-Jones (1993) considered *P. thomasi* and *P. hosei* to be subspecies of *P. comata*.

Head and body length is 420–610 mm, tail length is 500–850 mm, and weight is 5.0–8.1 kg (Brandon-Jones 1984). The upper parts generally are brownish, grayish, or blackish, and the underparts are paler. Some forms have light-colored markings on the head or stripes on the thighs. Reddish and whitish mutants also occur. The newborn are generally whitish. This genus is distinguished by poorly developed or absent brow ridges, prominent nasal bones, and a crest on the crown, which takes a variety of shapes in the different species (Brandon-Jones 1984; Napier and Napier 1967). The body is slender, the tail is long, the hands are long and slender, and the thumb is small but the other fingers are well developed and strong.

Langurs are diurnal, arboreal forest dwellers (Medway 1970). In Borneo *P. hosei* and *P. rubicunda* are found in tall and, less abundantly, secondary forests (Payne, Francis, and Phillipps 1985). They are active throughout the tree canopy, occasionally descending to the ground to visit natural mineral sources. Daily foraging range is about 500–800 meters; although the diet includes a substantial amount of foliage, the greater part consists of fruits and seeds (Bennett and Davies 1994).

MacKinnon (1986) listed population densities of 2–30/sq km for the various species in Indonesia. Ruhiyat (1983) found densities of *P. comata* to be 11–12/sq km in one part of Java and 35/sq km in another. Respective home range sizes were 35–40 ha. with little overlap and 14 ha. with much overlap. On the Malay Peninsula, Chivers (1973) found group territories of about 13 ha. for *P. femoralis*. According to Bennett and Davies (1994), a group of *P. rubicunda* occupied about 84 ha. of forest in Borneo, but ranges usually are smaller and mostly constitute actively defended territories. Groups usually include a single adult male and two or more adult females, but *P. potenziani* sometimes is found in monogamous pairs. Heterosexual group size averaged about 15 individuals for *P. melalophos* on the Malay Peninsula, 7 for *P. rubicunda* on Borneo, and 6 for *P. thomasi* on Sumatra; there also are lone males and

Stripe-crested langur *(Presbytis comata)*, photo by Russell A. Mittermeier.

small all-male units. Outside males may attempt to take over a heterosexual group or split off some of its females. Juvenile males disperse from their natal groups. According to Medway (1978), the alarm call of *P. femoralis* is a harsh rattle followed by a loud "chak-chak-chak-chak." There normally is a single young. A specimen of *P. melalophos* lived in captivity for 18 years and 4 months (Marvin L. Jones, Zoological Society of San Diego, pers. comm., 1995).

The IUCN classifies *P. comata* as endangered, noting that it has declined by at least 50 percent in the last decade because of habitat destruction and that fewer than 2,500 mature individuals survive. The IUCN also classifies *P. potenziani* as vulnerable and *P. femoralis* and *P. thomasi* as near threatened. *P. potenziani* is listed as threatened by the USDI and is on appendix 1 of the CITES. It is endemic to the Mentawai Islands, which are now being logged. It also is hunted extensively for food by the increasing human population. An estimated 45,000 individuals occupy the remaining 4,500 sq km of suitable habitat (MacKinnon 1986). All other species of *Presbytis* are on appendix 2 of the CITES. Most are suffering to some extent because of habitat loss. Khan (1978) estimated that populations of *P. femoralis* in peninsular Malaysia declined from 962,000 in 1958 to 554,000 in 1975. Eudey (1987) designated *P. comata* as highly endangered because of deforestation and fragmentation of remnant populations. It also now is listed as endangered by the IUCN.

PRIMATES; CERCOPITHECIDAE; **Genus**
SEMNOPITHECUS
Desmarest, 1822

Hanuman Langur

The single species, *S. entellus*, is found in extreme southern Tibet, Nepal, Sikkim, northern Pakistan, Kashmir, India, Bangladesh, and Sri Lanka (Corbet and Hill 1992; Eller-

man and Morrison-Scott 1966). Although this species often is considered to be simply a distinctive member of the genus *Presbytis*, Eudey (1987), based on studies by V. Weitzel and C. P. Groves, regarded it as the representative of a separate genus. Eudey did not accept Brandon-Jones's (1984) recognition of *S. hypoleucos* of southwestern India as a species distinct from *S. entellus*. *Semnopithecus* was treated as a monotypic genus by Groves (1989 and *in* Wilson and Reeder 1993) and Oates, Davies, and Delson (1994) but was considered by Corbet and Hill (1992) to include *Trachypithecus* (see account thereof).

According to Brandon-Jones (1984), head and body length is 410–780 mm, tail length is 690–1,080 mm, and weight is 5.4–23.6 kg. The upper parts are various shades of gray, brown, or buff, and the crown and underparts are white, orange-white, or yellowish. The coat of the newborn is blackish brown. From *Presbytis, Semnopithecus* is distinguished by its prominent, shelflike brow ridges. Napier and Napier (1967) stated also that the brow hairs are directed forward, and the crown hairs backward, by a whorl on the forepart of the crown. Postcranial morphology is like that of *Presbytis*.

The Hanuman langur is found in many environments, from desert edge to wet tropical forest and alpine scrub, and from sea level to more than 4,000 meters. In some areas where trees are scarce it has adapted well and spends most of its time on the ground (Oppenheimer 1977). On the ground it walks or runs on all four feet. In the trees it is remarkably agile; it can make horizontal leaps of 3–5 meters or cover up to 13 meters with some loss of height (Roonwal and Mohnot 1977). It is generally most active in the early morning and late afternoon, and it sleeps at midday. Like all langurs, it is almost entirely a vegetarian; leaves make up most of its diet, but fruits, flowers, and cultivated crops are also eaten. Groups may forage over several kilometers in the course of a day.

Population densities of up to 904/sq km have been reported (Roonwal and Mohnot 1977), but more usual density seems to range from about 3/sq km in grass and cropland to 130/sq km in forest (Oppenheimer 1977). Home range is about 0.05–13.00 sq km for groups made up of both males and females and 7–22 sq km for all-male groups. Bennett and Davies (1994) listed home range sizes of 2–12 sq km in the Himalayas but only 17–18 ha. in Sri Lanka. A portion of the home range is a core area where most time is spent, but there may be much overlap of entire ranges and there appears to be no strictly territorial defense. The most outstanding vocalization is a booming whoop, frequently given in a morning chorus and apparently functioning to space groups.

Social structure is variable (Bennett and Davies 1994; Oppenheimer 1977; Roonwal and Mohnot 1977). In areas where populations are below carrying capacity of the habitat, groups containing both sexes may include more than a single adult male. In areas where the environment is subject to at least seasonal stress heterosexual troops may have only a single adult male, and there may also be separate all-male groups. Reported size for heterosexual groups is 5–125 individuals, but perhaps the larger figures involve temporary aggregations of several social units. The more normal group size range is 13–37, and there are usually about 2 adult females for each adult male. Additional males may live alone or in groups of 2–32 individuals, including adults, subadults, and a few juveniles. Intragroup structure is generally stable once the males establish a dominance hierarchy through fighting. Both males and females sometimes migrate between heterosexual groups, but males do so more often. The leading male may eventually be ousted

Hanuman langur *(Semnopithecus entellus)*, photo from Zoological Society of San Diego.

from his position by a younger male, or the younger individual may succeed in splitting off some of the females to form a new group of his own. Females also have a rank order, but it is not pronounced, and a dominant male is the leader and defender of the group. One male is known to have remained in control of a group for 10 years. Relations between heterosexual troops are generally peaceful, but in some areas these troops are subject to attack from all-male groups. If an outside male succeeds in defeating the dominant male of a heterosexual group, his next action may be to kill the infants of the group. This measure will bring the females into estrus within two weeks and allow the new group leader to mate and father progeny of his own.

Births may occur at any time of the year but are concentrated in the dry season during April–May in northern India and in December–March in southern India. There is much intergroup movement of males during the breeding season, and birth periods in groups have been reported to shift or to cease altogether in the face of persistent attacks by outside males (Bennett and Davies 1994; Oppenheimer 1977). In western India there is a definite season of births, November–March, and in Sri Lanka the young are born from March to May and in September (Roonwal and Mohnot 1977). The estrous cycle is about 24 days long (Winkler, Loch, and Vogel 1984). Estrus lasts 5–7 days (Hayssen, Van Tienhoven, and Van Tienhoven 1993). If an infant is lost, cycles can resume within 8 days (Oppenheimer 1977). The normal interbirth interval is about 15–24 months, and the gestation period is about 190–210 days (Roonwal and Mohnot 1977). There usually is a single young, but twins are said to occur frequently in the Himalayas (Roonwal and Mohnot 1977). A birth weight of

500 grams was listed by Hayssen, Van Tienhoven, and Van Tienhoven (1993). Weaning is completed at about 10–12 months. Adulthood is reached by females at 3–4 years and by males at 6–7 years. Subadult males may be driven from a group by the dominant male. A captive specimen was still alive after 25 years (Marvin L. Jones, Zoological Society of San Diego, pers. comm., 1995).

The Hanuman langur is classified as endangered by the USDI and as near threatened by the IUCN and is on appendix 1 of the CITES. Although considered the most widely distributed nonhuman primate in India, it is declining through loss of habitat to agriculture and forest clearance. It is regarded as sacred to the Hindu religion, but its raids on crops may no longer be tolerated in the face of human population increases and food shortages. The total population in India is estimated at only 233,800 individuals (Southwick and Lindburg 1986).

PRIMATES; CERCOPITHECIDAE; **Genus**
TRACHYPITHECUS
Reichenbach, 1862

Brow-ridged Langurs, or Leaf Monkeys

There are two subgenera and nine species (Chasen 1940; Corbet and Hill 1992; Ellerman and Morrison-Scott 1966; Eudey 1987; Fooden 1976*a*; Lekagul and McNeely 1977; Medway 1977; Morris 1991; Weitzel and Groves 1985):

Brow-ridged langurs *(Trachypithecus obscurus)*, photo from Amsterdam Zoo.

subgenus *Kasi* Reichenbach, 1862

T. vetulus, Sri Lanka;
T. johnii, southwestern India;

subgenus *Trachypithecus* Reichenbach, 1862

T. geei, Nepal, Bhutan, northeastern India;
T. pileatus, Assam, Bangladesh, western and northern Burma;
T. phayrei, eastern Assam and Bangladesh, Burma, Thailand, Laos, northern Viet Nam, extreme western Yunnan (southern China);
T. francoisi, southern Guangxi (southeastern China), northern Viet Nam, west-central Laos;
T. cristatus, eastern Assam, parts of southeastern Burma and southern Thailand north of Malay Peninsula, southern Indochina, western mainland Malaysia, Riau and Lingga archipelagoes, Sumatra, Bangka, Java, Bali, Borneo, Belitung Island, South Natuna Islands;
T. auratus, Java, Bali, Lombok;
T. obscurus, Malay Peninsula (including extreme southern Burma and Thailand) and nearby islands.

Trachypithecus often is considered a subgenus or synonym of *Presbytis* (or of *Semnopithecus*) but was recognized as a separate genus by Eudey (1987), Groves (1989 and *in* Wilson and Reeder 1993), Oates, Davies, and Delson (1994), and Weitzel (1992). Corbet and Hill (1992) included *Trachypithecus* in *Semnopithecus*. All of those authorities accepted the species listed above and did not accept the following views set forth by Brandon-Jones (1984): that *T. leucocephalus* of southwestern Guangxi and *T. delacouri* of northern Viet Nam are species distinct from *T. francoisi;* that populations of *T. cristatus* and *T. phayrei* in southern China and northern Burma and Indochina actually represent another species, *T. barbei;* and that the remaining populations of *T. phayrei* are referable to *T. cristatus.* The species *T. vetulus* sometimes has been called *T. senex.*

Head and body length is 400–760 mm, tail length is 570–1,100 mm, and weight is 4.2–14.0 kg (Brandon-Jones 1984). The overall coloration is usually brown, dark gray, or black. Most species have various white, yellowish, or black markings on the head, shoulders, rump, limbs, or tail.

T. obscurus and *T. phayrei* have conspicuous white circles around the eyes. Some species have a pointed crest on top of the head. The coat of the newborn is bright yellow

Brow-ridged langur *(Trachypithecus cristatus)*, photo from New York Zoological Society.

or orange-red. From *Presbytis, Trachypithecus* is distinguished by its prominent brow ridges, which resemble raised eyebrows (Brandon-Jones 1984). The postcranial morphology is like that of *Presbytis*, but the thumb is particularly short (Napier and Napier 1985) and the hind limbs are relatively shorter (Oates, Davies, and Delson 1994).

The well-studied species *T. johnii* inhabits upland forests and is more arboreal than *Semnopithecus entellus*. It may shift its range in the course of a year in response to the fruiting and flowering seasons of various plants. It is also known to raid potato and cauliflower crops and occasionally to feed on insects (Poirier 1970; Roonwal and Mohnot 1977). The species *T. pileatus* is considered entirely arboreal and apparently never leaves the trees except when forced to find water during the dry season (Roonwal and Mohnot 1977). In a study of that species in Bangladesh, Stanford (1992) found groups to move an average of 325 meters per day in moist deciduous forest habitat and 485 meters in wet semievergreen forest; the respective population densities were 53/sq km and 13/sq km, and the diet consisted mostly of leaves and fruit. *T. phayrei* feeds almost entirely on leaves and also eats some flowers and fruits but never was observed to invade cultivated fields (Mukherjee 1982). Bennett and Davies (1994) reported that foliage amounts to about 60 percent of the diet for *Trachypithecus*.

Population densities of 116–327/sq km have been reported for *T. vetulus*, of Sri Lanka. According to MacKinnon (1986), about 2 million *T. cristatus* occur in 133,167 sq km of suitable habitat in Indonesia, at a density of 15/sq km. Group home range averages smaller in *T. johnii* than in *Semnopithecus*, about 5–50 ha., but territories are maintained through the vigilance, loud vocalizations, and visual displays of the adult males (Bennett and Davies 1994; Poirier 1970). No intergroup aggression has been noted in *T. geei*, but *T. vetulus* is highly territorial and adult males defend the entire home range of just 1–10 ha. through powerful calls and spectacular displays of running and leaping (Roonwal and Mohnot 1977). *T. vetulus* usually lives in cohesive groups of about 8 individuals, including a single adult male, but on average an outside male takes over the group every three years; at such times there is much violent behavior, with the new male driving out or killing all the young in the group (Bennett and Davies 1994). On the Malay Peninsula, Chivers (1973) found group territories of 5–12 ha. for *T. obscurus*.

In *T. johnii* reported heterosexual group size is 6–30 and most units contain only a single adult male. Additional males live alone or in small groups. There is a strong, stable dominance hierarchy among the females, who appear to be responsible for most group socialization. Intragroup relations are generally peaceful, but the leading male has frequent territorial encounters with the males of other groups and may eventually be ousted by an outside male (Bennett and Davies 1994; Poirier 1970; Roonwal and Mohnot 1977). Observations of most other species of *Trachypithecus* suggest a comparable group size and social structure (Bennett and Davies 1994; Chivers 1973; Medway 1970; Mukherjee 1982; Roonwal and Mohnot 1977; Stanford 1992).

No seasonality has been reported for *T. johnii* or *T. geei*. For *T. vetulus* Rudran (1973) reported that in one area of Sri Lanka high rainfall and abundant food stimulated a peak mating period in October–January that resulted in a peak birth season in May–August. In another area, where rainfall and food supplies were high all year, no reproductive peaks were observed. Lekagul and McNeely (1977) stated that births in wild *T. obscurus* usually occur from January to March and that the estrous cycle lasts about 3 weeks. The normal interbirth interval is 2 years. Gestation is 200–220 days in *T. vetulus* (Rudran 1973) and 140–50 days in *T. obscurus* (Lekagul and McNeely 1977). Litter size is usually 1 in the genus, but twins are born occasionally. Hayssen, Van Tienhoven, and Van Tienhoven (1993) listed a birth weight of 113 grams for *T. obscurus* and an age at sexual maturity of around 3–4 years for *T. vetulus*. A captive *T. cristatus* lived to 31 years and 1 month (Marvin L. Jones, Zoological Society of San Diego, pers. comm., 1995).

Some species of *Trachypithecus* are of serious concern from a conservation standpoint. The USDI lists *T. geei, T. pileatus,* and *T. francoisi* as endangered and *T. vetulus* as threatened. The IUCN classifies *T. delacouri* (recognized as a species distinct from *T. francoisi;* see above) as critically endangered, *T. poliocephalus* (recognized as a species distinct from *T. francoisi;* see below) as endangered, *T. vetulus, T. johnii, T. pileatus, T. francoisi,* and *T. auratus* as vulnerable, and *T. cristatus* as near threatened. In addition, *T. geei* and *T. pileatus* are on appendix 1 of the CITES and all other species are on appendix 2. *T. johnii* is restricted to a small part of southern India, where its range continues to shrink through deforestation and other environmental disturbances. Most other species of *Trachypithecus* also are suffering to some extent because of habitat loss (Wolfheim 1983). Khan (1978) estimated that populations in peninsular Malaysia declined from 1958 to 1975 as follows: *T. cristatus,* from 6,000 to 4,000 individuals, and *T. obscurus,* from 305,000 to 155,000. Eudey (1987) designated *T. francoisi* as endangered and noted that the subspecies *T. f. delacouri* of central Viet Nam may be the most endangered monkey in Asia. Fewer than 250 mature individuals are thought to survive. Weitzel (1992) noted that *T. f. poliocephalus* is now restricted to Cát Bà Island, just off Haiphong, and Canh and Campbell (1994) estimated that fewer than 200 individuals survived in a protected national park on the island. Tan (1985) reported that the total population of *T. francoisi leucocephalus* of Guangxi Province, southeastern China, had fallen to only about 400, at least in part because of hunting by people who believe its parts have medicinal value.

PRIMATES; CERCOPITHECIDAE; **Genus COLOBUS**
Illiger, 1811

Black-and-white Colobus Monkeys

There are five species (Dandelot *in* Meester and Setzer 1977; Oates and Trocco 1983; Rahm 1970*a*):

C. satanas (black colobus), southern Cameroon, Equatorial Guinea, Gabon, island of Bioko (Fernando Poo);
C. angolensis, northern Angola, Zaire, Uganda, Rwanda, Burundi, southern Kenya, Tanzania, northern Zambia;
C. polykomos, Gambia to Ivory Coast;
C. vellerosus, Ivory Coast to southwestern Nigeria;
C. guereza (guereza), eastern Nigeria to Ethiopia and Tanzania.

Some authorities regard *Piliocolobus* and/or *Procolobus* (see accounts thereof) as subgenera of *Colobus*. Most recent authorities, including Groves (*in* Wilson and Reeder 1993), have regarded *C. vellerosus* as a subspecies of *C. polykomos*, but Oates and Trocco (1983), on the basis of vocalizations and morphology, considered *C. vellerosus* to be a full species most closely related to *C. guereza*. Groves, Angst, and Westwood (1993) explained further that the supposed subspecies *dollmani* between the Sassandra and Bandama rivers in Ivory Coast, which has been reported to be morphologically intermediate to *C. polykomos* and *C. vellerosus*, actually represents a hybrid swarm formed when the latter species extended its range westward, genetically swamping the original population of *C. polykomos*.

Head and body length is 450–720 mm, tail length is 520–1,000 mm, and weight is 5.4–14.5 kg (Brandon-Jones 1984). *C. satanas* is entirely black; the other four species have some white markings. In *C. polykomos* the whiskers and chest are white, the rest of the body is black, and the tail is entirely white and not tufted. In *C. vellerosus* most of the body is black, but there are white thigh patches, the face is framed in a white mane, and the tail is entirely white and not tufted. In *C. guereza* there is a white beard, a conspicuous white mantle extending from the shoulders to the lower back, and a large white tuft on the end of the tail. And in *C. angolensis* there are long white hairs around the face and on the shoulders, and the tail ends in a small white tuft. Infants are pure white at birth but gain the full adult markings by about 3.5 months of age (Napier and Napier 1985).

Colobus has a slender body, a long tail, and prominent rump callosities. As in the langurs of Asia, a complex stomach is present and cheek pouches are absent. The thumb, which is small but present in the langurs, is suppressed and

Colobus monkeys *(Colobus guereza)*, photo from San Diego Zoological Garden. Inset: *C. guereza*, photo from *Bull. Amer. Mus. Nat. Hist.*

represented only by a tubercle in *Colobus*. The skull is somewhat prognathous, and the orbits are oval with narrow superciliary ridges (Napier and Napier 1967). The nostrils are more or less lengthened by an extension of the nasal skin and may nearly reach the mouth (Dandelot *in* Meester and Setzer 1977).

Colobus monkeys are generally diurnal, highly arboreal residents of deep forests, but there is some variation. *C. guereza* may be found in dry, moist, or riparian forest either in lowlands or up to 3,300 meters; it is most abundant in secondary forest or along rivers. Where trees are not densely packed it frequently feeds and travels on the ground (Oates 1977*b*, 1977*c*). *Colobus* spends much of the day resting, and groups generally have a daily foraging path of only about 500 meters (Oates 1994). The diet consists mostly of leaves, but fruits and flowers are at least seasonally important. Population densities of 100–500/sq km have been reported for *C. guereza* (Oates 1977*a*).

According to Oates (1994), most populations of *Colobus* live in relatively small social groups, normally with a single fully adult male and 2–6 adult females. Groups with more than one male tend to be larger. Temporary aggregations of more than 300 *C. angolensis* have been observed. Permanent groups use a home range of up to 84 ha., at least part of which is a defended territory.

Oates (1977*b*, 1977*c*) found *C. guereza* to live in small,

Guereza *(Colobus guereza)*, photo by Bernhard Grzimek.

highly cohesive social groups. There were usually 3–15 individuals per group, averaging about 9 in large forest blocks and 7 in riparian forest and small patches of forest. Most groups had a single fully adult male, but sometimes more than 1 male was present. There generally were also 3–4 adult females, plus young. The female membership seemed stable, but adult males sometimes were ousted by younger males that had either grown up within the group or moved in from the outside. Intragroup relations normally were friendly and were reinforced by considerable mutual grooming. Infants often were handled by individuals other than the mother. Relatively small home ranges were utilized, that of one group being about 15 ha. Either the entire home range or a core area therein was vigorously defended as a territory. Intergroup relations were usually agonistic, with most hostility being expressed by the adult males through gestures, vocalizations, displays of leaping, and occasional chasing and fighting. Some groups, however, did share water holes and other resources. There were loud nocturnal and dawn roaring choruses by adult males, probably as a means of spacing groups. McKey and Waterman (1982) reported an absence of early morning roaring in *C. satanas*. The annual home range in that species was found to be about 60 ha., and groups averaged about 15 individuals.

There seems to be little or no reproductive seasonality in most populations of *Colobus* that have been studied. Sabater Pi (1973), however, reported that in Equatorial Guinea there apparently was a birth season in *C. satanas* extending from December to early April, coincident with the period of greatest fruit production. Oates (1977*b*) stated that each adult female *C. guereza* produced one young every 20 months. Hayssen, Van Tienhoven, and Van Tienhoven (1993) listed an estrus of 1–3 days, a gestation period of about 175 days, a birth weight of 820 grams, a lactation period of 6 months, and an age at sexual maturity of about 2 years for *C. polykomos*. Oates (1977*c*) determined that the age of full sexual maturity in *C. guereza* was at

least 6 years for males and 4 years for females. C. Hill (1975) reported that a specimen of *C. polykomos* died after spending 23 years and 6 months in the San Diego Zoo and that the animal probably had been brought into captivity when it was 4–7 years old.

Oates (1977*b*) reported that all species of *Colobus* had declined drastically over the last 100 years, though *C. guereza* was not as seriously threatened as *C. polykomos* and *C. satanas*. The decline was caused initially by the international fur trade in the nineteenth century and subsequently by rapid human population growth and habitat destruction in Africa. All species are on appendix 2 of the CITES; the IUCN classifies *C. satanas*, *C. vellerosus*, and *C. angolensis ruwenzorii* (Rwanda, Uganda, eastern Zaire) as vulnerable and *C. polykomos* as near threatened; and the USDI lists *C. satanas* as endangered. This last species, according to Lee, Thornback, and Bennett (1988), has the most limited distribution of any species in the genus and is declining as a result of hunting for its meat and skin and loss of habitat to logging and agriculture. It seems unable to survive in secondary forest. According to Oates (1996), both *C. satanas* and *C. vellerosus* are among the most threatened primate species in Africa.

PRIMATES; CERCOPITHECIDAE; **Genus**
PILIOCOLOBUS
Rochebrune, 1887

Red Colobus Monkeys

Following Lee, Thornback, and Bennett (1988) and Oates (1985), five species are tentatively recognized here:

P. badius, Senegal to Ghana;
P. pennantii, southeastern Nigeria, Cameroon, Congo, Bioko (Fernando Poo);

Red colobus *(Piliocolobus badius)*, photo by Dawn Starin.

P. rufomitratus, northern and eastern Zaire, Uganda, mouth of Tana River in eastern Kenya, western Tanzania;
P. gordonorum, Uzungwa Mountains in south-central Tanzania;
P. kirkii, Zanzibar.

There is considerable disagreement regarding the taxonomy of red colobus monkeys at both the generic and the specific level. Some authorities, including Dandelot (*in* Meester and Setzer 1977), consider *Piliocolobus* to be a subgenus of *Colobus.* Others, including Brandon-Jones (1984), Lee, Thornback, and Bennett (1988), and Oates (1985), place *Piliocolobus* in the genus *Procolobus* (see account thereof). Groves (1989 and *in* Wilson and Reeder 1993) treated *Piliocolobus* as a subgenus of *Procolobus.*

Struhsaker (1981) noted general agreement that there are 14 named kinds of the genus *Piliocolobus,* with different authorities accepting from 1 to 6 of those kinds as full species. Oates, Davies, and Delson (1994), Rahm (1970a), Wolfheim (1983), and various other recent authors have recognized only the single widely distributed species *P. badius.* Oates (1985), followed by Lee, Thornback, and Bennett (1988), thought the best arrangement to be ranking of *P. badius* as a "superspecies" that includes the 5 allopatric species listed above. Dandelot (*in* Meester and Setzer 1977) listed *P. badius, P. pennantii, P. kirkii, P. rufomitratus* (including *P. gordonorum*), and also *P. tholloni* of east-central Zaire, which is here considered a subspecies of *P. rufomitratus.* Dandelot also listed three other "potential species" in *Piliocolobus: P. ellioti* of the Ituri Forest in Zaire, *P. preussi* of the Nigeria-Cameroon border area, and *P. waldroni* of the Ivory Coast–Ghana region. Groves (*in* Wilson and Reeder 1993) recognized *P. badius, P. pennantii* (including *kirkii, gordonorum,* and *tholloni*), *P. preussi* (here considered a subspecies of *P. pennantii*), and *P. rufomitratus.* Corbet and Hill (1991) and Napier and Napier (1985)

regarded only *P. kirkii* as a species distinct from *P. badius.*

Head and body length is 450–670 mm, tail length is 520–800 mm, and weight is 5.1–11.3 kg (Brandon-Jones 1984). These monkeys have black, slaty, or brownish upper parts and red or chestnut brown arms, legs, and head; tail tufts and long fringes of hair are lacking. According to Napier and Napier (1967), the body form of *Piliocolobus* is similar to that of *Colobus,* but the fingers are relatively longer and the hallux shorter. The thumb tubercle, apparent in *Colobus* and *Procolobus,* is totally absent. The orbits of the skull are angular and have thick superciliary ridges with a distinct supraorbital groove or foramen, which is lacking in *Colobus* and *Procolobus.* Dandelot (*in* Meester and Setzer 1977) noted also that in contrast to *Colobus, Piliocolobus* has nostrils that are V-shaped and not lengthened by an extension of the nasal skin.

Most populations of red colobus are found in rainforests, but in Senegal and Gambia they often inhabit savannah woodland (Struhsaker 1975). Some inhabit mangroves and floodplains. They are diurnal and arboreal. Red colobus are generally considered to be clumsy climbers, but they only infrequently descend to the ground and seem to favor heights of 16.5–27.0 meters. They appear to lack a precise pattern of activity, and their day alternates between periods of feeding and resting. During the course of 54 days one group of *P. badius* had an average daily movement of 648.9 (222.5–1,185.0) meters (Struhsaker 1975). The diet consists principally of leaves and also includes some fruits and seeds. A population density of 80–100/sq km has been reported for *P. badius* (Nishida 1972a).

Apparently there is a pronounced difference in social structure between the genera *Colobus* and *Piliocolobus.* According to information collected or cited by Struhsaker (1975), the red colobus, like the black-and-white, form stable groups, but these range in size from 12 to 82 members, averaging about 50. There are usually several adult males and about 1.5–3 times as many adult females. One group of 75–80 individuals included 10 adult males and 21 adult females. There were also solitary animals, mostly immature males driven out of groups by older males. There is a pronounced dominance hierarchy within a group, expressed by priority access to food, space, and grooming. Most mating is done by the highest-ranking male. Rank order is maintained by aggressive interaction, but actual physical fighting is rare. Infants are handled by the mother alone until they are 1–3.5 months old. Red colobus have a great variety of vocalizations, perhaps the most outstanding of which is the alarm bark of the adult males. Reported group home range size has varied from about 8.5 to 132.1 ha., there evidently usually being about 1–2 ha. for each individual in the group. There is no territoriality, but the area being used at a given time may be defended through aggressive displays and vocalizations.

Reproductive seasonality is not generally evident but has been reported in Sierra Leone, where most births occur in the early dry season, from October to December (Oates 1994). Struhsaker (1975) gave the estimated gestation period of red colobus as 4.5–5.5 months, and Oates (1994) cited interbirth intervals of around 26 months. A single offspring is born. Studies of *P. badius* in Gambia (Starin 1981) indicate that the young of both sexes leave their natal group prior to maturity. Females depart voluntarily and are readily accepted into other troops, but males are forcibly expelled and then may again encounter hostility, and even be killed, when they try to join another group.

Struhsaker (1975) warned that red colobus monkeys were being adversely affected by timber exploitation, not only through loss of required forest habitat but also because logging made the animals more accessible to hunters. He considered red colobus to be the easiest monkeys in Africa to hunt. The IUCN includes *Piliocolobus* in the genus *Procolobus* and recognizes only the single species *P. badius*, which it designates generally as near threatened. It also assigns the following additional designations (technical names adjusted in accordance with the taxonomic classification used above): *P. badius waldroni*, Ivory Coast and Ghana, critically endangered; *P. b. temmincki*, Senegal to northwestern Guinea, endangered; *P. pennantii pennantii*, island of Bioko (Fernando Poo), endangered; *P. p. epieni*, Nigeria, endangered; *P. p. preussi*, Cameroon, endangered; *P. p. bouvieri*, Congo, endangered; *P. rufomitratus rufomitratus*, floodplains of Tana River in eastern Kenya, endangered; *P. gordonorum*, endangered; and *P. kirkii*, endangered. *P. r. rufomitratus* and *P. kirkii* also are listed as endangered by the USDI and are on appendix 1 of the CITES; all other species and subspecies are on appendix 2. Oates (1996) reported that *P. badius waldroni* appeared to be on the verge of extinction because of intensive hunting and habitat destruction. Lee, Thornback, and Bennett (1988) expressed particular concern for the following: *P. r. rufomitratus*, numbers declined from about 1,860 in 1972 to 200–300 and range greatly reduced; *P. gordonorum*, extremely rare with a patchy distribution in a restricted range; *P. pennantii bouvieri*, probably found only along the Lefini River in southern Congo, numbers perilously low; *P. pennantii preussi*, eliminated from southeastern Nigeria and now confined to a small part of adjacent Cameroon, fewer than 8,000 survive; and *P. kirkii*, about 1,500 individuals remaining in isolated patches of forest, all of which are subject to cutting. More recently, Butynski and Mwangi (1995) reported that numbers of *P. r. rufomitratus* were somewhat higher, about 1,100–1,300, but that the total occupied habitat was considerably less than 13 sq km and was immediately jeopardized by planned construction of hydroelectric dams. The one population of *P. gordonorum* with a reasonable chance for long-term survival, comprising about 544

individuals, inhabits a 6-sq-km forested area, legal ownership of which is currently being disputed between commercial interests and the Selous Game Reserve (Decker 1994).

PRIMATES; CERCOPITHECIDAE; **Genus PROCOLOBUS**
Rochebrune, 1887

Olive Colobus Monkey

The single species, *P. verus*, is found from Sierra Leone and southeastern Guinea to Benin; there also is an isolated population to the east of the Niger River in southeastern Nigeria, which suggests that once the overall range once was more widespread (Napier and Napier 1985). *Procolobus* sometimes has been treated as a subgenus of *Colobus* but was regarded as a separate monotypic genus by Corbet and Hill (1991) and Napier and Napier (1985). In contrast, Brandon-Jones (1984), Oates (1985), and Lee, Thornback, and Bennett (1988) also placed the red colobus monkeys (here considered to belong to the genus *Piliocolobus*) in *Procolobus*. Oates, Davies, and Delson (1994), while recognizing "that the olive colobus is a very distinctive animal," treated *Piliocolobus* as a subgenus of *Procolobus*. Groves (1989 and *in* Wilson and Reeder 1993) also accepted the genus *Procolobus* with the two subgenera *Procolobus* and *Piliocolobus*.

Except as noted, the information for the remainder of this account was taken from Napier and Napier (1967, 1985). Head and body length is 430–90 mm, tail length is 570–640 mm, and weight is 2.9–4.4 kg. The upper parts are olive gray or olive brown and the underparts and limbs are grayish. There is a small longitudinal crest on the crown. The morphological characters of *Procolobus* are similar to those of *Colobus*, but in the former there are five cusps on the lower third molar tooth, while in the latter there are six. Unique among primates, the glans penis of *Procolobus* bears minute horny papillae (Oates, Davies, and Delson 1994). Dandelot (*in* Meester and Setzer 1977) noted also that like *Piliocolobus*, but in contrast to *Colobus*, *Procolobus* has V-shaped nostrils that are not lengthened by an extension of the nasal skin. Unlike either of the other genera, *Procolobus* has two partings of the frontal hairs separated by a median crest.

The olive colobus is restricted to rainforest and is diurnal and arboreal. It inhabits thickets and the low stratum of the canopy, often near riverbanks and swamps. It moves into the middle stratum to sleep but never ascends to the upper stratum. The diet is strictly vegetarian and probably consists mainly of leaves. A population density of 21/sq km has been reported (Davies 1994). One series of observations indicates that young foliage is eaten most often and that fruit and seeds are also important (Oates 1994). Groups contain 5–20 individuals, commonly 10–15, and there may be more than a single adult male. One group used a home range of about 28 ha. but did not appear to engage in active territorial defense (Oates 1994). The voice is little used, but there is a characteristic complex call, alto in pitch and ending in a scream. In Sierra Leone most matings have been seen to occur from March to August and very small infants were seen only from November to April, suggesting a distinct breeding season and a gestation period of about 6 months (Oates 1994). A newborn infant is carried in its mother's mouth for several months, a practice unique among monkeys and apes.

The olive colobus is classified as near threatened by the IUCN and is on appendix 2 of the CITES. Its presence in

Nigeria was unknown until a single specimen was taken in 1967. Oates (1982) observed a few living individuals there but reported the population to be threatened by intensive hunting and habitat destruction. Anadu and Oates (1988) received reports of the possible presence of *Procolobus* in the Niger River Delta.

PRIMATES; Family HYLOBATIDAE; Genus HYLOBATES
Illiger, 1811

Gibbons, or Lesser Apes

The single living genus, *Hylobates*, contains 4 subgenera and 11 species (Chivers 1977; Chivers and Gittins 1978; Corbet and Hill 1992; Geissmann 1994; Groves 1972b, 1984, 1993; Groves and Wang 1990; Haimoff et al. 1982; Ma and Wang 1986; Marshall 1981a; Marshall and Sugardjito 1986; Prouty et al. 1983):

subgenus *Nomascus* Miller, 1933

H. concolor (crested gibbon), southeastern China, Hainan, extreme northwestern corner of Laos east of Mekong River, northern Viet Nam northeast of Song Bo and Song Ma rivers;
H. leucogenys, extreme southern Yunnan (southern China), northern Laos, northern Viet Nam southwest of the Song Bo and Song Ma rivers;
H. gabriellae, southern Laos, eastern Cambodia, central and southern Viet Nam;

subgenus *Symphalangus* Gloger, 1841

H. syndactylus (siamang), mainland Malaysia, Sumatra;

subgenus *Bunopithecus* Matthew and Granger, 1923

H. hoolock (white-browed gibbon), east of the Brahmaputra River in Assam and Bangladesh, Burma and adjacent border area of Yunnan (China);

subgenus *Hylobates* Illiger, 1811

H. lar (white-handed gibbon), extreme southern Yunnan (southern China), eastern and southern Burma, Thailand south to below the Isthmus of Kra, eastern and southern mainland Malaysia, northern Sumatra;
H. pileatus (capped gibbon), southeastern Thailand, extreme southwestern Laos, Cambodia;
H. agilis (dark-handed gibbon), extreme southern peninsular Thailand, northwestern mainland Malaysia, Sumatra except north, southwestern Borneo;
H. moloch (silvery gibbon), Java;
H. muelleri (gray gibbon), Borneo except southwest;
H. klossii (Kloss's gibbon), Siberut, Sipura, North Pagai, and South Pagai in the Mentawai Islands off western Sumatra.

This family often has been considered to be only a subfamily of the Pongidae, but anatomical, immunological, karyological, paleontological, and behavioral evidence now support its recognition as a separate though closely related family (Andrews and Groves 1976; Chivers and Gittins 1978). *Symphalangus* often is treated as a distinct genus, and Lekagul and McNeely (1977) treated both *Symphalangus* and *Nomascus* as full genera. All of the above species of *Hylobates* are allopatric, except that the range of *H. syndactylus* overlaps parts of those of *H. lar* and *H. agilis* (Corbet and Hill 1992), there evidently was an area of sympatry between *H. lar* and *H. pileatus* to the east of Bangkok (Geissmann 1991b), and *H. concolor* and *H. leucogenys* seem to occur at the same localities in parts of extreme southern Yunnan, northern Viet Nam, and northwestern Laos (Groves 1993; Groves and Wang 1990). Fooden, Quan, and Luo (1987) considered *H. leucogenys* to be only a subspecies of *H. concolor*, but both it and *H. gabriellae* were recognized as distinct species by Corbet and Hill (1992) and Groves and Wang (1990). The species of the subgenus *Hylobates* readily interbreed in captivity, and small hybrid populations now are known at points of contact between some of these species in the wild, though there is no evidence of massive introgression (Brockelman and Gittins 1984; Marshall and Sugardjito 1986).

H. syndactylus is by far the largest species, with an armspread of as much as 1.5 meters. Its head and body length is 750–900 mm, there is no tail, and weight is 8–13 kg. The coat is longer and less dense than in the other species. The body and limbs are black, and there is a large gray or pink throat sac that is inflated during calls. This species also is distinguished from most other gibbons by a webbing uniting the second and third toes.

In the other species head and body length is 440–635 mm, there is no tail, and weight is usually 4–8 kg. Chivers and Gittins (1978) listed the following diagnostic features: *H. concolor*, spidery form, conical shape to crown, males black, females golden or gray brown; *H. hoolock*, longer hair than in species of subgenus *Hylobates*, males black with white eyebrows; *H. klossii*, both sexes completely black, like *H. syndactylus*, and sometimes with interdigital webbing, but much smaller and with shorter and denser hair; *H. pileatus*, males black with white hands and feet, females ash blond with black cap and chest; *H. muelleri*, coloration varied, from mouse gray to brown, cap and chest darker; *H. moloch*, both sexes blue gray, cap and chest darker; *H. agilis*, coloration varied, from very dark brown to light buff, often with reddish tinge, males with white brows and cheeks, females with white eyebrows; and *H. lar*, coloration varied, from black to light buff, hands and feet white. Groves and Wang (1990) added that in *H. leucogenys* the pelage is longer and coarser than in *H. concolor*, the males being black with some silvery hairs intermixed and with white patches on the cheeks and the females tending to be more richly colored than those of *H. concolor* and having no conical tuft on the crown. Corbet and Hill (1992) indicated that the pelage of *H. gabriellae* is fine, like that of *H. concolor*, but the males have pinkish cheeks and the females have only a short crown patch. Many gibbons are almost white at birth and do not attain their final color until two to four years old.

Prouty et al. (1983) characterized the four subgenera as follows: *Nomascus*, diploid chromosome number 52, late maturation color change in female, nasal bones flat, prominent laryngeal sac not in both sexes, male component of the "great call" vocalization concentrated at end; *Symphalangus*, diploid number 50, no late color change in female, nasal bones flat, prominent laryngeal sac in both sexes, male and female parts of great call interspersed; *Bunopithecus*, diploid number 38, late color change in female, nasal bones convex, no prominent laryngeal sac, male and female parts of great call interspersed; and *Hylobates*, diploid number 44, no late color change in female, nasal bones flat, no prominent laryngeal sac, male component of great call concentrated at end (unless call is solo).

Gibbons resemble the Pongidae in lacking a tail and in having the same dental formula but differ in their much

Gibbon (*Hylobates* sp.), photo by Bernhard Grzimek. Inset: *Hylobates agilis*, photo from West Berlin Zoo (Gerhard Budich).

smaller size, more slender form, relatively longer arms, longer canine teeth, and the presence of buttock pads. Gibbons do not have the pronounced sexual dimorphism in size of body, skull, and teeth found in the great apes, but as noted above, several species of gibbons are sexually dichromatic. The thumb of *Hylobates* is unique among higher primates in having the basal part freed from the palm and extending from near the wrist, allowing a wide range of movement (Lekagul and McNeely 1977).

Hylobates means "dweller in the trees," and gibbons fully justify the name. They exceed all other animals in agility. They move primarily by brachiation, in which the long arms are extended above the head to suspend and propel the body. Gibbons do most of their traveling through the trees by swinging alternately from branch to branch, using the hands like hooks, not grasping the limb, and often making long swings, like leaps, in which neither arm supports them. By this method of rapid locomotion gibbons may cover 3 meters in a single swing. When traveling on a large limb or on the ground they usually walk upright with their arms held high for balance. Gibbons can leap from branch to branch for a distance of 9 meters or more. Roonwal and Mohnot (1977) reported that *H. hoolock* can swim well, but Marshall and Sugardjito (1986) stated that gibbons do not swim, avoid open water, and do not walk long distances on the ground. These characteristics were factors in the speciation and largely allopatric distribution of gibbons.

Gibbons occur throughout the deciduous monsoon and evergreen rainforests of the islands and mainland of Southeast Asia (Chivers 1977, 1984). Most inhabit lowlands and hills, though *H. syndactylus* is sometimes found in sub-

montane forests at altitudes of up to 1,800 meters. All species are primarily diurnal and almost exclusively arboreal. *H. syndactylus* is usually found at a height of 25–30 meters in the trees. It is active for about 10.5 hours a day, during which it alternates periods of feeding and rest. Its average daily movement is nearly 1 km, being more during the dry season and less in the wet season. At around 1600–1800 hours groups settle in a regularly used sleeping tree for the night (Chivers 1972, 1974, 1977). *H. lar* generally is seen higher in the trees than the siamang and moves faster and farther during the day. Srikosamatara (1984) listed the following mean distances (in meters) traveled per day: *H. lar*, 1,490; *H. agilis*, 1,150; *H. muelleri*, 890; *H. klossii*, 1,514; and *H. pileatus*, 833. About 50 percent of the diet of *H. syndactylus* consists of leaves, 40 percent is fruits, and the remainder consists of flowers, buds, and insects (Chivers 1977). The other species are mostly frugivorous but also eat leaves, other vegetable matter, a variety of insects, and perhaps eggs and small vertebrates. Raemaekers (1984) suggested that the difference in dietary requirements between *H. syndactylus* and the smaller gibbons has allowed sympatry but that the members of the latter group exclude one another geographically through resource competition.

The following account of social structure is based on the studies of Chivers (1972, 1974, 1977, 1984), Ellefson (1974), Tilson (1981), and Whitten (1982). All species of gibbons are monogamous. A mated adult pair and their offspring occupy a small, stable home range, most or all of which is defended as a territory. The usual group size observed is three or four individuals, but solitary animals are

Siamang gibbon *(Hylobates syndactylus):* A. Photo by Ernest P. Walker; B. Head showing throat sac inflated while calling, photo by Marcel Langer. Hands and foot photos by Ernest P. Walker.

Crested gibbon *(Hylobates concolor)*, photo by B. E. Joseph.

often seen. Apparently, the lone individuals are subadults that have been forced out of the family group and have not yet established a territory of their own.

There is a density of 0.8 groups of *H. syndactylus* per sq km, and reported densities of the smaller species range from about 0.7 to 6.5 groups per sq km. Surprisingly, densities are highest in the drier and more seasonal forests of the northern part of the range of *Hylobates,* where there are usually at least 4 groups per sq km; on Sumatra, Java, and Borneo there tend to be only 3 per sq km. Where two species are sympatric, as on the Malay Peninsula, there may be only 1 group of each species per sq km.

Groups of *H. syndactylus* include up to three immature animals, separated in age by two or three years. These groups are very cohesive, with the animals being an average of only eight meters apart during the day. Social bonds are reinforced by considerable mutual grooming. Departure of young siamangs from the group after they reach adolescence at around five or six years is brought about both by their motivation to mate and by the increasing intolerance of the adult pair. Young of both sexes begin to call and move separately from their parents and may spend several years searching for a mate. Social behavior in *H. lar* is much the same, except that groups seem somewhat less cohesive and may include up to four offspring. Adults of *H. klossii* may assist their offspring to obtain territories by accompanying them into unoccupied areas and intimidating potential rivals or by forcefully usurping land belonging to other families.

All species of *Hylobates* are strongly territorial. In most

species about 75 percent of the group home range is defended. The overall average home range size for the genus is 34.2 ha., and modal territory size is about 25 ha. Home ranges are largest on the Malay Peninsula, being about 50 ha. there for *H. lar.* The smallest reported ranges are 22 ha. for *H. hoolock* in Assam and Bangladesh; 17 ha. for *H. moloch* in Java; and about 10–20 ha. for *H. klossii.* Average size for *H. syndactylus* is about 30 ha. Defense rarely involves physical contact, but groups range near the borders of their territories and call at, display to, and chase intruders. Displays include acrobatics and breaking of branches in the trees. Spacing of groups also apparently is accomplished through loud vocalizations given even when groups are not in contact. Such calls, which may carry for several kilometers to the human ear, are emitted on about 30 percent of days by *H. syndactylus* and on 80–90 percent of days by *H. lar.*

The calls or songs of gibbons have been the subject of considerable study, and pertinent information was summarized by Chivers and Gittins (1978). The vocalizations vary from one species to another, being valuable in identification, and also usually differ markedly between sexes. In some species males alone give a morning chorus, but group calls generally are composed of a duet between the male and female, with some accompaniment from their offspring. The female emits the longest and most distinctive sounds. The call of *H. klossii* is the most spectacular, with a long series of slowly ascending pure notes leading into a bubbling trill. In *H. syndactylus* there are alternate booms and barks, made louder by resonance across the greatly inflated throat sac. Cowlishaw (1992) hypothesized that the male-female duets are unrelated to pair bonding but are mainly for intimidation of neighbors; the female's song is associated strongly with territorial defense, the male's with keeping potential rivals away.

Roonwal and Mohnot (1977) stated that the young of *H. hoolock* are probably born from November to March, but otherwise there seems to be little information on reproductive seasonality in *Hylobates.* The interbirth interval is 2–3 years in *H. syndactylus* (Chivers 1977) and 2–2.5 years in *H. lar* (Ellefson 1974). The estrous cycle averages 28 days in *H. hoolock* and 30 days in *H. lar* (Roonwal and Mohnot 1977). Reported gestation periods are 230–35 days in *H. syndactylus* (Hill 1967), 7–7.5 months in *H. lar* (Roonwal and Mohnot 1977), and 7.5 months in *H. pileatus* (Lekagul and McNeely 1977). Normally a single young is born, and it initially clings around the mother's body like a belt. Weaning is gradual, not being completed in *H. lar* until the young is about 1 year and 8 months old (Ellefson 1974). In *H. syndactylus* early in the young animal's second year the father takes over most of its care. The age of full sexual maturity is about 8–9 years, though Geissmann (1991a) reported cases in which both sexes bred successfully in captivity at around 4–6 years. The animals in one pair of wild *H. syndactylus* were known to be nearly 25 years old (Chivers 1977). A captive specimen of *H. syndactylus* lived to about 40 years, and individuals of both *H. agilis* and *H. leucogenys* are reported to have lived in zoos for 44 years (Marvin L. Jones, Zoological Society of San Diego, pers. comm., 1995).

The following quotation from Rumbaugh (1972) on maintaining *Hylobates* in captivity may apply to a number of other primates as well: "Infant gibbons are a delight and can be tractable, loving and friendly. But not always! Intermixed with these treasured attributes are episodes of tantrums and blind fury that render them quite unpredictable. This unpredictability almost invariably increases with maturity and the day comes when as adults, equipped with formidable canines, they attack without either warning or obvious cause even their most trusted human affiliates."

Siamang gibbon *(Hylobates syndactylus)*, photo by Bernhard Grzimek.

Wild gibbons have suffered severe losses through the increase of people and consequent hunting and destruction of forest habitat. About 1,000 years ago gibbons, probably of the species *H. concolor,* could be found as far north as the Yellow River in eastern China. *Hylobates* now occurs in China only in the extreme southern border region and on the island of Hainan. The subspecies in the latter area, *H. concolor hainanus,* is restricted to a reserve of less than 2,000 ha. and may be on the verge of extinction (Haimoff 1984). Because of poaching and deforestation within the reserve the population declined from perhaps 30–40 individuals in the 1980s to only 15 in 1993 (Fooden, Quan, and Luo 1987; Zhang and Sheeran 1994). Bleisch and Nan (1990) estimated that 1,750–5,350 *H. concolor* survived in reserves on the mainland of China and noted that the species was otherwise extremely rare. Even the so-called reserves were subject to deforestation, mining activity, and hunting.

Populations in several other countries have been greatly reduced and fragmented. In his excellent summary of data on the natural history and conservation of gibbons Chivers (1977) listed the following estimated numbers: *H. concolor* (including *H. leucogenys* and *H. gabriellae*), 228,000; *H. syndactylus,* 167,000; *H. hoolock,* 532,000; *H. klossii,* 84,000; *H. pileatus,* 100,000; *H. muelleri,* 1,825,000; *H. moloch,* 20,000; *H. agilis,* 744,000; *H. lar,* 250,000; total, 3,950,000. He observed that while some species might remain numerous, present levels of logging and agriculture and of taking for food and commerce would lead within 15–20 years to the overall destruction of 80–90 percent of

then current gibbon populations, with the extinction of *H. moloch* and possibly of *H. klossii, H. pileatus,* and *H. concolor.*

Brockelman and Chivers (1984) reported that the anticipated declines were being realized as forests were destroyed and opened to human access and that probably all species were headed eventually for relict status. Indeed, Kool (1992) thought that the habitat of *H. moloch,* of Java, had been reduced by 96 percent and that only 4,800 individuals survived, 1,800 of which were in protected areas. In peninsular Malaysia alone deforestation is resulting in an annual loss of 31,000 *H. agilis* and *H. lar* and 3,400 *H. syndactylus* (Chivers 1986). MacKinnon and MacKinnon (1987) provided the following revised estimates for some other species: *H. concolor* (including *H. leucogenys* and *H. gabriellae*), 131,000; *H. hoolock,* 169,134; and *H. pileatus,* 33,600. More recently, however, Eames and Robson (1993) calculated that the total world population of *H. concolor* (including *H. leucogenys* and *H. gabriellae*) could be as low as 10,500–14,000. All species of *Hylobates* are listed as endangered by the USDI and are on appendix 1 of the CITES. The IUCN now classifies *H. moloch* as critically endangered and *H. concolor* (including *H. leucogenys* and *H. gabriellae*) as endangered and estimates the surviving number of mature individuals at fewer than 250 for the former and 2,500 for the latter. The IUCN also designates *H. pileatus* and *H. klossii* as vulnerable and *H. syndactylus, H. lar, H. agilis,* and *H. muelleri* as near threatened.

Fossil evidence suggests that the Hylobatidae diverged

from the Pongidae prior to the separation of the Pongidae and the Hominidae. Gibbonlike remains are known from the Oligocene and Miocene of Africa, the Miocene of Europe, and the Miocene, upper Pliocene, and Pleistocene of Asia. Possibly, however, the similarity of the earlier material represents parallel development rather than direct relationship (Chivers 1977; Fleagle 1984; Lekagul and McNeely 1977). The family Pliopithecidae of the Oligocene and Miocene sometimes is considered to have given rise to the Hylobatidae. Both Groves (1989) and Tyler (1991*b*) rejected that view but indicated that there are fossil genera from the same period in India and China that appear to be ancestral hylobatids.

PRIMATES; Family PONGIDAE

Great Apes

This family of three living genera and four species is found in equatorial Africa and on Sumatra and Borneo. Some authorities, such as E. R. Hall (1981), place the great apes in the family Hominidae, along with humans. Andrews and Cronin (1982) suggested an alternative classification by which the living Pongidae would be restricted to *Pongo* and the Hominidae would comprise two subfamilies: the Gorillinae, with *Pan* and *Gorilla;* and the Homininae, with *Homo* (and the fossil *Australopithecus*). In the arrangement of Groves (1986*a*, 1989, and *in* Wilson and Reeder 1993) the family Hominidae comprises the subfamily Ponginae, with *Pongo*, and the subfamily Homininae, with *Gorilla, Pan,* and *Homo*. The phylogeny thus indicated is supported by a variety of genetic data showing that *Gorilla, Pan,* and *Homo* are more similar to one another than any is to *Pongo* (Marks 1988). Schwartz (1986) divided the involved taxa into three full families: the Panidae, with *Pan* and *Gorilla;* the Pongidae, with *Pongo* (and the fossil *Gigantopithecus* and *Sivapithecus*); and the Hominidae, with *Homo* (and *Australopithecus*). Martin (1990) argued for continuing to combine all three great apes in the Pongidae and to place humans in the distinctive family Hominidae (see account thereof).

All three genera of great apes contain many individuals that exceed most human beings in size. The gorilla may be 1.75 meters tall when standing on two feet and may weigh well over 200 kg. Like humans, the Pongidae lack a tail. The pelage is short, shaggy, and coarse. The face is nearly naked, and the ears are round and naked. Cheek pouches are not present. Unlike in humans, the arms are longer than the legs. The second to fifth fingers are long and the thumb is short. The big toe is opposable, allowing at least some individuals to carry objects with their feet.

The dental formula is: (i 2/2, c 1/1, pm 2/2, m 3/3) $\times$ 2 = 32. The incisors are nearly equal in size, the canines are large, the premolars resemble the molars, and the molars are wide-crowned and supplied with tubercles. The form of the skull is variable, but it is always longer than broad. In general form and structure the brain closely resembles that of *Homo*. Average brain capacities in adult Pongidae are: *Pongo*, 411 cu cm; *Pan*, 394 cu cm; and *Gorilla*, 506 cu cm. In *Hylobates* this capacity is 95 cu cm; in an 11-month-old *Homo* it is 850 cu cm; and in an adult human it is 1,350 cu cm (Rumbaugh 1970).

The great apes generally support themselves on all four limbs, but they have a limited bipedal capability. They do not take naturally to water and are not known to be able to swim. They can distinguish colors, and their best-developed senses are vision and hearing. They have a wide range of vocalizations, and some have laryngeal sacs that can be inflated for resonance. A variety of facial expressions are evidenced, especially in the chimpanzee.

The great apes are intelligent, at least from the human viewpoint. Chimpanzees are used extensively in laboratory research (often with little consideration of the needs of the individual or the conservation of the species) and are capable of learning many complicated tricks. Orangutans may be as intelligent as chimpanzees, but since they tend to be introverts by nature, testing their intelligence is more difficult than testing that of the extrovert chimpanzees, who love to show off. D. P. Erdbrink (pers. comm.) reported that a newly captured young male orangutan almost succeeded in unscrewing the bars of his travel cage; having observed how workers used a shifting spanner, he grasped the spanner through the bars when the workers left. In the zoo at Bandung, Java, before World War II an orangutan, seeing tame chickens wandering near his cage, captured one by throwing a path of cornseeds (part of his food) toward

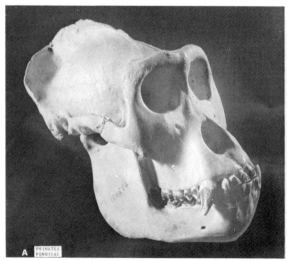
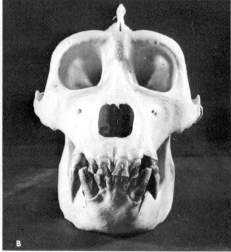

Skull of gorilla *(Gorilla gorilla),* photos by P. F. Wright of specimen in U.S. National Museum of Natural History.

Chimpanzee *(Pan troglodytes)*, photo by Bernhard Grzimek.

the cage. The size, strength, and relative rarity of gorillas has precluded intensive training and mental analysis, though some recent work suggests a remarkable ability to learn from and communicate with people (Patterson 1986).

According to Andrews (1978), the family Pongidae originated in the Oligocene, at which time the genera *Propliopithecus* and *Oligopithecus* were present in Africa. By the early Miocene, about 22 million years ago, the genus *Proconsul* had developed in Africa, and this genus is thought to represent the ancestral line of both the modern great apes and humans. The last common ancestor of all living Pongidae and Hominidae is probably the genus *Kenyapithecus*, which lived about 15 million years ago in the middle Miocene of eastern Africa (Andrews and Martin 1987; McCrossin and Benefit 1993). There once was a general belief that a somewhat later Miocene descendant, *Sivapithecus*, represented the start of the evolutionary line leading to *Homo*. More recent studies, however, have indicated that *Sivapithecus* is not a direct ancestor of *Homo* and that the divergence between the lines leading to modern humans and apes did not occur until the early Pliocene (Kelley and Pilbeam 1986; Zihlman and Lowenstein 1979). Andrews and Cronin (1982) suggested that *Sivapithecus* and *Pongo* are closely related and in an evolutionary line completely separate from that leading to *Pan, Gorilla,* and *Homo*. The latter view has now been widely accepted, but Schwartz (1986, 1987, 1988*a*) argued that the closest living relative of *Homo* is *Pongo*.

Andrews (1978) also considered that in the late Miocene

Sivapithecus gave rise to *Gigantopithecus*, a huge ape that survived at least until the middle Pleistocene in Asia. This ape may have been 2.5–3.0 meters tall and probably was the largest primate ever to exist. It was first discovered by von Koenigswald, the Dutch geologist, who located its teeth during his travels in China and Java from 1934 to 1939 while wandering through Chinese pharmacies to study fossil bones used as medicines (see also account of the family Hominidae).

PRIMATES; PONGIDAE; **Genus PONGO**
Lacépède, 1799

Orangutan

The single species, *P. pygmaeus*, is now found only on the islands of Sumatra and Borneo. The two subspecies, *P. p. abelii*, of Sumatra, and *P. p. pygmaeus*, of Borneo, have many morphological and behavioral differences (Courtenay, Groves, and Andrews 1988; Groves 1986*a*; Groves and Holthius 1985). The population in southwestern Borneo may also warrant subspecific designation (Groves, Westwood, and Shea 1992). Pleistocene fossils from southern China, northern Viet Nam, Laos, and Java indicate that the genus once had a much greater distribution. Additional remains discovered in northern Viet Nam, as well as traditional beliefs in peninsular Malaysia, suggest that the

Nest of a Bornean orangutan *(Pongo pygmaeus)* accommodating two babies, photo by Barbara Harrison.

orangutan still occurred on the mainland of Southeast Asia in early Recent times (Groves 1986*a*; Groves and Holthius 1985; Kahlke 1973).

Head and body length is about 1.25–1.50 meters (according to D. P. Erdbrink, an unusually large Bornean male measured 1.80 meters) and the sitting height is about 0.70–0.90 meters. The arms, which reach to the ankles when the animal is erect, have a spread of about 2.25 meters. Rijksen (1978) estimated adult weight at 30–50 kg for females and 50–90 kg for males. Most of those that have survived in captivity have become fatter and heavier than those in the wild. Fooden and Izor (1983) reported full adult weight in captivity to average 65 kg for females and 144 kg for males and indicated that several of the latter had reached about 200 kg. Coloration is dark rufous or reddish brown. The hairy coat is rather thin and shaggy.

The forehead is high, and the snout bulging. The skull profile is more sloping than that of *Pan* and *Gorilla*, and the skull exhibits little of the brow ridging so prominent in those genera. Cheek pads are present in adults, especially in old males. These represent localized deposits of subcutaneous fat bound by connective tissue and covered by unmodified skin with scant and irregularly distributed hairs. The mouth projects and the lips are thin. The small ears are devoid of hair. The legs are short and relatively weak, but the hands and arms are powerful. According to Courtenay, Groves, and Andrews (1988), adult males from Sumatra are characterized by cheek pads that lie flat against the face, giving it a very wide appearance; the pads are thickly covered with hair, and there also is a pronounced beard. In males from Borneo the pads bulge outward and have little or no hair and the beard is short and scruffy.

A number of recent field studies have helped to clarify our understanding of *Pongo* (Horr 1975; MacKinnon 1971, 1973, 1974; Rijksen 1978; Rodman 1973, 1977, 1988). The orangutan appears to be well adapted to several different types of primary forest, ranging from swamps and other areas near sea level to mountainous forests at around 1,500

meters. The species is primarily arboreal and diurnal. It uses vegetation to construct a large platform nest in a tree, in which it sleeps at night. Usually a new nest is made for each night, but occasionally the same one is reused. Smaller nests may be made for rests during the day. The orangutan also shelters itself from rain and sun by holding leafy branches over its head or draping large leaves around its head and neck.

Tool use by fully wild orangutans is much more limited than has been reported for the chimpanzee, but released captive individuals use sticks for digging, fighting, prying, eating, scratching, and many other purposes. Some of these ex-captives learned (without human assistance) to untie complex knots securing boats and rafts and then to shove off, board, and ride the vessels across rivers (Galdikas 1982).

Normal movements are mainly by climbing and walking through large trees and swinging from branch to branch. Terrestrial movement is rare, cautious, and usually only to get from one tree to another. Surface locomotion is mainly quadrupedal, with the clenched fist, but not the knuckles, being placed on the ground. Movements generally are much less hurried than those of gibbons. In a lowland area of Borneo, Galdikas (1988) found daily movement usually to be 200–1,000 meters. There are morning and late afternoon peaks of activity and a rest period around the middle of the day. The diet is largely frugivorous and seems to include a high proportion of wild figs. Various species of these figs ripen at different times of the year, and this sequence is responsible for considerable local movement by the orangutan. Many other kinds of vegetation, as well as mineral-rich soil, insects, and perhaps small vertebrates and birds' eggs, are also eaten. Captive orangutans have accepted meat readily, and Sugardjito and Nurhuda (1981) observed a wild individual to consume a gibbon carcass.

Reported population densities have ranged from 0.2 to 5.0 individuals per sq km. Local variation has been reported in the social structure of the orangutan. Observations generally indicate, however, that individuals usually occur in very small groups or alone, that most animals in a given area maintain a loose relationship, that temporary aggregations are sometimes formed, and that adult males occupy larger home ranges than adult females and are hostile to one another. Perhaps some of the reported variation results from inconsistent interpretation by different investigators.

Studies in Borneo (Galdikas 1981, 1985*a*, 1985*b*; Horr 1975; MacKinnon 1971, 1973; Rodman 1973, 1977) showed that the number of orangutans seen together averaged about 1.8. Adult females, usually with dependent young, occupied overlapping home ranges of about 0.65 sq km or less. Adult males were generally alone and used home ranges of about 2–6 sq km that overlapped the ranges of several adult females. In Sabah (Malaysian Borneo) male home ranges also were reported to overlap one another, but in Kalimantan (Indonesian Borneo) these ranges were considered discrete. Although infrequent, contact between mature males invariably involved aggression or marked avoidance. Adult females, however, frequently paired and traveled and fed together for up to three days. Several immature animals sometimes associated with one another or with an adult of either sex. Adults of opposite sex came together only for a brief period of courtship. For purposes of mating, males preferred fully adult females and females preferred dominant adult males. Choice of sexual partners was very much a prerogative of females, who could not be monopolized by a particular male, though "rapes" did occur.

For Sumatra, MacKinnon (1973, 1974) reported a closer

Orangutan *(Pongo pygmaeus)*, photo from New York Zoological Society. A. Adult male showing great width of face produced by deposits of tissue beneath the skin and heavy beard and mustache, photo by Bernhard Grzimek. B. Photo from San Diego Zoological Garden. C. Photo by Bernhard Grzimek.

relationship than in Borneo among the various orangutans in a particular area. Although he thus recognized dispersed groups of up to 17 individuals apparently centered on the leadership of an adult male, he again found the number of animals seen together to average fewer than 2; adult males, however, were sometimes observed in the presence of females with young. Rijksen (1978), also working in Sumatra, determined group size to average only 1.47; adult fe-

males were usually accompanied by offspring, and adult males were usually solitary. Both sexes lived in home ranges of 2–10 sq km that overlapped considerably. Upon reaching adolescence, at about four or five years of age, young animals became increasingly independent of their mothers and formed small groups of their own. There also was temporary association of adults of either sex with immature individuals.

Orangutan *(Pongo pygmaeus):* Top, photo by Polly McCann; Bottom, photo by Bernhard Grzimek.

Rijksen (1978) described 13 vocalizations, and MacKinnon (1971) listed 15. The orangutan is generally quieter than other higher primates. The most notable sound is the long call of the adult male, a series of groans that may be heard up to 1 km away by a human, which probably serves as a spacing mechanism.

The following reproductive information was taken largely from Graham (1988), MacKinnon (1971, 1974), Nadler (1988), Rijksen (1978), and Rodman (1988). The estrous cycle of *Pongo* averages about 28 days, receptivity usually lasts 5–6 days, gestation averages 245 (227–75) days, and a single young is normal, but twins occur rarely. Mean birth weight is 1.56 kg for females and 1.93 kg for male when born singly (Fooden and Izor 1983). The young clings to the ventral surface of the female until it is nearly 1 year old and may still ride on the mother at 2.5 years. Weaning is usually completed at about 3.5 years. The interbirth interval is generally about 3–4 years but has been reported to be 8 years in an area subject to considerable human disturbance. Young individuals become increasingly independent of the mother after a second young is born but may still seek protection from the female until about 7–8 years. Physiological sexual maturity occurs at around 7 years in females and shortly thereafter in males; the females, however, usually do not give birth in the wild until at least 12 years, and the males do not attain full physical and social maturity, and hence reproductive capability, until 13–15 years or more. Young females generally remain in the vicinity of their birth, but males emigrate to other areas. According to Marvin L. Jones (Zoological Society of San Diego, pers. comm., 1995), a pair of orangutans that died in 1976 and 1977, after 48 and 49 years, respectively, in captivity, were estimated to have been 58 and 59 years old. These two animals hold the known record longevity for nonhuman primates.

The range of the orangutan has been shrinking since the Pleistocene, largely because of excessive hunting and environmental disruption by people. Distribution in Sumatra has declined by 20–30 percent since 1935–38, and the species is now found only in the northern part of that island. Both MacKinnon (1971) and Rijksen (1978) suggested that the orangutan's arboreal mode of life and dispersed social structure may have resulted in part from many thousands of years of human hunting pressure. The other large primates *(Papio, Mandrillus, Gorilla, Pan, Homo)* are mainly terrestrial and highly gregarious. Pleistocene remains from China and Sumatra suggest that some orangutans were considerably larger than those of today, possibly even exceeding the gorilla in size. Galdikas (1988), however, considered the solitary nature of *Pongo* to result mainly from the association of its large size and frugivorous diet and the consequent competition at opportunistic food sources; she doubted that Pleistocene orangutans had been terrestrial.

The killing of the orangutan may have been stimulated in part by its resemblance to humans and may have served as an alternative to headhunting. Through much of the twentieth century a major threat to surviving populations was considered to be the collection of young animals, inevitably requiring the killing of the mother, for exhibition in zoos and circuses. The species has long been protected by the laws of the countries where it lives, but it is still sometimes taken and exported illegally. Indeed, in the late 1980s as many as 1,000 young orangutans were being smuggled from Indonesia to Taiwan (Eudey 1991).

By far the greatest current problem is the logging of the forests upon which the orangutan depends. Reports had circulated that fewer than 5,000 individuals survived, but apparently there are still more than that. MacKinnon (1986) calculated that there could be as many as 179,000 in the Indonesian portion of the range. Rijksen (1986), after considering the large amounts of unsuitable and destroyed habitat, estimated populations to number about 6,000 on Sumatra and 37,000 on Borneo. More recent and precise estimates are that Sumatra has about 11,700 sq km of remaining habitat with 9,200 orangutans and that Borneo has about 22,400 sq km of habitat with 10,282–15,546 animals; there are another 901 in captivity (Perkins 1993; Tilson and Eudey 1993). The orangutan is classified as vulnerable by the IUCN and as endangered by the USDI and is on appendix 1 of the CITES. Recent conservation efforts have included the live capture of individuals isolated by logging and their relocation to more suitable areas of forest (Andau, Hiong, and Sale 1994).

PRIMATES; PONGIDAE; Genus GORILLA
I. Geoffroy St.-Hilaire, 1852

Gorilla

The single species, *G. gorilla*, is found mainly in two widely separated regions of equatorial Africa (Dandelot *in* Meester and Setzer 1977; Goodall and Groves 1977; Groves 1970*b*, 1971*b*; Schaller 1963; Vedder 1987). The subspecies *G. g. gorilla* (western lowland gorilla) occurs in southeastern Nigeria, west-central and southern Cameroon, extreme southwestern Central African Republic, mainland Equatorial Guinea, Gabon, Congo, and the extreme western tip of Zaire, as well as in the Cabinda enclave of Angola at the mouth of the Congo River. An isolated population of this subspecies may also have existed in the Bondo district of north-central Zaire. The subspecies *G. g. graueri* (eastern lowland gorilla) is separated by about 1,000 km from the western lowland gorilla. *G. g. graueri* occurs in east-central Zaire to the southeast of Kisangani (Stanleyville), between the Lualaba (Congo) River, Lake Edward, and the northern part of Lake Tanganyika. A third subspecies, *G. g. beringei* (mountain gorilla), is restricted to the six extinct volcanoes of the Virunga Range straddling the Zaire-Rwanda-Uganda border and may also be represented by a small population in the Bwindi (Impenetrable) Forest of southwestern Uganda. However, a recent morphological analysis indicates that this population is subspecifically distinct from that of the Virungas and may be referable either to *G. g. graueri* or to an undescribed taxon (Sarmiento, Butynski, and Kalina 1995). A population in the Mount Kahuzi vicinity, just west of Lake Kivu, in Zaire, also sometimes has been assigned to *G. g. beringei*, but Casimir (1975) suggested that this population actually represents *G. g. graueri*, and most subsequent authorities have agreed. A recent study of mitochondrial DNA indicates a striking divergence between *G. g. gorilla*, on the one hand, and *G. g. graueri* and *G. g. beringei*, on the other, and has raised the possibility that the western and eastern populations may constitute separate species (Morell 1994*b*). There also have been reports of a pygmy species of *Gorilla*, but Groves (1985*a*) showed that such reports were based on misunderstanding. Goodall and Groves (1977), Tuttle (1986), and some other authors have taken the position that *Gorilla* is only a subgenus of *Pan*, but Groves (1986*a*) diagnosed the two as being generically distinct.

When standing on two feet the gorilla has a height of 1.25–1.75 meters; the total length is somewhat greater since the animal normally stands with its knees slightly bent. There is no tail. The span of the outstretched arms, about 2.00–2.75 meters, is far greater than the standing

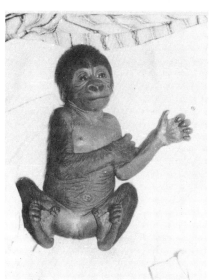

Gorillas *(Gorilla gorilla):* A & C. Photos from San Diego Zoological Society; B. Baby, not more than 36 hours old, photo from Zoologischer Garten, Basel, through E. M. Lang.

Gorilla *(Gorilla gorilla)*, photo by Bernhard Grzimek.

height. The shoulder breadth of this powerful primate is nearly twice that of the chimpanzee, and the chest is as much as 508 mm across. The circumference of the chest of male *Gorilla* is 1.25–1.75 meters. According to Grzimek (1975), weight is 70–140 kg in females and 135–275 kg in males, but occasionally it reaches 350 kg in individuals that become fat in captivity. The face, ears, hands, and feet are bare, and the chest in old males lacks hair. There is no beard. A pad of skin and connective tissue relatively dense and fibrous in nature is present on the crown. According to Groves (1970b), the coloration of both the skin and hair of *G. g. beringei* and *G. g. graueri* is jet black, and the adult male develops a silvery white saddle on the back between the shoulders and rump; *G. g. beringei* can easily be distinguished, however, by its much longer, silkier fur, especially on the arms. In *G. g. gorilla* the hair is short as in *graueri* but more grayish or brownish, and the male's saddle of whitish hair extends onto the thighs and grades more into the body color.

The gorilla has a short muzzle and an extremely stocky body. The nostrils are large, the eyes are small, and the small ears lie close to the head. The forearm is much shorter than the upper arm, and the hand is very large. The thumbs are larger than the fingers. The only animal that resembles the gorilla at all is the chimpanzee, which is smaller, usually has large, conspicuous ears, and is quicker in its movements. According to Napier and Napier (1967), the skull of *Gorilla* is characterized by a markedly prognathic face, rectangular and widely separated orbits, sagittal and nuchal crests in males and sometimes in females, and lower molar teeth, which increase in size toward the rear.

Most gorillas inhabit lowland tropical rainforests, but in the Cross River region of southeastern Nigeria and southern Cameroon and in extreme eastern Zaire and adjacent parts of Uganda and Rwanda some populations are found in montane rainforest between 1,500 and 3,500 meters and

in bamboo forest from about 2,500 to 3,000 meters. The gorilla is primarily terrestrial but is fully capable of climbing. Goodall and Groves (1977) reported that some animals in the Mount Kahuzi area of Zaire fed in trees at heights of 40 meters and that even the largest male, weighing at least 200 kg, frequently climbed to 20 meters. Surface locomotion is by a quadrupedal walk, with the soles of the feet and the middle phalanges of the fingers placed on the ground. A few bipedal steps are often taken, but Schaller (1963) never observed an animal to progress in this manner for more than about 18 meters. The collection and preparation of food is done almost entirely with the hands. The gorilla does not seem to be able to swim and, indeed, is sometimes reluctant to wade across shallow streams; this factor may partially explain the unusual distribution of the species. Every adult and juvenile builds its own crude platform nest in which to sleep at night and occasionally for rest during daylight. Construction time is less than five minutes. In some areas more than 90 percent of nests are made on the ground, but in other areas, apparently because of availability of materials, most nests are built in trees. A nest is not used for more than a single night (Goodall and Groves 1977; Schaller 1963, 1965).

According to Schaller (1963), the gorilla is diurnal, with nearly all activity occurring between 0600 and 1800 hours. After awakening, the animals feed intensively for a while, then rest from about 1000 to 1400 hours, and then travel and feed until bedding down for the night. Groups observed by Schaller moved 90–1,800 meters per day. Other observations (Casimir and Butenandt 1973; Fossey and Harcourt 1977; Jones and Sabater Pi 1971) suggest that daily movement is usually about 400–1,000 meters. Goodall and Groves (1977) stated that during the wet season (April–June) in eastern Zaire groups moved an average of 596 meters per day but that in the dry months (July–August) they covered 1,240 meters per day. A. G. Goodall (1977) determined that movements in the course of a year were not random but followed an established migratory pattern with respect to availability of food. Casimir and Butenandt (1973) found that one gorilla group in the Mount Kahuzi area spent October and November in bamboo forest, because of the abundance then of young bamboo shoots, and the rest of the year in other kinds of forest. Harcourt and Stewart (1984) determined that the mountain gorilla spends about 45 percent of the day feeding. The gorilla is almost entirely a vegetarian in the wild, though captives have been known to eat meat. Fossey and Harcourt (1977) reported that 85.8 percent of the diet of *G. g. beringei* consists of leaves, shoots, and stems. Small amounts of wood, roots, flowers, fruits, and grubs also are eaten. Harcourt and Harcourt (1984) noted that many small insects are eaten inadvertently while feeding on plants. Tutin and Fernandez (1983) found that in Gabon the gorilla frequently breaks into termite nests to feed on the insects therein. A. G. Goodall (1977) observed that gorillas had ample opportunity to eat eggs, helpless young birds, and the honey of stingless bees but never did.

Population densities of 0.35–0.75/sq km have been recorded in various parts of Africa where gorillas occupy areas of at least 100 sq km; if unfavorable habitat within such areas is not considered, maximum density would be nearly 1/sq km (Harcourt, Fossey, and Sabater Pi 1981). Annual group home range probably varies from 4 sq km to 25 sq km and averages about 8 sq km (Fossey 1974; Fossey and Harcourt 1977). A. G. Goodall (1977), however, reported the home range of a group to be 34 sq km, and Casimir and Butenandt (1973) stated that over a period of years 40–50 sq km probably would be used by one group. Schaller

(1963) found much overlap of home range in the Virunga Volcanoes region, with several groups frequenting the same parts of the forest. Fossey and Harcourt (1977) reported that groups in the same region had core areas that were not entered by outsiders but that there was no overt territorial defense. A. G. Goodall (1977) stated that some territoriality apparently did exist in the Mount Kahuzi area.

When two groups approach, they may ignore each other, temporarily associate, or express hostility through vocalizations and displays (Fossey 1974; Schaller 1963). It is possible that there is a larger community structure beyond the group level and that certain groups interact peacefully among themselves but not with other groups (Goodall and Groves 1977). A group may seemingly go far out of its way to avoid contact with another group. Fossey (1974) thought that the main determinant in overall group movement was the location of other groups. A. G. Goodall (1977), however, considered food availability to be the chief determinant.

Intragroup relations seem to be much more cohesive in the gorilla than in the chimpanzees. Schaller (1963) found the splitting of groups into smaller units to be infrequent and temporary. In his study area in the Virunga Volcanoes region group size ranged from 2 to 30 individuals. On the average there were 16.9 animals, including 1.7 fully adult males (silverbacks), 1.5 subadult males (blackbacks), 6.2 adult or subadult females, 2.9 juveniles (3–6 years old), and 4.6 infants (under 3 years). In other regions average group size has varied from 6 to 13 (Goodall and Groves 1977). Median group composition throughout the range of *Gorilla* is 1 silverback, 1 blackback, 3 adult females, and 2 or 3 immatures (Harcourt, Fossey, and Sabater Pi 1981). The smallest groups, averaging 4–8 members, are found in the western lowland subspecies (Dixson 1981). If there is more than a single adult male, one is dominant and is the leader of the group, and only that one normally breeds. There is a general rank order for the group based mainly on size, with silverback males being dominant over all other animals.

Groups are highly stable, and the dominant male retains leadership for years. Certain of the other males are only temporarily associated with the group and eventually leave to live alone or to join other groups (Goodall and Groves 1977). The adult males that do remain are probably the sons of the dominant male and eventually will take over leadership (Harcourt 1978; Harcourt, Stewart, and Fossey 1976). Males that leave a group wander alone for a number of years, then sometimes establish a range adjacent to or overlapping that of their former group, and may eventually take one or more females to begin a new group (Caro 1976; Fossey 1974). Unlike the situation in most primates, nearly all female gorillas emigrate from their parental group at maturity. Departure usually seems to be of the female's own volition, and she almost immediately joins with another group or a lone young silverback male (Harcourt 1978, 1979; Harcourt, Fossey, and Sabater Pi 1981; Harcourt, Stewart, and Fossey 1976; Watts 1992). A female may join, and reproduce within, several groups in succession (Watts 1991). However, the long-term female residents of a large group sometimes form cliques that may try to prevent the arrival of, or remain hostile to, immigrant females (Watts 1994).

Fossey (1972) identified 16 vocalizations emitted by *Gorilla,* and Schaller (1963) described 22. These include roars and growls of aggression, grunts and barks for group coordination, and loud hoots that can be heard 1 km away by people and may serve to space different groups. Schaller also described the elaborate chest-beating display that may

be elicited by the presence of outsiders, either other gorillas or humans. It begins with a series of hoots, after which the animal stands on two legs and slaps its chest with slightly cupped hands, first one, then the other, 2–20 times. The gorilla then runs about and breaks vegetation.

According to Schaller (1963), there is no evidence of a particular breeding season in the wild. Females give birth every 3.5–4.5 years unless the infant dies. The estrous cycle lasts about 27–28 days, and estrus 1–3 days (Harcourt, Stewart, and Fossey 1981; Watts 1991). Estimates of gestation range from 251 to 295 days, though Dixson (1981) gave the average as precisely 257.6 days. There is normally a single young, but twins occur rarely. The weight at birth is about 2 kg. Young are weaned at 3–4 years (Watts 1991). The age of physiological sexual maturity is about 6–8 years in females and 10 years in males, but breeding in the wild normally does not begin until females are 9–10 and males 15 years old, about the time when emigration occurs. Mortality in stable populations is 42 percent for immatures, mostly in the first year of life, and 5 percent for adults. Females generally give birth to only 2–3 surviving young during their reproductive life (Harcourt, Fossey, and Sabater Pi 1981; Watts 1991). Life span in the wild may be up to 50 years. A captive gorilla lived for exactly 54 years (*Washington Post,* 1 January 1985). Although once considered impossible, breeding of *Gorilla* in captivity now is accomplished regularly; Olney, Ellis, and Fisken (1994) indicated that 362 of the 633 individuals held by the world's zoos had been bred in captivity.

The gorilla is classified as endangered by the USDI and is on appendix 1 of the CITES. The IUCN designates the species in general as endangered but designates the subspecies *G. g. beringei* as critically endangered. That subspecies is found in the Virunga Volcanoes region at the point where Zaire, Uganda, and Rwanda come together. Most of this region is included within national parks or reserves, but there have been encroachments for agriculture, grazing, and wood gathering. During the 1970s poaching intensified in the Virunga region, especially to obtain gorilla heads for commercial sale (Harcourt and Curry-Lindahl 1978). The number of gorillas in the region fell from about 450 in 1960 to about 225 two decades later (Harcourt and Fossey 1981). A conservation program implemented in Rwanda in 1978 and funded in part by tourism may have helped to stabilize the population and reduce threats to the habitat (A. H. Harcourt 1986; Weber and Vedder 1983).

G. g. beringei also may be represented by the gorilla population of the Bwindi (Impenetrable) Forest in southwestern Uganda. As noted above, however, the most recent assessment (Sarmiento, Butynski, and Kalina 1995) indicates that the Bwindi population is referable to *G. g. graueri* or to an undescribed subspecies. Although this area is a reserve, the gorilla population was thought to have been reduced to only about 100 animals because of intense human activity (A. H. Harcourt 1981). Recent surveys indicate, however, that there are approximately 320 gorillas in the Bwindi Forest (Butynski, Werikhe, and Kalina 1990). Apparently based on the earlier figure, Vedder (1987) estimated total numbers of *G. g. beringei* as 370–440. By 1995 there was recognition that the Virunga and Bwindi gorilla populations each had grown to about 320 animals, but protection had become more difficult and poaching had increased in association with the terrible social unrest sweeping Rwanda and adjoining areas (Morris 1995).

Comparatively little is known about the overall status of *G. g. graueri* and *G. g. gorilla.* Schaller (1963) estimated the entire population of eastern gorillas as 5,000–15,000 individuals. More recent estimates were only 2,500–4,500

(Vedder 1987) and 3,000–5,000 (Lee, Thornback, and Bennett 1988). A population just west of Lake Kivu in the vicinity of Mount Kahuzi, Zaire, has been estimated at about 275 animals (Yamagiwa et al. 1993). The western subspecies has been estimated to number 40,000 animals (A. H. Harcourt 1986), of which about 35,000 are located in the nation of Gabon (Tutin and Fernandez 1984). There also may be a large number in Congo (Fay and Agnagna 1992). In contrast, the species had not been recorded at all in Nigeria for 30 years, but a few isolated groups totaling perhaps 150 individuals were located there recently by Harcourt, Stewart, and Inahoro (1989). This Nigerian population is classified as critically endangered by the IUCN. Jones and Sabater Pi (1971) considered the western lowland gorilla to be declining because of habitat destruction and killing by people for food and allegedly to protect crops. A small population of *G. g. gorilla* that once may have existed in the Bondo district of north-central Zaire has not been reported since 1908 (Goodall and Groves 1977).

PRIMATES; PONGIDAE; Genus PAN
Oken, 1816

Chimpanzees

There are two species (Dandelot *in* Meester and Setzer 1977; Kano 1992; Lee, Thornback, and Bennett 1988):

P. troglodytes (chimpanzee), originally found in the forest zone from Gambia to Uganda, and south to Lake Tanganyika, but not including the central forests of Zaire south and west of the Congo–Lualaba River;

P. paniscus (pygmy chimpanzee, bonobo), central Zaire to the south and west of the Congo–Lualaba River and north of the Kasai–Sankuru River.

There has been considerable controversy regarding the technically proper generic name for the chimpanzees. *Pan* is used here because of its familiarity and in the hope of promoting nomenclatural stability. Van Gelder (*Bull. Zool. Nomencl.* 32 [1975]: 69–73) presented a detailed argument in favor of using the name *Chimpansee* Voigt, 1831. *P. paniscus* sometimes is considered a subspecies of *P. troglodytes* (see, e.g., Horn 1979), but recent evidence suggests not only that the two are distinct but that *P. paniscus* bears some resemblance to the ancestral stock that gave rise to the Pongidae and Hominidae (see, e.g., papers *in* Susman 1984*b*). Three other subspecies of *P. troglodytes* usually are recognized: *P. t. verus*, found from Gambia to the Niger River in central Nigeria; *P. t. troglodytes*, from the Niger River to the Congo–Ubanghi River along the eastern border of Congo; and *P. t. schweinfurthi*, from the northwestern corner of Zaire to western Uganda and Tanzania (Lee, Thornback, and Bennett 1988; Shea, Leigh, and Groves 1993). Tuttle (1986) followed some older authorities in listing a fourth, *P. t. koolokamba*, a large, "gorilla-like" animal. Shea (1984), however, explained that *koolokamba* is based in part on local folklore and that purported specimens thereof have generally turned out to be either large male chimpanzees or small female gorillas. Recent analyses of DNA (Morell 1994*a*; Morin et al. 1994) have suggested such a great genetic distance between *P. t. verus*, on the one hand, and *P. t. troglodytes* and *P. t. schweinfurthi*, on the other, that *verus* may warrant elevation to full species rank. This work also indicates the following times of divergence: between *P. t. troglodytes* and *P. t. schweinfurthi*, 440,000 years ago; between *P. t. verus* and the other two subspecies, 1,580,000 years ago; between the species *P. troglodytes* and *P. paniscus*, 2,500,000 years ago; and between the genera *Pan* and *Homo*, 4,700,000 years ago.

According to Napier and Napier (1967), *Pan* is characterized by prominent ears, protrusive lips, arms that are longer than the legs, a long hand but short thumb, and no tail. The facial skeleton is moderately prognathic. The orbits are frontally directed and surmounted by prominent supraorbital crests. A small sagittal crest is seen only occasionally in large males and females, and there is no nuchal crest. The molar teeth decrease in size toward the rear. Additional information is provided separately for each species, and the account of *P. troglodytes* was prepared with the assistance of Geza Teleki, of the Committee for the Conservation and Care of Chimpanzees.

Pan troglodytes (chimpanzee)

Head and body length is about 635–940 mm, and the height when erect is about 1.0–1.7 meters. The arms, which extend below the knees when the animal is standing, have a spread that is about 50 percent greater than the animal's height. Weight in the wild is 34–70 kg for males and 26–50 kg for females (Jungers and Susman 1984), but respective figures for captives are as much as 80 kg and 68 kg. The face is usually bare and generally black in color. Younger animals have flesh-colored ears, nose, hands, and feet and a white patch near the rump. At maturity the overall skin color is dark and the pelage varies from deep black to light brown. The hair on the head may grow in any direction, and both sexes are prone to partial baldness early in maturity.

The distribution of *Pan troglodytes* is centered in tropical rainforest but also extends into forest-savannah mosaic and montane forest up to 3,000 meters (Dandelot *in* Meester and Setzer 1977). During daylight hours most time is spent in trees, and chimpanzees, especially young individuals, sometimes brachiate from branch to branch. Movement through the trees, however, seldom proceeds very far, and most travel takes place on the ground (Reynolds and Reynolds 1965; Van Lawick–Goodall 1968). The most common means of locomotion is a quadrupedal walk with the hind legs slightly flexed, the body inclined forward, and the arms straight. The soles of the feet and the backs of the middle phalanges of the fingers are placed on the ground. In this position chimpanzees may proceed at a rapid gallop. They also are capable of a bipedal walk, but with the toes turned inward, and may proceed for more than 1 km in this manner. The hands are used to eat and may also be used to throw objects at enemies. For sleeping at night, each individual, except infants, constructs a nest of vegetation in a tree, usually at a height of 9–12 meters. Nests normally are occupied for only one night but occasionally are reused. Nests also may be constructed for rest during daytime in the rainy season (Van Lawick–Goodall 1968).

The chimpanzee has achieved some renown for its manufacture and utilization of tools. It frequently feeds by carefully poking a stick or vine, which it may have modified for the purpose, into an entrance of a termite nest and then withdrawing the tool, which has become covered with the insects (McGrew and Collins 1985). Sticks also have been used as hooks to pull down fruit-laden branches, as weapons in inter- and intraspecific fighting, and for various other purposes (Goodall 1986). Stones are used as hammers and anvils to crack nuts (Kortlandt and Holzhaus 1987). The spread of this nut-cracking ability through a group, with one animal after another learning, was documented by Hannah and McGrew (1987). Chimpanzees also have been observed to use leaves as sponges to soak up drinking

Chimpanzees *(Pan troglodytes):* A. Mother and baby, photo from San Diego Zoological Garden; B. Young chimpanzee, photo by Ernest P. Walker. Inset: photo from *Bull. Amer. Mus. Nat. Hist.*

water and as tissues for cleaning the body. Apparently, tool use and nest construction are not instinctive skills but are culturally acquired by younger animals observing more experienced ones (Van Lawick–Goodall 1968). Tool use has been found in all populations of *P. troglodytes,* but there are regional differences; it is especially extensive in far West Africa and nut-cracking seems to be confined mainly to that region (Boesch and Boesch 1990; McGrew 1992).

Chimpanzees are largely diurnal but sometimes move about by night. During the day they feed for 6–8 hours and forage over a distance of 1.5–15.0 km. Peaks of activity occur in the early morning and at around 1530–1830 hours (Van Lawick–Goodall 1968). They may shift their activity seasonally depending on the fruiting times of various kinds of plants (Izawa 1970). The diet includes fruits, leaves, blossoms, seeds, stems, bark, resin, honey, insects, eggs, and meat. Food intake varies by season, consisting on an annual basis of about 60 percent fruits, 30 percent other vegetation, and 10 percent animal matter. Termites are the most important insect. Chimpanzees occasionally stalk, kill, and eat young artiodactyls, baboons, and other monkeys. Individuals can hunt some prey successfully, but cooperative hunting seems essential in taking baboons and the large young of bush pigs. Cannibalism on infant chimpanzees has been observed, and there have been a few reports of the carrying off of young human children (J. Goodall 1977, 1986; Nishida, Uehara, and Nyundo 1979; Sugiyama 1973; Suzuki 1971; Van Lawick–Goodall 1968).

Reported population densities for *P. troglodytes* are 0.05–26.00/sq km, usually about 1–10/sq km (IUCN 1972; Jones and Sabater Pi 1971). Populations are divided into "communities," loose and flexible associations of males and

females with a shared home range (Tutin, McGrew, and Baldwin 1983). Subgroups of a community vary from solitary individuals to large mixed parties with all ages of both sexes. Only rarely, if ever, do all members of a community congregate in one place. Subgroups are generally unstable, except for females with immature offspring. There is little difference in community range size between forest and woodland; eight studies in such habitats have produced estimates of 5–40 sq km, with an average of 12 sq km. Savannah ranges are much larger, size estimates being 120–560 sq km. Communities of about 50 individuals each have been reported in all three kinds of habitat, but overall size range is around 15–80.

A community of 28 animals investigated by Tutin, McGrew, and Baldwin (1983) occurred in dry savannah, and this marginal habitat apparently resulted in a large home range of 228 sq km and the formation of large, relatively long-lasting traveling parties. Izawa (1970) investigated two communities of 40 animals each. Both groups used a nomadic home range of about 120 sq km, but there was only a 20 percent overlap in these ranges and the groups avoided one another. Sugiyama (1973) reported that a group of 70–80 animals maintained itself in a home range of 7–8 sq km for several years. He added that another group of 50–60 had a total range of 7–8 sq km and that a group of 60–80 used 17 sq km.

The chimpanzee is thought to be territorial, and intercommunity relationships sometimes appear to resemble those between enemy human tribes (Goodall 1986). Borders are patrolled, and outsiders may be attacked and killed. Males even have been seen to attack nonestrous females of another community. Long-term studies have revealed that

Chimpanzee *(Pan troglodytes)*, photo from West Berlin Zoo.

considerable greeting behavior, and large adult males sometimes embrace and groom one another (Sugiyama 1973). Wrangham (1977) observed that when an adult male arrived at a major food source, it would give a "food call" that tended to attract other individuals of both sexes.

Of 498 parties seen by Van Lawick–Goodall (1968), 44 percent had 2–4 animals each, and only 1 percent had more than 20. However, there was a recognized dominance hierarchy, and the highest-ranking individual present would lead the party. Because associations were continuously changing, most mature and adolescent animals had an opportunity to be dominant. Rank depended partly on size and age but also on relationship. A young animal might show dominance over an older one if its mother was high-ranking and present. Aggressive behavior within communities was frequent but seldom violent, and even serious fights resulted in no evident injury. Communication was complex and involved constantly varying facial expression, numerous gestures, and 24 identified vocalizations. The most common adult call is the "pant-hoot," which is given in many different situations and often seems to identify the caller and solicit information from other animals. Mitani and Nishida (1993) suggested that pant-hooting serves to space potential rivals and bring together allies. Grooming, particularly the removal of external parasites from one individual by another, is an important factor in maintaining friendly relationships but also can be used as a means of exchange to gain favors from other individuals (Goodall 1986).

The chimpanzee once was thought to be largely promiscuous, with several males, showing little aggression toward one another, following and mating with a female in estrus. Such activity does occur, but a male may enter a short-term relationship with a receptive female and prevent lower-ranking males from mating with her. An adult pair may also establish a temporary consortship, leaving the vicinity of other animals or even moving beyond the community range and mating exclusively with one another (Tutin and McGinnis 1981). Consortships may last up to three months. Older females are more popular with the males of their community than are younger females. Females, especially those that have not reached full adulthood, may leave their natal community either temporarily or permanently and mate with the males of neighboring communities. According to Goodall (1986), the presence or absence of an estrous female is the single most significant factor in the overall patterning of a chimpanzee community. Such an individual, unless involved in a consortship, may be surrounded by most or all of the community males and many of the females.

The following information on reproduction and life history was taken in part from Goodall (1986) and Van Lawick–Goodall (1968, 1973). Breeding occurs throughout the year. The estrous cycle lasts about 36 days, and females are receptive for about 6.5 days during a period of maximum genital swelling. There is a minimum interbirth interval of 3 years in the wild, but the period usually is 5–6 years if the first young survives. A female can give birth every year if the young is removed when only a few months old. The gestation period is 202–61 days, averaging about 230 days. The normal number of young is one; twins are rare, though perhaps more common than in *Homo sapiens*. Average birth weight is about 1.9 kg. For the first 3 months of life the infant is cradled by the mother when she is sitting. Until around 6 months the young clings ventrally when the female moves. For the next several years the young rides on the mother's back. Weaning takes place at 3.5–4.5 years, but the young remains dependent on the mother for a longer period and may still travel with her at 10 years.

an entire community may be destroyed as its range is gradually occupied by a stronger neighboring group and its members are systematically killed or driven away. Manson and Wrangham (1991) noted that "lethal raiding," in which several individuals make unprovoked aggressive invasions of the range of another group, where they deliberately seem to hunt and kill members of that group, is unknown in primates except for *Pan* and *Homo*. According to Goodall (1986:534), "The chimpanzee, as a result of a unique combination of strong affiliative bonds between adult males on the one hand and an unusually hostile and violently aggressive attitude toward nongroup individuals on the other, has clearly reached a stage where he stands at the very threshold of human achievement in destruction, cruelty, and planned intergroup conflict."

Goodall (1986) described the chimpanzee community as a "fusion-fission" society. Individuals of either sex have almost complete freedom to come and go as they wish. They associate with one another for various lengths of time, depending on the intensity of the relationship, reproductive status, and resource distribution. Each animal has its own network of social contacts. A community has a dominant male leader, and there also are power-wielding coalitions among both males and females. The ability of a male to enlist support during conflict is perhaps the most crucial factor in attaining and maintaining his rank. Subgroups of a community are temporary and may change in composition within a matter of hours or days. They can include any category of animals—all adult males, all adult females, both sexes, adults and young, or all young. Friendly relationships, even between adult males, appear to predominate within these groupings, though animals can suddenly turn violent. When two parties meet at a fruiting tree there is

Bonds between young, especially females, and the mother sometimes persist throughout life. Puberty in both sexes occurs at about 7 years, but females do not usually give birth until 13–14 years and males are not fully integrated into the social hierarchy until 15–16 years. Reproductive capability in the female apparently lasts until at least age 40, and maximum life span in the wild may be 60 years. According to Marvin L. Jones (Zoological Society of San Diego, pers. comm., 1995), a female died in 1992 at the Yerkes Primate Center at an age of about 59 years, another female was still living in 1976 at the Mysore Zoo in India at an age of 53 years, and a male ("Cheeta" of the *Tarzan* movies) was still living in 1990 at 56 years.

There may have been several million *P. troglodytes* in the early twentieth century, but current estimates generally range from 150,000 to 235,000, and the species has disappeared entirely from large areas (Lee, Thornback, and Bennett 1988; Luoma 1989). The largest numbers remaining are found in the Central African forests of Cameroon, Gabon, Congo, and Zaire. A survey during the early 1980s suggested the presence of about 64,000 chimpanzees in Gabon alone (Tutin and Fernandez 1984). More recently, however, Blom et al. (1992) indicated that numbers may be declining in Gabon, and J. Goodall (1994) warned that alarming declines were occurring even in Central Africa. Populations have been reduced and fragmented through loss of forest habitat to agriculture and logging, hunting by people for food and to protect crops, and commercial exportation for the animal trade. In 1976 the USDI classified *P. troglodytes* as threatened, and consequent regulations reduced legal importation to the United States, though illicit trade continued. Recently there has been widespread concern that a demand for animals for biomedical research, especially involving work related to Acquired Immune Deficiency Syndrome, or AIDS, would lead to large-scale collection of young animals in the wild and consequent devastation of remaining populations. There are several thousand animals in captivity, many of them in colonies that are being managed with the objective of their becoming self-sustaining. In 1990 the USDI reclassified *P. troglodytes* as endangered in the wild and also strengthened regulations covering animals in captivity. *P. troglodytes* is on appendix 1 of the CITES. The IUCN also now classifies the species as endangered.

Pan paniscus (pygmy chimpanzee, bonobo)

Despite the common name, *P. paniscus* is often equal in size to *P. troglodytes*. Head and body length is 700–828 mm and weight is 37–61 kg for males and 27–38 kg for females (Jungers and Susman 1984). Coloration of *P. paniscus* is about the same as in *P. troglodytes*. *P. paniscus* has longer limbs than *P. troglodytes*, and the chest girth of the former is smaller relative to its height (Zihlman 1984). *P. paniscus* has long hair on the sides of the head, relatively small ears, and a relatively high forehead (Kano 1992).

According to Kano and Mulavwa (1984), *P. paniscus* has a narrower ecological range than *P. troglodytes*, being found exclusively in lowland, especially primary, forest. Activity is largely diurnal; the animals feed in trees shortly after waking, rest, move leisurely on the ground to the next food trees, become less active toward the middle of the day, and then repeat the pattern in the afternoon. Average daily movement is about 2.4 km. Nishida (1972b) reported *P. paniscus* to build sleeping nests in trees in the same manner as *P. troglodytes*. Tool use in the wild has not been observed, but Jordan (1982) reported captives to construct ropes to swing from, to wipe themselves with leaves, to use sticks to probe, rake, and pole-vault over water. Movements in the wild include quadrupedal knuckle walking on

Pygmy chimpanzee *(Pan paniscus)*, photo by Bernhard Grzimek.

the ground, arm swinging through the trees, and leaping from branches (Susman 1984a). The natural diet consists mostly of fruits and also includes leaves, seeds, and, rarely, invertebrates and small vertebrates. Badrian and Malenky (1984) reported predation on young duikers *(Cephalophus)*.

The following information on population structure and social life was taken from Badrian and Badrian (1984), Kano (1992), and Kano and Mulavwa (1984). *P. paniscus* has a "fusion-fission" community organization comparable to that of *P. troglodytes* but seems to be more gregarious, more mutually tolerant, and less agonistic. Each known community or "unit group" contains about 40–120 individuals occupying a home range of about 22–68 sq km, at a density of approximately 2/sq km. Membership is fluid, mainly because young, nulliparous females transfer freely from group to group, but is more stable than in groups of *P. troglodytes*. Subgroups, or "parties," usually contain 2–15 individuals but may contain as many as 40 and are commonly based on a female and her immature male offspring, as well as on adult female associations, and also usually include adult males. Larger subgroups sometimes are found in the vicinity of rich food sources. Upon discovery of such a site, individuals may emit a loud call, perhaps both to alert members of their community and to warn other groups. There is extensive overlap between community ranges, but parties of one community avoid those of the other. When there is an encounter, there may be loud vocalizing, agonistic displays, and occasionally serious fighting, but the deliberate "lethal raids" of *P. troglodytes* have not been observed.

Adolescent female *P. paniscus* emigrate from their natal groups and have no further ties with their mothers. Adult females, however, appear to have a greater tendency to come together in strongly bonded units than do female *P. troglodytes*. Such female association seems to form the primary basis of the community, and it is the females that generally have the leadership role. In contrast to young adult male *P. troglodytes*, which are incorporated into a male hierarchy and then have little to do with their mothers, male *P. paniscus* stay in their mothers' units and maintain a permanent association with them. Adult male *P. paniscus* seem to lack the dominant role of male *P. troglodytes* and are about equal in rank to the females. The males are sometimes aggressive toward one another but are only rarely aggressive toward the females.

Reproductive seasonality is not known. The estrous cycle lasts about 46 days (Thompson-Handler, Malenky, and Badrian 1984). The period of receptivity is about twice as long as that of *P. troglodytes*, but interbirth interval usually is about the same. There normally is a single young and interaction with the mother is similar to that seen in *P. troglodytes* (Kano 1992). Based on observations in captivity, the gestation period is 220–30 days, birth weight is 1–2 kg, and females reach sexual maturity at about 9 years (Neugebauer 1980). There are far fewer captive *P. paniscus* than captive *P. troglodytes*, the number of the former being about 100 (Reinartz and Van Puijenbroeck *in* Bowdoin et al. 1994).

The pygmy chimpanzee is classified as endangered by the IUCN and the USDI and is on appendix 1 of the CITES. Although it occurs over a large region of dense jungle, its distribution is discontinuous because of both natural ecological requirements and fragmentation of habitat through increasing human populations, agricultural expansion, and logging. It also is hunted extensively by people for meat and for parts used in religious rituals. Its numbers are unknown, but extrapolation of estimates of known populations to potential remaining habitat suggests an overall figure of about 15,000 (Lee, Thornback, and Bennett 1988). Assessments of continued habitat deterioration and expansion of human activity in Zaire suggest that numbers in the wild may be fewer than 5,000 (Thompson-Handler, Malenky, and Reinartz 1995). Recently, Kortlandt (1995) suggested that the pygmy chimpanzee may still inhabit a vast area of relatively undisturbed habitat and could number more than 100,000. Susman (1995) agreed that remaining distribution and population size may be large but cautioned that human activity could quickly jeopardize survival of the species. Both authorities emphasized the need for more intensive and careful field surveys.

PRIMATES; **Family HOMINIDAE; Genus HOMO**
Linnaeus, 1758

People, or Human Beings

The single living genus and species, *Homo sapiens*, maintains permanent residence in nearly all terrestrial parts of the earth's surface. Exceptions include extremely arid regions, such as certain sections of Australia and the central Sahara Desert, and extensive ice-covered regions, especially most of Greenland and Antarctica. Their knowledge and technical skills allow humans to temporarily occupy even those inhospitable areas, move freely on and under most water-covered portions of the world, and fly through and orbit above the atmosphere. A Russian space station has been maintained about 350 km above the earth since February 1986, usually occupied by several persons, and in 1995–98 was visited by American crews. From December 1968 to December 1972 a series of rocket-powered spacecraft carried people from the United States to the vicinity of the moon and back to earth; three of these expeditions merely circled the moon, but six landed teams of two men each on the lunar surface.

There is controversy regarding the content of the family Hominidae. Although it long was common practice to treat *Homo sapiens* as the only living species, some authorities have suggested that the great apes (here considered to compose the Pongidae) should be placed within the same family as humans. Groves (1986*a*, 1989, and *in* Wilson and Reeder 1993), for example, divided the Hominidae into two living subfamilies: the Homininae, with the living genera *Pan, Gorilla,* and *Homo;* and the Ponginae, with *Pongo.* Based largely on DNA evidence, Goodman (1992) went even further, including the gibbons in the Hominidae as the subfamily Hylobatinae and placing the great apes and humans in a second subfamily, the Homininae, with *Pan* and *Homo* united at the subtribal level. The relegation of humans to a common family with the apes is partly a consequence of a principle of the taxonomic school known as cladistics. This view holds (1) that all components of one systematic grouping must be considered to have descended from a common ancestor and (2) that the grouping must include all the descendants of that ancestor. With regard to the first point, there is general agreement that apes and humans did have a common ancestor. Nonetheless, a problem develops relative to the currently favored version of subsequent phylogenetic history, as discussed below. If *Pongo* diverged from the ancestral stock prior to the split between the two African apes and *Homo,* and if *Pongo* is placed in the same family as the African apes, then in accordance with the second point of cladistics *Homo* must also be included in that family. Not all taxonomists accept such an arrangement even if they agree with its evolutionary basis. Martin (1990) discussed this matter in some detail and concluded that the great morphological distinction between the three living great apes, on the one hand, and *Homo,* on the other, warrants placing the latter in a separate family. That position has been followed here.

Until about 1970 there was a general view, based primarily on fossil evidence, that the evolutionary line leading to *Homo* diverged from that leading to the great apes in the middle of the Miocene epoch, about 15 million years ago. It was thought that the first clearly hominid genera were *Sivapithecus* and *Ramapithecus,* of southern Asia and Africa. Subsequently, based in large part on biochemical analyses of humans and apes, there has been a growing consensus that *Sivapithecus* (which may include *Ramapithecus*) is close to the ancestral line of only *Pongo.* That line is thought to have split off in the middle Miocene from the line leading to humans and the two African apes. The latter line, however, did not begin to branch until the late Miocene, around 8 million years ago, with the split between the ancestors of *Pan* and *Homo* taking place in the Pliocene, perhaps only 4–5 million years ago (Andrews and Cronin 1982; Hasegawa 1992; Johanson and White 1979; Pilbeam 1984; Simons 1989; Zihlman and Lowenstein 1979). It is commonly thought that *Pan* is the closest living relative of *Homo* and that the ancestral stock of *Gorilla* split off considerably earlier, but Rogers (1993) suggested that the lines leading to all three genera diverged over a relatively brief period. There is a minority view, supported by certain morphological evidence, that *Pongo* is the closest living relative of *Homo* (Schwartz 1987, 1988*a*). There also have been some suggestions that *Gigantopithecus,* the huge ape of the Pliocene and Pleistocene of southeastern Asia, which usually is thought to have been a relative of *Pongo,* may actually represent an aberrant branch of the line leading to *Homo* (Ciochon 1988; Gelvin 1980; Oxnard 1985).

If the possible affinities of *Pongo* and *Gigantopithecus* are disregarded, and if the Hominidae are restricted to creatures that developed subsequent to their divergence from the ancestors of the modern African apes, then the current consensus is that the family comprises only *Homo,* the extinct *Australopithecus,* and perhaps one or two other extinct genera closely related to the latter; it also is thought that all major developments in the evolution of the family took place in Africa (Lewin 1993; Martin 1990; Pilbeam 1984; Simons 1989; Wood 1992). Some fragmentary remains, possibly referable to *Australopithecus,* date from around 5 million years ago, but the earliest described

Homo sapiens. Astronaut Edward H. White II makes America's first "space walk" during the *Gemini IV* mission, 3 June 1965. He was accompanied by James A. McDivitt. On 27 January 1967, White, along with astronauts Virgil I. Grissom and Roger B. Chaffee, died in a fire during a test of an Apollo spacecraft. Photo courtesy of National Aeronautics and Space Administration.

hominid species is based mostly on dentition from sites about 4.4 million years old in Ethiopia. This species, *A. ramidus,* originally was assigned to *Australopithecus* (White, Suwa, and Asfaw 1994; Wood 1994) but subsequently was placed in an entirely new genus, *Ardipithecus* (White, Suwa, and Asfaw 1995). In several critical morphological characters it is closer to *Pan* than to *Australopithecus,* but it evidently represents a basal line leading to the latter.

Fossilized bones and bipedal footprints suggest that *Australopithecus* itself arose about 4 million years ago. The oldest relatively complete specimen, popularly known as "Lucy" and representing a species designated *A. afarensis,* was found in Ethiopia. There is some controversy about the content of this taxon, with Ferguson (1992) arguing that it is a composite species. In any case, Lucy is thought to have been about 1 meter in height and 30 kg in weight and to have had a brain capacity of 400 cu cm, about equal to that of a modern chimpanzee. The males of her species may have been up to 1.5 meters tall and averaged 45 kg in weight. Later fossil material from eastern and southern Africa suggests that *A. afarensis,* either directly or after having evolved into another species, *A. africanus* (Skelton and McHenry 1992), gave rise to two evolutionary lines. One seems to have consisted of relatively lightly built creatures that probably were ancestral to *Homo.*

The other line is represented by a specimen known as the "black skull," which was found in Kenya in deposits about 2.5 million years old. This line apparently gave rise to two heavily built species, *A. robustus* and *A. boisei,* which some authorities, including Groves (1989) and Wood (1992), place in the separate genus *Paranthropus.* These

species survived for a substantial period, were contemporary in Africa with the earliest known populations of *Homo,* and did not become extinct until about 1.2 million years ago (Ward 1991). The remains of both *Australopithecus* and *Homo* have been found in the lowest strata of Olduvai Gorge in Tanzania, so these two genera of hominids may have occurred together in this region. Both may also have used stone implements and tools, since some of these have been found in association with the skeletal remains.

The earliest definitely known specimens of *Homo* itself are from Kenya and Malawi and have been dated to about 2.4 million years ago (Hill et al. 1992; Schrenk et al. 1993). These occurrences suggest that the genus arose in tropical Africa during a period of climatic cooling and subsequently spread southward. There is some question about the number of early species, with certain authorities recognizing only one, *H. habilis,* and others accepting at least one contemporary, *H. rudolfensis.* The brain capacity of *H. habilis* was about 700 cu cm, considerably greater than that of any modern ape. These early forms of *Homo* disappeared, but from among them evolved *H. erectus.* The latter is first represented by specimens about 1.6 million years old from eastern Africa, which had a brain capacity of around 900 cu cm. At least 1 million years ago *H. erectus* expanded its range from Africa to Eurasia. From among the resulting widespread populations there seems to have gradually evolved an archaic form of *Homo sapiens.* In Europe and southwestern Asia this form became established about 150,000 years ago as the subspecies *H. sapiens neanderthalensis.* Some genetic studies suggest that modern *H. sapiens* meanwhile originated about 100,000–200,000

The first record of *Homo sapiens*, or of any mammal, on the surface of a world other than the Earth. Astronaut Edwin E. Aldrin stands on the moon during the *Apollo XI* mission, 20 July 1969. He was photographed by the mission commander, Neil A. Armstrong, while pilot Michael Collins remained in lunar orbit. Photo courtesy of National Aeronautics and Space Administration.

years ago in Africa and eventually spread out to replace other human populations. There also is substantial evidence that the replacement of *H. s. neanderthalensis* by modern human beings occurred 30,000–40,000 years ago. Strahan (1983) noted that people first entered Australia at least 40,000 years ago. According to Owen (1984), there is no substantial evidence that *Homo* occupied the Americas, except possibly the area immediately east of the Bering Strait, until about 12,000 years ago.

There is substantial disagreement about the phylogeny, biogeography, and nomenclature of the populations involved in the development of *Homo* (Sussman 1993). Some authorities, for example, have argued that *H. erectus* represents an early offshoot that developed in Asia and was not directly involved in the evolution of *H. sapiens*. Instead, they say, another species, *H. ergaster*, would have arisen in Africa subsequent to the emigration of the forebears of *H. erectus* and represented the ancestral stock of modern humans (Groves 1989; Harrison 1993; Turner and Chamberlain 1989; Wood 1992). In contrast, Kramer (1993) concluded that all the variation demonstrated by *Homo* in the period from 1.7 million to 0.5 million years ago can be

accommodated by the single species *H. erectus*. There are questions concerning how distinct archaic *H. sapiens* was from the modern populations and whether the disappearance of the former involved interbreeding with, or total replacement by, the latter. Tattersall (1992) argued that *H. s. neanderthalensis* should be regarded as a distinctive species. A highly publicized analysis of mitochondrial DNA suggesting that all modern humans can trace their ancestry to a single female that lived in Africa about 200,000 years ago was refuted by Pickford (1991).

The family Hominidae, as accepted here, is distinguished from the Pongidae by skeletal modifications that adapt it to erect bipedalism; such characters in the Pongidae are directed primarily to arboreal brachiation. Thus, in the Hominidae there is a proportional lengthening of the lower extremities, as well as modifications of the pelvis, femur, and musculature to reflect erect posture, whereas in the Pongidae there is lengthening of the upper extremities and modification of morphological details to reflect brachiation. The two families also differ in a number of other minor skeletal and dental details (Buettner-Janusch 1966).

The genus *Australopithecus* has been described as having, among others, the following characters: (1) a large masticatory apparatus reflecting an adaptation to a diet that required heavy chewing; (2) a small cranial capacity (350–530 cu cm); (3) strong supraorbital ridges; (4) a low sagittal crest; (5) a skeleton resembling that in *Homo* but differing in detail (a forward prolongation of the region of the anterior superior spine of the ilium and a relatively small sacroiliac surface); (6) large, long-fingered hands; and (7) arms shorter in relation to torso length than in apes of comparable size (Johanson and White 1979; Le Gros Clark 1955; Simons 1989). The primary characters of the genus *Homo* have been described as (1) no specialization relating to a heavily masticated diet; (2) a skeleton modified for habitual erect posture; and (3) a large cranial capacity (700 to almost 2,000 cu cm) (Johanson and White 1979; Leakey, Tobias, and Napier 1964; Le Gros Clark 1955). Mayr (1951) perceived the genus *Homo* as being characterized by progressive brain enlargement associated with increasing cultural elaboration. The species *H. sapiens* presents the ultimate in the tendencies toward brain expansion and reduction of the masticatory apparatus characteristic of earlier forms of hominids.

There is considerable variation in size of *Homo*. Recently there were announcements of the death of both the world's smallest man, who was about 74 cm tall, and the world's tallest man, 243 cm tall (*Washington Post*, 13 February 1988, C-4, and 24 January 1990, B-7, respectively). According to Thorington and Anderson (1984), the standing height of adult males averages 163–72 cm; weight averages about 75 kg in males and 52 kg in females; pelage is sparse, usually of one color, black to blond or reddish; and dental formula is: (i 2/2, c 1/1, pm 2/2, m 3/3) × 2 = 32.

Living populations of *H. sapiens* differ from one another only in minor details of skin coloration, hair texture and color, facial features, and so on. Using standard criteria as applied to other mammals, four or five possible geographic races of people can be recognized, with a number of minor variants.

1. Negroid—originally distributed in tropical parts of Africa and Southeast Asia; now also widely found in North, Central, and South America and in the Caribbean area. This race contains some of the shortest human beings (Ituri Forest pygmies in Zaire, whose height is usually under 1.4 meters) as well as some of the tallest (the Tutsi of Burundi and Rwanda, whose height is as great as 2 meters and over). Negroids are

The transcendency and diversity of *Homo sapiens*, as represented by the crew of the space shuttle *Challenger*. Front row, left to right: Michael J. Smith, Francis R. Scobee, Ronald E. McNair; Back row, left to right: Ellison S. Onizuka, Sharon Christa McAuliffe, Gregory Jarvis, Judith A. Resnik. All died when their spacecraft exploded during launch on 28 January 1986. Photo courtesy of National Aeronautics and Space Administration.

characterized by dark skin coloration and curly, stiff hair that is only sparsely distributed on the body.

2. Caucasoid—originally distributed in Europe, North Africa, and extreme western Asia; now widely scattered throughout much of the world. The subspecies is characterized by light or dark skin coloration and wavy, soft hair that forms a heavy beard in the males and is usually densely distributed over the body.

3. Mongoloid—originally distributed in eastern Asia; Mongoloid stock also migrated in late Pleistocene and early Recent times to the Americas by way of the Bering Land Bridge or some other route and gave rise to the American Indians. Mongoloids tend to be intermediate in skin color to Negroids and Caucasoids, though many are very pale. They are characterized by straight, very dark hair sparsely distributed over the body, a flat face with protruding cheekbones, and, in some varieties at least, the Mongolian eye folds.

4. Australoid and Khoisanic—possibly distinct from the three major races listed above. The Australoids are represented by the Australian aborigines (and perhaps by the Ainus of Japan, as well as others) and are characterized in very general terms as having dark skin coloration, wavy hair, which may be thick over the body and face, and heavy brow ridges. The true Khoisanics are represented today by only a few Bushman tribes living in the deserts of southern Africa but may at one time have been very widely distributed across the African continent. In general, they can be described as having yellowish skin coloration; sparse,

wiry hair that grows in clumps over the head ("peppercorn hair"); steatopygia, especially marked in females; short stature; and delicate facial features with little development of a brow ridge. Both Australoids and Khoisanics have suffered greatly from persecution by other peoples and today are reduced in numbers and distribution.

It must be emphasized that at least in recent years *Homo sapiens* has become a case apart from all other mammals. The extreme mobility of the species has thoroughly mixed the races enumerated above in many areas and has distributed at least the three major races widely throughout the world. Even before modern means of rapid transportation, humans were wide-ranging animals and there was extensive interbreeding between the races. Because of the movements of human populations, zones of intergradation tended to cover vast areas, thus further confusing the picture.

Three calculations of average length of the menstrual cycle in *Homo sapiens* are 28.7, 30.4, and 33.9 days (Hayssen, Van Tienhoven, and Van Tienhoven 1993). The gestation period of the species varies from 243 to 298 days, usually being about 280 days. Generally a single baby is born, with twins occurring once in 88 births, triplets once in 7,600 births, and quadruplets once in 670,000 births. Very rarely there are five, six, or more offspring. In the United States, apparently in association with a societal trend toward delaying childbearing, the rate of twins recently has increased to once in 43 births, triplets to once in 1,341 (*Washington Post Health*, 26 July 1994, 5). The weight of single babies averages 3.25 kg and ranges from 2.5 to 6 kg. Humans have one of the longest periods of development following birth of any living creature. Human females are fertile from about 13 to 49 years of age. Maximum longevity may be as great as 100 years or more, but average life expectancy is much lower and differs widely from population to population, depending mainly on such factors as the availability of modern medical expertise, equipment, and drugs and also, apparently, on the lifestyle of the people. The average life expectancy at birth of people in some of the developing countries may be as low as 30 years, mainly because of high infant mortality, but it is well above 70 years in some of the more developed countries. In the United States, life expectancy at birth is 78.9 years for females and 72.0 years for males (*Washington Post*, 1 September 1993, A-3).

All three of the principal racial groups of people (Negroid, Caucasoid, and Mongoloid) have populations that are simple hunters/gatherers, farmers, and herders, and all three have produced populations that have advanced civilizations, that is, having written languages, highly developed arts and sciences, and complex social and cultural developments. Hunters and gatherers perhaps represent the basic form of human society, and the one that comes closest to representing the type of life lived by early human beings. According to Hayden (1981), existing hunter/gatherer societies still occupy many habitats, including tropical forests, semideserts, temperate woodlands, and arctic tundra. The diet is omnivorous but may include more meat than is generally thought. Reported population densities range from about 0.4/sq km to 75.0/sq km. Group size varies from 9 to 1,500 but usually is about 25–50. For food gathering, members of such groups forage up to 8 km per day during a period of 2–5 hours, and hunting for meat involves movements as great as 16–24 km. The maximum yearly home range for groups that lack transportation aids is about 2,500 sq km.

The Negroid Mbuti pygmies of the Ituri Forest in Zaire are an example of a hunting and gathering people and have been well studied. According to Turnbull (1976), their economy requires a minimal technology and is still at the Stone Age level, though Harako (1981) explained that they have been substantially influenced by contact with village peoples. They domesticate neither plants nor animals. They gather mushrooms, roots, fruits, berries, and nuts, but hunting is the primary influence on their society. They live in small bands of 3–14 family units, and each band is territorial, with a large no man's land in the center of the forest. Harako (1981) indicated that originally each band had an exclusive hunting area of 100–200 sq km. When a band grows to a size of more than 30 individuals, it subdivides, since no band can grow in size beyond what local game and vegetable supplies can feed. To prevent excessive consumption in any one area, the bands move from camp to camp, never staying more than a month in one place.

The Mbuti family is the basic unit, but in a sense the entire band considers itself a single family. A band does not necessarily consist of families related to each other in one line or another, and the composition may change with every monthly move. The families within a band are extremely cooperative with one another, and great affection and intimacy are usually evident between all members. The entire band will participate in a hunt, with men setting up nets and guarding with spears, youths standing farther back to shoot with bow and arrow any game that escapes, and women and children forming an opposing semicircle at some distance to drive the game into the net. Usually the band can obtain more than enough game with several casts of the nets, whereupon everyone returns to camp, and by

noon each family is cooking its meal over an open fire in front of the small leaf hut in which it lives. Everyone, even the children, has a role to play in this sort of society.

The pygmies apparently have a belief in God as a universal creator; although they enjoy meat as food, they still somehow regard it as wrong to take life. At an early age, children are introduced to a concept of dependence on and trust in their forest world and made to feel a part of it. They personalize it and refer to it as father and mother and say that it provides them with all that they need, with life itself. Thus, they adapt to their environment instead of trying to control it.

Since pygmies are highly mobile, their social organization must be fluid. Because bands are constantly changing in size and composition, there can be no chiefs or individual leaders, for they would be as likely to move somewhere else as anyone, leaving the band without a leader. Elders, however, act as arbitrators and make decisions on major issues facing the band, and they are highly respected by all; there is no lineage system among the Mbuti.

The pygmies' technology is simple but more than adequate for their needs. Their clothing consists only of a barkcloth pubic covering, and their houses and furniture can be made within a matter of minutes from saplings and leaves (they obtain metal machetes and knives from neighboring village peoples). They place themselves completely under the forest's protection, and in times of trouble they sing songs to "awaken the forest" and draw its attention to the plight of its children; then all will be well.

The Mbuti are a society of human beings who live in harmony with their environment. The lifestyle of these people may be as close to that of early *H. sapiens* as can be found in the world today. They and other hunting and gathering societies have been under great pressure in recent years as vast areas of forest and woodland have been cut over for agricultural and pastoral purposes; nevertheless, such cultures still exist widely in all subspecies of humans.

Agricultural and herding peoples represent the next step in the ladder leading to the "technologically advanced" societies that control much of the world today. It was a revolutionary step when people first learned that they could plant crops and reap a steady harvest year after year and not have to depend on the availability and abundance of wild nuts, fruits, and tubers. It was also a major achievement when people learned to raise their own meat in the form of cattle, pigs, and sheep, so that they no longer had to be constantly on the move searching for game. Then, according to Turnbull (1976), there was the dramatic development of true civilization. Apart from the trend toward industrialization, however, herding and cultivating cultures still characterize most human societies today.

With crops and herds providing a ready and dependable source of food, people were able to devote more of their time and energies to other affairs. Humans thus began to control nature rather than simply live in harmony with it. As better methods of herding and farming were developed in certain areas, fewer and fewer people could supply food for more and more people, so that the mass of humanity no longer needed to be tied closely to the soil. Cities arose and people devoted their minds to such matters as art, literature, medicine, and science. Great technologies developed that enabled humans to change and control more and more of their natural environment. Rapid means of transportation were developed, roads built, and diseases conquered, and populations grew and expanded. Except for some of the social insects, no other animal has a society as complex and varied as humans', and none has been able to exert such control over the environment.

Along with these advances have come some serious problems, not only for humans themselves but for the earth as a whole. Until recently, human population grew slowly. It is thought that there were only 250 million people on earth in the first year A.D. but 1 billion by 1830. It took a century to add the second billion, 30 years to add the third, and only 15 years to add the fourth. The rate of increase recently has declined dramatically in some areas, particularly because women in developing countries have become better educated and have been able to reduce their number of births. Nonetheless, the world population now is 5.6 billion people, and it is expected to reach at least 10 billion by the year 2050 (*Washington Post*, 4 September 1994, A-1). To provide for the needs of these people, forests are being rapidly cleared, coal and other minerals are being torn from the earth, oil resources are being exploited and probably depleted, industrialization and development are spreading into even remote regions, fisheries are being overharvested, and other animals and plants are being exterminated at an unprecedented rate. The waste and incidental products of human activity are poisoning freshwater and even the oceans, devastating the natural and cultural environment through pollution and acid rain, and threatening the integrity of the atmosphere and climate.

In a recent overview, Vitousek et al. (1997) pointed out that between one-third and one-half of the earth's land surface has been transformed by human action, the carbon dioxide concentration in the atmosphere has increased by nearly 30 percent since the beginning of the Industrial Revolution, more atmospheric nitrogen is fixed by humanity than by all natural terrestrial sources combined, more than half of all accessible surface fresh water is put to human use, two-thirds of major marine fisheries are fully or overexploited or depleted, and the rate of species extinction is now 100 to 1,000 times that before human dominance of the earth. Never has a mammal been so destructive to the habitat upon which it depends for survival. That there is still hope, however, may be seen in the accompanying photographs illustrating the remarkable capabilities and courage of the human species.

World Distribution of Primates

For maximum usefulness, it has been necessary to devise the simplest practicable outline of the approximate distribution of the genera in the sequence used in the text. The tabulation should be regarded as an index guide to groups of primates or to geographic regions. At the same time it gives a good overall picture of the general distribution of primates.

The major geographic distribution of the genera of Recent primates that appears in the tabulation is designed to show their natural distribution at the present time or within comparatively recent times. It should be noted that most of the animals occupy only a portion of the geographic region that appears at the head of the column. Some are limited to the tropical regions, others to temperate zones. Mountain ranges and streams sometimes have been natural barriers preventing the spread of animals on the lands that are a part of the same continent. Also, many restricted ranges cannot be designated either by letters to show the general area or by footnotes because of limited space on the tabulation. *It therefore should not be assumed that a mark indicating that an animal occurs within a geographic region implies that it inhabits that entire area.* For more detailed outlines of the ranges of the respective genera, it is necessary to consult the generic texts.

Explanation of Geographic Column Headings

Europe and Asia constitute a single land mass, but this land mass comprises widely different types of zoogeographic areas created by high mountain ranges, plateaus, latitudes, and prevailing winds. The general distribution of Recent primates can be shown much more accurately by two columns, headed "Europe" and "Asia," than by a single column headed "Eurasia."

Most islands are included with the major land masses nearby unless otherwise specified, though in many instances some of the primates indicated for the continental mass do not occur on the islands.

With Europe are included the British Isles and other adjacent islands, including those in the Arctic.

With Asia are included the Japanese Islands, Taiwan, Hainan, Sri Lanka, and other adjacent islands, including those in the Arctic.

With North America are included Mexico and Central America south to Panama, adjacent islands, the Aleutian chain, the islands in the arctic region, and Greenland but not the West Indies.

With South America are included Trinidad, the Netherlands Antilles, and other small adjacent islands but not the Falkland and Galapagos Islands unless named in footnotes.

With Africa are included only Zanzibar Island and small islands close to the continent but not the Cape Verde or Canary Islands.

The island groups treated separately are:

Southeastern Asian islands, in which are included the Andamans, the Nicobars, the Mentawais, Sumatra, Java, the Lesser Sundas, Borneo, Sulawesi, the Moluccas, and the many other adjacent small islands;

New Guinea and small adjacent islands;

the Australian region, in which are included Australia, Tasmania, and adjacent small islands;

the Philippine Islands and small adjacent islands;

the West Indies;

Madagascar and small adjacent islands;

other islands that have only one or a few forms of primates and are named in footnotes.

Footnotes indicate the major easily definable deviations from the distribution indicated in the tables.

Symbols

†	The primates are extinct.
■	The primates occur on the land area.
N	Northern portion
S	Southern portion
E	Eastern portion
W	Western portion
Ne	Northeastern portion
Se	Southeastern portion
Sw	Southwestern portion
Nw	Northwestern portion
C	Central portion

Examples: "N, C" = northern and central; "Nc" = north-central. Numerals refer to footnotes indicating clearly defined limited ranges within the general area.

Genera of Recent Primates	page	North America	West Indies	South America	Madagascar	Africa	Europe	Asia	Southeast Asia Islands	Philippine Islands	New Guinea	Australian Region	Antarctic Region	Arctic Region	Atlantic Ocean	Indian Ocean	Pacific Ocean
PRIMATES LORISIDAE																	
Pseudopotto	54					■C											
Arctocebus	55					■Wc											
Loris	56							■Sc									
Nycticebus	57							■Se	■	■S							
Perodicticus	58					■W,C											
Euoticus	60					■Wc											
Galago	60					■											
Otolemur	62					■E,S											
Galagoides	63					■											
PRIMATES CHEIROGALEIDAE																	
Microcebus	66				■												
Mirza	67				■W												
Cheirogaleus	68				■												
Allocebus	69				■E												
Phaner	70				■												
PRIMATES LEMURIDAE																	
Hapalemur	72				■												
Lemur	73				■S												
Varecia	75				■												
Pachylemur†	77				■												
Eulemur	77				■1												
PRIMATES MEGALADAPIDAE																	
Lepilemur	82				■												
Megaladapis†	83				■												
PRIMATES INDRIIDAE																	
Indri	84				■Ne												
Avahi	85				■												
Propithecus	86				■												
PRIMATES PALAEOPROPITHECIDAE																	
Mesopropithecus†	89				■												
Babakotia†	89				■N												
Palaeopropithecus†	90				■												
Archaeoindris†	91				■												
PRIMATES ARCHAEOLEMURIDAE																	
Archaeolemur†	91				■												
Hadropithecus†	92				■												
PRIMATES DAUBENTONIIDAE																	
Daubentonia	92				■												
PRIMATES TARSIIDAE																	
Tarsius	94								■	■S							
PRIMATES CEBIDAE																	
Lagothrix	98			■Nw													
Ateles	100	■S		■N,C													
Brachyteles	101			■E													

1. And Comoro Islands.

Genera of Recent Primates	page	North America	West Indies	South America	Madagascar	Africa	Europe	Asia	Southeast Asia Islands	Philippine Islands	New Guinea	Australian Region	Antarctic Region	Arctic Region	Atlantic Ocean	Indian Ocean	Pacific Ocean
PRIMATES CEBIDAE Continued																	
Alouatta	103	■S		■													
Pithecia	105			■N,C													
Chiropotes	106			■C													
Cacajao	107			■Nc													
Xenothrix†	108		■1														
Paralouatta†	109		■2														
Antillothrix†	109		■3														
Callicebus	109			■													
Aotus	111	■4		■													
Cebus	113	■S		■													
Saimiri	115	■S		■													
PRIMATES CALLITRICHIDAE																	
Callimico	119			■Wc													
Saguinus	120	■S		■N,C													
Callithrix	123			■C													
Cebuella	126			■Wc													
Leontopithecus	127			■E													
PRIMATES CERCOPITHECIDAE																	
Erythrocebus	130					■											
Chlorocebus	132					■											
Cercopithecus	133					■											
Miopithecus	135					■Wc											
Allenopithecus	136					■C											
Cercocebus	137					■											
Lophocebus	139					■W,C											
Macaca	140					■Nw	■5	■S	■	■							
Papio	148					■		■Sw									
Mandrillus	151					■Wc											
Theropithecus	152					■6											
Nasalis	154								■7								
Simias	156								■8								
Pygathrix	157							■Se									
Rhinopithecus	158							■E									
Presbytis	159							■9	■								
Semnopithecus	160							■Sc									
Trachypithecus	161							■S	■								
Colobus	164					■											
Piliocolobus	165					■											
Procolobus	167					■W											
PRIMATES HYLOBATIDAE																	
Hylobates	168							■Se	■								
PRIMATES PONGIDAE																	
Pongo	174								■10								
Gorilla	178					■C											
Pan	182					■W,C											
PRIMATES HOMINIDAE																	
Homo	186	■	■	■	■	■	■	■	■	■	■	■					

1. Jamaica only.   2. Cuba only.   3. Hispaniola only.   4. Panama only.   5. Gibraltar only.   6. Ethiopia only.   7. Borneo only.
8. Mentawai Islands only.   9. Malay Peninsula only.   10. Sumatra and Borneo only.

Appendix

GEOLOGICAL TIME

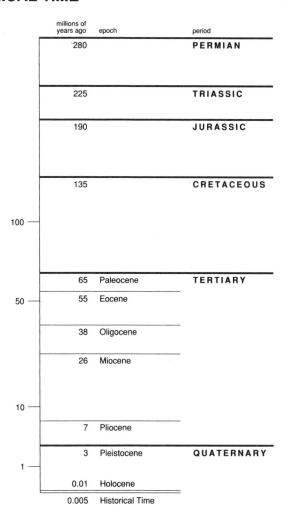

millions of years ago	epoch	period
280		**PERMIAN**
225		**TRIASSIC**
190		**JURASSIC**
135		**CRETACEOUS**
100		
65	Paleocene	**TERTIARY**
55	Eocene	
38	Oligocene	
26	Miocene	
10		
7	Pliocene	
3	Pleistocene	**QUATERNARY**
1		
0.01	Holocene	
0.005	Historical Time	

LENGTH

scales for comparison of metric and U.S. units of measurement

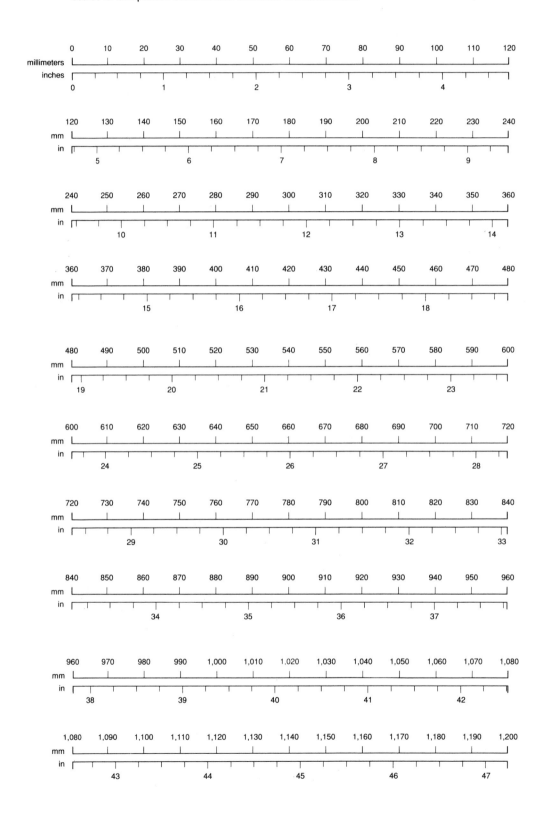

LENGTH

scales for comparison of metric and U.S. units of measurement

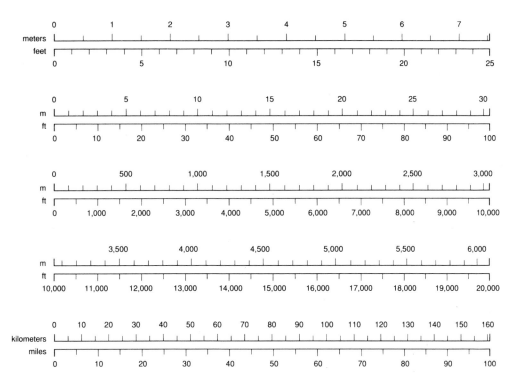

CONVERSION TABLES

Abbreviations

	U.S. to Metric		Metric to U.S.	
	to convert	*multiply by*	*to convert*	*multiply by*
LENGTH	in. to mm.	25.4	mm. to in.	0.039
	in. to cm.	2.54	cm. to in.	0.394
	ft. to m.	0.305	m. to ft.	3.281
	yd. to m.	0.914	m. to yd.	1.094
	mi. to km.	1.609	km. to mi.	0.621
AREA	sq. in. to sq. cm.	6.452	sq. cm. to sq. in.	0.155
	sq. ft. to sq. m.	0.093	sq. m. to sq. ft.	10.764
	sq. yd. to sq. m.	0.836	sq. m. to sq. yd.	1.196
	sq. mi. to ha.	258.999	ha. to sq. mi.	0.004
VOLUME	cu. in. to cc.	16.387	cc. to cu. in.	0.061
	cu. ft. to cu. m.	0.028	cu. m. to cu. ft.	35.315
	cu. yd. to cu. m.	0.765	cu. m. to cu. yd.	1.308
CAPACITY (liquid)	fl. oz. to liter	0.03	liter to fl. oz.	33.815
	qt. to liter	0.946	liter to qt.	1.057
	gal. to liter	3.785	liter to gal.	0.264
MASS (weight)	oz. avdp. to g.	28.35	g. to oz. avdp.	0.035
	lb. avdp. to kg.	0.454	kg. to lb. avdp.	2.205
	ton to t.	0.907	t. to ton	1.102
	l. t. to t.	1.016	t. to l. t.	0.984

Abbreviations

avdp.	avoirdupois
cc.	cubic centimeter(s)
cm.	centimeter(s)
cu.	cubic
ft.	foot, feet
g.	gram(s)
gal.	gallon(s)
ha.	hectare(s)
in.	inch(es)
kg.	kilogram(s)
lb.	pound(s)
l. t.	long ton(s)
m.	meter(s)
mi.	mile(s)
mm.	millimeter(s)
oz.	ounce(s)
qt.	quart(s)
sq.	square
t.	metric ton(s)
yd.	yard(s)

WEIGHT
scales for comparison of metric and U.S. units of measurement

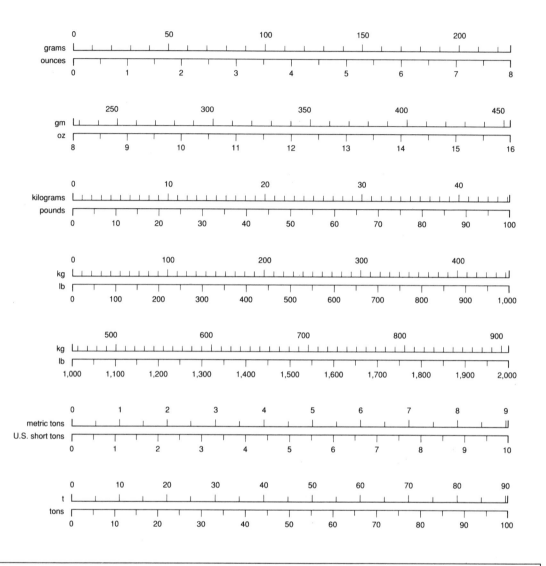

TEMPERATURE
scales for comparison of metric and U.S. units of measurement

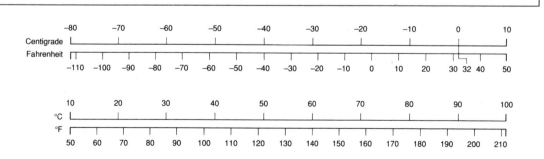

AREA

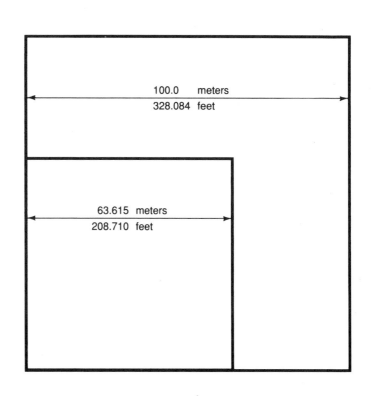

HECTARE
10,000.0	square meters
107,639.1	square feet

ACRE
4,046.86	square meters
43,560.0	square feet

100.0 meters
328.084 feet

63.615 meters
208.710 feet

CONVERSION TABLES

U.S. to Metric / Metric to U.S.

	to convert	multiply by	to convert	multiply by
LENGTH	in. to mm.	25.4	mm. to in.	0.039
	in. to cm.	2.54	cm. to in.	0.394
	ft. to m.	0.305	m. to ft.	3.281
	yd. to m.	0.914	m. to yd.	1.094
	mi. to km.	1.609	km. to mi.	0.621
AREA	sq. in. to sq. cm.	6.452	sq. cm. to sq. in.	0.155
	sq. ft. to sq. m.	0.093	sq. m. to sq. ft.	10.764
	sq. yd. to sq. m.	0.836	sq. m. to sq. yd.	1.196
	sq. mi. to ha.	258.999	ha. to sq. mi.	0.004
VOLUME	cu. in. to cc.	16.387	cc. to cu. in.	0.061
	cu. ft. to cu. m.	0.028	cu. m. to cu. ft.	35.315
	cu. yd. to cu. m.	0.765	cu. m. to cu. yd.	1.308
CAPACITY (liquid)	fl. oz. to liter	0.03	liter to fl. oz.	33.815
	qt. to liter	0.946	liter to qt.	1.057
	gal. to liter	3.785	liter to gal.	0.264
MASS (weight)	oz. avdp. to g.	28.35	g. to oz. avdp.	0.035
	lb. avdp. to kg.	0.454	kg. to lb. avdp.	2.205
	ton to t.	0.907	t. to ton	1.102
	l. t. to t.	1.016	t. to l. t.	0.984

Abbreviations

avdp.	avoirdupois
cc.	cubic centimeter(s)
cm.	centimeter(s)
cu.	cubic
ft.	foot, feet
g.	gram(s)
gal.	gallon(s)
ha.	hectare(s)
in.	inch(es)
kg.	kilogram(s)
lb.	pound(s)
l. t.	long ton(s)
m.	meter(s)
mi.	mile(s)
mm.	millimeter(s)
oz.	ounce(s)
qt.	quart(s)
sq.	square
t.	metric ton(s)
yd.	yard(s)

Literature Cited

A

Abbott, D. H., J. Barrett, and L. M. George. 1993. Comparative aspects of the social suppression of reproduction in female marmosets and tamarins. *In* Rylands (1993*b*), 152–63.

Aiello, L. C. 1993. The origin of the New World monkeys. *In* George, W., and R. Lavocat, eds., The Africa–South America connection, Clarendon Press, Oxford, 100–118.

Aimi, M., and A. Bakar. 1992. Taxonomy and distribution of *Presbytis melalophos* group in Sumatra, Indonesia. Primates 33: 191–206.

Albignac, R. 1987. Status of the aye-aye in Madagascar. Primate Conserv. 8:44–45.

Ali, R. 1985. An overview of the status and distribution of the lion-tailed macaque. Monogr. Primatol. 7:13–25.

Altmann, J., M. Warneke, and J. Ramer. 1988. Twinning among *Callimico goeldii*. Internatl. J. Primatol. 9:165–68.

Altmann, S. A., and J. Altmann. 1970. Baboon ecology. Univ. Chicago Press, vii + 220 pp.

Anadu, P. A., and J. F. Oates. 1988. The olive colobus monkey in Nigeria. Nigerian Field 53:31–34.

Anadu, P. M., L. K. Hiong, and J. B. Sale. 1994. Translocation of pocketed orangutans in Sabah. Oryx 28:263–68.

Andrews, P. 1978. Taxonomy and relationships of fossil apes. *In* Chivers and Joysey (1978), 43–56.

Andrews, P., and J. E. Cronin. 1982. The relationships of *Sivapithecus* and *Ramapithecus* and the evolution of the orang-utan. Nature 297:541–46.

Andrews, P., and C. P. Groves. 1976. Gibbons and brachiation. *In* Rumbaugh, D. M., ed., Gibbon and siamang, IV, S. Karger, Basel, 167–218.

Andrews, P., and L. Martin. 1987. Cladistic relationships of extant and fossil hominoids. J. Human Evol. 16:101–18.

Angst, W. 1975. Basic data and concepts on the social organization of *Macaca fascicularis*. Primate Behav. 4:325–88.

Antinucci, F., and E. Visalberghi. 1986. Tool use in *Cebus apella*: a case study. Internatl. J. Primatol. 7:351–63.

Aquino, R., and F. Encarnación. 1986. Population structure of *Aotus nancymai* (Cebidae: Primates) in Peruvian Amazon lowland forest. Amer. J. Primatol. 11:1–7.

———. 1994. Owl monkey populations in Latin America: field work and conservation. *In* Baer, Weller, and Kakoma (1994), 59–95.

Ayres, J. M. 1985. On a new species of squirrel monkey, genus *Saimiri*, from Brazilian Amazona (Primates: Cebidae). Papéis Avulsos de Zoologia, Mus. Zool. Univ. Sao Paulo, 36:147–64.

Ayres, J. M., and A. D. Johns. 1987. Conservation of white uacaris in Amazonian várzea. Oryx 21:74–80.

Ayres, J. M., and J. L. Nessiman. 1982. Evidence for insectivory in *Chiropotes satanas*. Primates 23:458–59.

B

Badrian, A., and N. Badrian. 1984. Social organization of *Pan paniscus* in the Lomako Forest, Zaire. *In* Susman (1984*b*), 325–46.

Badrian, N., and R. K. Malenky. 1984. Feeding ecology of *Pan paniscus* in the Lomako Forest, Zaire. *In* Susman (1984*b*), 275–99.

Baer, J. F. 1994. Husbandry and medical management of the owl monkey. *In* Baer, Weller, and Kakoma (1994), 133–65.

Baldwin, J. D. 1970. Reproductive synchronization in squirrel monkeys *(Saimiri)*. Primates 11:317–26.

Baldwin, J. D., and J. I. Baldwin. 1971. Squirrel monkeys *(Saimiri)* in natural habitats in Panama, Colombia, Brazil, and Peru. Primates 12:45–61.

———. 1972. Population density and use of space in howling monkeys *(Alouatta villosa)*

in south-western Panama. Primates 13: 371–79.

———. 1976. Primate populations in Chiriqui, Panama. *In* Thorington and Heltne (1976), 20–31.

Barnett, A., and A. C. Da Cunha. 1991. The golden-backed uacari on the upper Rio Negro, Brazil. Oryx 25:80–88.

Beard, K. C., and B. Wang. 1991. Phylogenetic and biogeographic significance of the tarsiiform primate *Asiomomys changbaicus* from the Eocene of Jilin Province, People's Republic of China. Amer. J. Phys. Anthropol. 85:159–66.

Beattie, J. C., A. T. C. Feistner, N. M. O. Adams, P. Barker, and J. B. Carroll. 1992. First captive breeding of the aye-aye *Daubentonia madagascariensis*. Dodo 28:23–30.

Beck, B. B., D. Anderson, J. Ogden, B. Rettberg, C. Brejla, R. Scola, and M. Warneke. 1982. Breeding the Goeldi's monkey *Callimico goeldii* at Brookfield Zoo, Chicago. Internatl. Zoo Yearbook 22:106–14.

Beck, B. B., D. G. Kleiman, J. M. Dietz, I. Castro, C. Carvalho, A. Martins, and B. Rettberg-Beck. 1991. Losses and reproduction in reintroduced golden lion tamarins *Leontopithecus rosalia*. Dodo 27:50–61.

Bennett, E. L. 1987. Big noses of Borneo. Anim. Kingdom 90(2):9–15.

Bennett, E. L., and A. G. Davies. 1994. The ecology of Asian colobines. *In* Davies and Oates (1994), 129–71.

Bennett, E. L., and A. C. Sebastian. 1988. Social organization and ecology of proboscis monkeys *(Nasalis larvatus)* in mixed coastal forest in Sarawak. Internatl. J. Primatol. 9:233–55.

Bernstein, I. S., and T. P. Gordon. 1980. Mixed taxa introductions, hybrids, and macaque systematics. *In* Lindburg (1980), 125–47.

Biquand, S., V. Biquand-Guyot, A. Bourg, and J.-P. Gautier. 1992*a*. The distribution of *Papio hamadryas* in Saudi Arabia: ecological correlates and human influence. Internatl. J. Primatol. 13:223–43.

———. 1992b. Group composition in wild and commensal hamadryas baboons: a comparative study in Saudi Arabia. Internatl. J. Primatol. 13:533–43.

Bleisch, W. 1991. Preliminary comments on the Guizhou snub-nosed monkey (Rhinopithecus brelichi). Asian Primates 1(3):4.

Bleisch, W., and Chen Nan. 1990. Conservation of the black-crested gibbon in China. Oryx 24:147–56.

Blom, A., M. P. T. Alers, A. T. C. Feistner, R. F. W. Barnes, and K. L. Barnes. 1992. Primates in Gabon—current status and distribution. Oryx 26:223–34.

Bodini, R., and R. Pérez-Hernández. 1987. Distribution of the species and subspecies of cebids in Venezuela. Fieldiana Zool., n.s., 39:231–44.

Boesch, C., and H. Boesch. 1990. Tool use and tool making in wild chimpanzees. Folia Primatol. 54:86–99.

Bogart, M. H., R. W. Cooper, and K. Benirschke. 1977. Reproductive studies of black and ruffed lemurs Lemur macaco macaco and L. variegatus. Internatl. Zoo Yearbook 17:177–82.

Boinski, S. 1985. Status of the squirrel monkey Saimiri oerstedi in Costa Rica. Primate Conserv. 6:15–16.

Boskoff, K. J. 1977. Aspects of reproduction in ruffed lemurs (Lemur variegatus). Folia Primatol. 28:241–50.

Boubli, J. P. 1993. Southern expansion of the geographical distribution of Cacajao melanocephalus melanocephalus. Internatl. J. Primatol. 14:933–37.

Bouchardet da Fonseca, G. A., and T. E. Lacher, Jr. 1984. Exudate-feeding by Callithrix jacchus penicillata in semideciduous woodland (cerradao) in central Brazil. Primates 25:441–50.

Bourliere, F., C. Hunkeler, and M. Bertrand. 1970. Ecology and behavior of Lowe's guenon (Cercopithecus campbelli lowei) in the Ivory Coast. In Napier and Napier (1970), 297–350.

Bowdoin, J., R. J. Wiese, K. Willis, and M. Hutchins, eds. 1994. AZA Annual Report on Conservation and Science, 1993–94. American Association of Zoological Parks and Aquariums, Bethesda, Maryland, 375 pp.

Brandon-Jones, D. 1984. Colobus and leaf monkeys. In Macdonald (1984), 398–409.

———. 1993. The taxonomic affinities of the Mentawai Islands sureli, Presbytis potenziani (Bonaparte, 1856) (Mammalia: Primata: Cercopithecidae). Raffles Bull. Zool. 41:331–57.

Brennan, E. J. 1985. De Brazza's monkeys (Cercopithecus neglectus) in Kenya: census, distribution, and conservation. Amer. J. Primatol. 8:269–77.

Brockelman, W. Y., and D. J. Chivers. 1984. Gibbon conservation: looking to the future. In Preuschoft et al. (1984), 3–12.

Brockelman, W. Y., and S. P. Gittins. 1984. Natural hybridization in the Hylobates lar species group: implications for speciation in gibbons. In Preuschoft et al. (1984), 498–532.

Brumback, R. A. 1974. A third species of the owl monkey (Aotus). J. Hered. 65:321–23.

Buchanan-Smith, H. M. 1991. A field study on the red-bellied tamarin, Saguinus l. labiatus, in Bolivia. Internatl. J. Primatol. 12:259–76.

Budnitz, N., and K. Dainis. 1975. Lemur catta: ecology and behavior. In Tattersall and Sussman (1975), 219–35.

Buettner-Janusch, J. 1966. Origins of man; physical anthropology. John Wiley & Sons, New York, 674 pp.

Buettner-Janusch, J., and I. Tattersall. 1985. An annotated catalogue of Malagasy primates (families Lemuridae, Indriidae, Daubentoniidae, Megaladapidae, Cheirogaleidae) in the collections of the American Museum of Natural History. Amer. Mus. Novit., no. 2834, 45 pp.

Butchart, S. H. M., R. Barnes, C. W. N. Davies, M. Fernandez, and N. Seddon. 1995. Threatened mammals of the Cordillera de Colán, Peru. Oryx 29:275–81.

Butynski, T. M. 1988. Guenon birth seasons and correlates with rainfall and food. In Gautier-Hion et al. (1988), 284–322.

Butynski, T. M., and G. Mwangi. 1995. Census of Kenya's endangered red colobus and crested mangabey. African Primates 1(1): 8–10.

Butynski, T. M., S. E. Werikhe, and J. Kalina. 1990. Status, distribution, and conservation of the mountain gorilla in the Gorilla Game Reserve, Uganda. Primate Conserv. 11:31–41.

C

Cabrera, A. 1957, 1961. Catalogo de los mamíferos de América del Sur. Rev. Mus. Argentino Cien. Nat. "Bernardo Rivadavia," 4:1–732.

Caine, N. G. 1993. Flexibility and cooperation as unifying themes in Saguinus social organization and behaviour: the role of predation pressures. In Rylands (1993b), 200–219.

Caldecott, J. O. 1986. Mating patterns, societies, and the ecogeography of macaques. Anim. Behav. 34:208–20.

Cameron, R., C. Wiltshire, C. Foley, N. Dougherty, X. Aramayo, and L. Rea. 1989. Goeldi's monkey and other primates in northern Bolivia. Primate Conserv. 10:62–70.

Camperio Ciani, A., R. Stanyon, W. Scheffrahn, and B. Sampurno. 1989. Evidence of gene flow between Sulawesi macaques. Amer. J. Primatol. 17:257–70.

Canh, L. X., and B. Campbell. 1994. Population status of Trachypithecus francoisi poliocephalus in Cat Ba National Park. Asian Primates 3(3–4):16–20.

Carman, M. 1979. The gestation and rearing periods of the mandrill Mandrillus sphinx. Internatl. Zoo Yearbook 19:159–60.

Caro, T. M. 1976. Observations on the ranging behaviour and daily activity of lone silverback mountain gorillas (Gorilla gorilla beringei). Anim. Behav. 24:889–97.

Carpenter, C. R. 1965. The howlers of Barro Colorado Island. In De Vore (1965), 250–91.

Carroll, J. B., and J. C. Beattie. 1993. Maintenance and breeding of the aye-aye Daubentonia madagascariensis at the Jersey Wildlife Preservation Trust. Dodo 29:45–54.

Casimir, M. J. 1975. Some data on the systematic position of the eastern gorilla population of the Mt. Kahuzi region (République du Zaire). Z. Morph. Anthrop. 66:188–201.

Casimir, M. J., and E. Butenandt. 1973. Migration and core area shifting in relation to some ecological factors in a mountain gorilla group (Gorilla gorilla beringei) in the Mt. Kahuzi region (République du Zaire). Z. Tierpsychol. 33:514–22.

Castro, R., and P. Soini. 1977. Field studies on Saguinus mystax and other callitrichids in Amazonian Peru. In Kleiman (1977a), 73–78.

Chalmers, N. R. 1968a. Group composition, ecology, and daily activities of free living mangabeys in Uganda. Folia Primatol. 8: 247–62.

———. 1968b. The social behaviour of free living mangabeys in Uganda. Folia Primatol. 8:263–81.

———. 1973. Differences in behavior between some arboreal and terrestrial species of African monkeys. In Michael and Crook (1973), 69–100.

Charles-Dominique, P. 1974. Aggression and territoriality in nocturnal prosimians. In Holloway (1974), 31–48.

———. 1977. Ecology and behaviour of nocturnal primates: prosimians of Equatorial West Africa. Duckworth, London, x + 277 pp.

Charles-Dominique, P., and S. K. Bearder. 1979. Field studies of lorisid behavior: methodological aspects. In Doyle and Martin (1979), 567–629.

Charles-Dominique, P., and J.-J. Petter. 1980. Ecology and social life of Phaner furcifer. In Charles-Dominique et al. (1980), 75–95.

Chasen, F. N. 1940. A handlist of Malaysian mammals. Bull. Raffles Mus., Singapore, no. 15, xx + 209 pp.

Chen Fuguan, Min Zhilan, Luo Shiyou, and Xie Wenzhi. 1983. An observation on the behaviour and some ecological habits of the golden monkey *(Rhinopithecus roxellanae)* in Qing Mountains. Acta Theriol. Sinica 3:141–46.

Cheverud, J. M., S. C. Jacobs, and A. J. Moore. 1993. Genetic differences among subspecies of the saddle-back tamarin *(Saguinus fuscicollis):* evidence from hybrids. Amer. J. Primatol. 31:23–39.

Cheverud, J. M., and A. J. Moore. 1990. Subspecific morphological variation in the saddle-back tamarin *(Saguinus fuscicollis).* Amer. J. Primatol. 21:1–15.

Chiarelli, A. B. 1972. Taxonomic atlas of living primates. Academic Press, London, vii + 363 pp.

Chism, J., and T. E. Rowell. 1988. The natural history of patas monkeys. *In* Gautier-Hion et al. (1988), 412–38.

Chivers, D. J. 1972. The siamang and the gibbon in the Malay Peninsula. *In* Rumbaugh (1972), 103–35.

———. 1973. An introduction to the socioecology of Malayan forest primates. *In* Michael and Crook (1973), 101–46.

———. 1974. The siamang in Malaya. Contrib. Primatol., no. 4, xiii + 335 pp.

———. 1977. The lesser apes. *In* Rainier III and Bourne (1977), 539–98.

———. 1984. Feeding and ranging in gibbons: a summary. *In* Preuschoft et al. (1984), 267–81.

———. 1986. Southeast Asian primates. *In* Benirschke (1986), 127–51.

Chivers, D. J., and S. P. Gittins. 1978. Diagnostic features of gibbon species. Internatl. Zoo Yearbook 18:157–64.

Christen, A., and T. Geissmann. 1994. A primate survey in northern Bolivia, with special reference to Goeldi's monkey, *Callimico goeldii.* Internatl. J. Primatol. 15:239–74.

Ciochon, R. L. 1988. *Gigantopithecus:* the king of all apes. Anim. Kingdom 91(2):32–37.

Ciochon, R. L., and A. B. Chiarelli, eds. 1980. Evolutionary biology of the New World monkeys and continental drift. Plenum Press, New York, xvi + 528 pp.

Coimbra-Filho, A. F., and R. A. Mittermeier. 1973a. Distribution and ecology of the genus *Leontopithecus* Lesson, 1840 in Brazil. Primates 14:47–66.

———. 1973b. New data on the taxonomy of the Brazilian marmosets of the genus *Callithrix* Erxleben, 1777. Folia Primatol. 20:241–64.

———. 1977a. Conservation of the Brazilian lion tamarins *(Leontopithecus rosalia).* In Rainier III and Bourne (1977), 59–94.

———. 1977b. Tree-gouging, exudate eating, and the "short-tusked" condition in *Callithrix* and *Cebuella. In* Kleiman (1977a), 105–15.

Colquhoun, I. C. 1993. The socioecology of *Eulemur macaco:* a preliminary report. *In* Kappeler and Ganzhorn (1993), 11–23.

Colyn, M., A. Gautier-Hion, and D. Thys van den Audenaerde. 1991. *Cercopithecus dryas* Schwarz 1932 and *C. salongo* Thys van den Audenaerde 1977 are the same species with an age-related coat pattern. Folia Primatol. 56:167–70.

Constable, I. D., J. I. Pollock, J. Ratsirarson, and H. Simons. 1985. Sightings of aye-ayes and red ruffed lemurs on Nosy Mangabe and the Masoala Peninsula. Primate Conserv. 5:59–62.

Corbet, G. B. 1978. The mammals of the Palaearctic Region: a taxonomic review. British Mus. (Nat. Hist.), London, 314 pp.

Corbet, G. B., and J. E. Hill. 1991. A world list of mammalian species. Natural History Museum Publ., London, and Oxford Univ. Press, viii + 243 pp.

———. 1992. The mammals of the Indomalayan region: a systematic review. Oxford Univ. Press, viii + 488 pp.

Cords, M. 1988. Mating systems of forest guenons: a preliminary review. *In* Gautier-Hion et al. (1988), 323–39.

Costello, R. K., C. Dickinson, A. L. Rosenberger, S. Boinski, and F. S. Szalay. 1993. Squirrel monkey (genus *Saimiri*) taxonomy: a multidisciplinary study of the biology of species. *In* Kimbel and Martin (1993), 177–210.

Courtenay, J., C. P. Groves, and P. Andrews. 1988. Inter- or intra-island variation? An assessment of the differences between Bornean and Sumatran orang-utans. *In* Schwartz (1988b), 19–29.

Cowgill, U. M., S. J. States, and K. J. States. 1989. A twenty-five-year chronicle of a group of captive nocturnal prosimians *(Perodicticus potto).* Mamm. Rev. 19:83–89.

Cowlishaw, G. 1992. Song function in gibbons. Behaviour 121:131–53.

Crockett, C. M., and J. F. Eisenberg. 1987. Howlers: variations in group size and demography. *In* Smuts, B. B., D. L. Cheney, R. M. Seyfarth, R. W. Wrangham, and T. T. Struhsaker, eds., Primate societies, Univ. Chicago Press, 54–68.

Crockett, C. M., and R. Rudran. 1987a. Red howler monkey birth data I: seasonal variation. Amer. J. Primatol. 13:347–68.

———. 1987b. Red howler monkey birth data II: interannual, habitat, and sex comparisons. Amer. J. Primatol. 13:369–84.

Crockett, C. M., and W. L. Wilson. 1980. The ecological separation of *Macaca nemestrina* and *M. fascicularis* in Sumatra. *In* Lindburg (1980), 148–81.

Cronin, J. E., R. Cann, and V. M. Sarich. 1980. Molecular evolution and systematics of the genus *Macaca. In* Lindburg (1980), 31–51.

Crovella, S., D. Montagnon, and Y. Rumpler. 1993. Highly repeated DNA analysis and systematics of the Lemuridae, a family of Malagasy prosimians. Primates 34:61–69.

Culotta, E. 1995. Many suspects to blame in Madagascar extinctions. Science 268:1568–69.

D

Daschbach, N. J., M. W. Schein, and D. E. Haines. 1981. Vocalizations of the slow loris, *Nycticebus coucang* (Primates, Lorisidae). Internatl. J. Primatol. 2:71–80.

Davidge, C. 1978. Ecology of baboons *(Papio ursinus)* at Cape Point. Zool. Afr. 13:329–50.

Davies, A. G. 1994. Colobine populations. *In* Davies and Oates (1994), 285–310.

Dawson, G. A. 1977. Composition and stability of social groups of the tamarin, *Saguinus oedipus geoffroyi,* in Panama: ecological and behavioral considerations. *In* Kleiman (1977a), 23–37.

Deag, J. M. 1977. The status of the Barbary macaque *Macaca sylvanus* in captivity and factors influencing its distribution in the wild. *In* Rainier III and Bourne (1977), 267–87.

De Bois, H. 1994. Progress report on the captive population of golden-headed lion tamarins, *Leontopithecus chrysomelas*—May 1994. Neotropical Primates 2(suppl.):28–29.

Decker, B. S. 1994. Endangered primates in the Selous Game Reserve and an imminent threat to their habitat. Oryx 28:183–90.

Defler, T. R. 1979a. On the ecology and behavior of *Cebus albifrons* in eastern Colombia: I. Ecology. Primates 20:475–90.

———. 1979b. On the ecology and behavior of *Cebus albifrons* in eastern Colombia: II. Behavior. Primates 20:491–502.

———. 1982. A comparison of intergroup behavior in *Cebus albifrons* and *C. apella.* Primates 23:385–92.

Delson, E. 1980. Fossil macaques, phyletic relationships and a scenario of deployment. *In* Lindburg (1980), 10–30.

———. 1994. Evolutionary history of the colobine monkeys in palaeoenvironmental perspective. *In* Davies and Oates (1994), 11–43.

Dene, H., M. Goodman, and W. Prychodko. 1980. Immunodiffusion systematics of the Primates. IV: Lemuriformes. Mammalia 44:211–23.

Denham, W. W. 1982. History of green monkeys in the West Indies. Part II. Population dynamics of Barbadian monkeys. J. Barbados Mus. and Hist. Soc. 36:353–71.

De Vore, I., and K. R. L. Hall. 1965. Baboon ecology. *In* De Vore (1965), 20–52.

DeVos, A., and A. Omar. 1971. Territories and movements of Sykes monkeys (*Cercopithecus mitis kolbi* Neuman) in Kenya. Folia Primatol. 16:196–205.

Dewar, R. E. 1984. Extinctions in Madagascar: the loss of the subfossil fauna. *In* Martin and Klein (1984), 574–93.

DiBitetti, M. S., G. Placci, A. D. Brown, and D. I. Rode. 1994. Conservation and population status of the brown howling monkey (*Alouatta fusca clamitans*) in Argentina. Neotropical Primates 2(4):1–4.

Dietz, J. M., and A. J. Baker. 1993. Polygyny and female reproductive success in golden lion tamarins *Leontopithecus rosalia*. Anim. Behav. 46:1067–78.

Dietz, J. M., N. F. De Sousa, and J. R. O. Da Silva. 1994. Population structure and territory size in golden-headed lion tamarins, *Leontopithecus chrysomelas*. Neotropical Primates 2(suppl.):21–23.

Digby, L. J., and C. E. Barreto. 1994. Social organization in a wild population of *Callithrix jacchus*. I. Group composition and dynamics. Folia Primatol. 61:123–34.

Digby, L. J., and S. F. Ferrari. 1994. Multiple breeding females in free-ranging groups of *Callithrix jacchus*. Internatl. J. Primatol. 15:389–97.

Disotell, T. R. 1994. Generic level relationships of the Papionini (Cercopithecoidea). Amer. J. Phys. Anthropol. 94:47–57.

Dittus, W. 1977. The socioecological basis for the conservation of the toque monkey (*Macaca sinica*) of Sri Lanka (Ceylon). *In* Rainier III and Bourne (1977), 237–65.

Dixson, A. F. 1977. Observations on the displays, menstrual cycles, and sexual behaviour of the "black ape" of Celebes (*Macaca nigra*). J. Zool. 182:63–84.

———. 1981. The natural history of the gorilla. Columbia Univ. Press, New York, xviii + 202 pp.

———. 1994. Reproductive biology of the owl monkey. *In* Baer, Weller, and Kakoma (1994), 113–32.

Dixson, A. F., T. Bossi, and E. J. Wickings. 1993. Male dominance and genetically determined reproductive success in the mandrill (*Mandrillus sphinx*). Primates 34:525–32.

Dixson, A. F., D. M. Scruton, and J. Herbert. 1975. Behaviour of the talapoin monkey (*Miopithecus talapoin*) studied in groups, in the laboratory. J. Zool. 176:177–210.

Doyle, G. A. 1979. Development of behavior in prosimians with special reference to the lesser bushbaby, *Galago senegalensis moholi*. *In* Doyle and Martin (1979), 157–206.

Doyle, G. A., and S. K. Bearder. 1977. The galagines of South Africa. *In* Rainier III and Bourne (1977), 1–35.

Du Mond, F. V. 1968. The squirrel monkey in a seminatural environment. *In* Rosenblum and Cooper (1968), 87–145.

Dunbar, R. I. M. 1974. Observations on the ecology and social organization of the green monkey, *Cercopithecus sabaeas*, in Senegal. Primates 15:341–50.

———. 1977a. Feeding ecology of gelada baboons: a preliminary report. *In* Clutton-Brock (1977), 251–73.

———. 1977b. The gelada baboon: status and conservation. *In* Rainier III and Bourne (1977), 363–83.

———. 1983a. Structure of gelada baboon reproductive units. II. Social relationships between reproductive females. Anim. Behav. 31:556–64.

———. 1983b. Structure of gelada baboon reproductive units. III. The male's relationship with his females. Anim. Behav. 31:565–75.

———. 1993a. Conservation status of the gelada. *In* Jablonski (1993b), 527–31.

———. 1993b. Social organization of the gelada. *In* Jablonski (1993b), 425–39.

Dunbar, R. I. M., and P. Dunbar. 1974a. On hybridization between *Theropithecus gelada* and *Papio anubis* in the wild. J. Human Evol. 3:187–92.

———. 1975. Social dynamics of gelada baboons. Contrib. Primatol., no. 6, vii + 157 pp.

Durham, N. M. 1971. Effects of altitude differences on group organization of wild black spider monkeys (*Ateles paniscus*). Proc. 3rd Internatl. Congr. Primatol. 3:32–40.

Dutrillaux, B., A.-M. Dutrillaux, M. Lombard, J.-P. Gautier, R. Cooper, F. Moysan, and J. M. Lernould. 1988. The karyotype of *Cercopithecus solatus* Harrison 1988, a new species belonging to *C. lhoesti*, and its phylogenetic relationships with other guenons. J. Zool. 215:611–17.

E

Eames, J. C., and C. R. Robson. 1993. Threatened primates in southern Vietnam. Oryx 27:146–54.

Eisenberg, J. F. 1977. Comparative ecology and reproduction of New World monkeys. *In* Kleiman (1977a), 13–22.

———. 1989. Mammals of the neotropics: the northern neotropics. Univ. Chicago Press, x + 449 pp.

Ellefson, J. O. 1974. A natural history of white-handed gibbons in the Malayan peninsula. *In* Rumbaugh, D. M., ed., Gibbon and siamang, III, S. Karger, Basel, 1–136.

Ellerman, J. R., and T. C. S. Morrison-Scott. 1966. Checklist of Palaearctic and Indian mammals. British Mus. (Nat. Hist.), London, 810 pp.

Epple, G., A. M. Belcher, I. Küderling, U. Zeller, L. Scolnick, K. L. Greenfield, and A. B. Smith, III. 1993. Making sense out of scents: species differences in scent glands, scent-marking behaviour, and scent-mark composition in the Callitrichidae. *In* Rylands (1993b), 123–51.

Erickson, C. J. 1991. Percussive foraging in the aye-aye, *Daubentonia madagascariensis*. Anim. Behav. 41:793–801.

———. 1994. Tap-scanning and extractive foraging in aye-ayes, *Daubentonia madagascariensis*. Folia Primatol. 62:125–35.

Estrada, A., and R. Coates-Estrada. 1984. Some observations on the present distribution and conservation of *Alouatta* and *Ateles* in southern Mexico. Amer. J. Primatol. 7:133–37.

———. 1988. Tropical rain forest conversion and perspectives in the conservation of wild primates (*Alouatta* and *Ateles*) in Mexico. Amer. J. Primatol. 14:315–27.

Eudey, A. A. 1987. Action plan for Asian primate conservation: 1987–1991. Internatl. Union Conserv. Nat., Gland, Switzerland, 65 pp.

———. 1991. IUCN resolution on orang utans. Asian Primates 1(1):1–3.

F

Fa, J. E. 1989. The genus *Macaca*: a review of taxonomy and evolution. Mamm. Rev. 19:45–81.

Fa, J. E., D. M. Taub, N. Menard, and P. J. Stewart. 1984. The distribution and current status of the Barbary macaque in North Africa. *In* Fa (1984), 79–111.

Fay, J. M., and M. Agnagna. 1992. Census of gorillas in northern Republic of Congo. Amer. J. Primatol. 27:275–84.

Fedigan, L., and L. M. Fedigan. 1988. *Cercopithecus aethiops*: a review of field studies. *In* Gautier-Hion et al. (1988), 389–411.

Feistner, A. T. C. 1992. Aspects of reproduction of female mandrills. Internatl. Zoo Yearbook 31:170–78.

Feistner, A. T. C., and M. Rakotoarinosy. 1993. Conservation of gentle lemur *Hapalemur griseus alaotrensis* at Lac Alaotra, Madagascar: local knowledge. Dodo 29:54–65.

Ferguson, W. W. 1992. "*Australopithecus afarensis*": a composite species. Primates 33:273–79.

Ferrari, S. F., and M. A. Lopes. 1992. A note on the behaviour of the weasel *Mustela* cf. *africana* (Carnivora, Mustelidae), from Amazonas, Brazil. Mammalia 56:482–83.

Ferrari, S. F., and S. L. Mendes. 1991. Buffy-headed marmosets ten years on. Oryx 25:105–9.

Ferrari, S. F., and H. L. Queiroz. 1994. Two new Brazilian primates discovered, endangered. Oryx 26:31–36.

Fittinghoff, N. A., Jr., and D. G. Lindburg. 1980. Riverine refuging in east Bornean *Macaca fascicularis. In* Lindburg (1980), 182–214.

Flannery, T. F. 1995. Mammals of the southwest Pacific and Moluccan Islands. Comstock/ Cornell Univ. Press, Ithaca, 464 pp.

Fleagle, J. G. 1984. Are there any fossil gibbons? *In* Preuschoft et al. (1984), 431–47.

Foerg, R. 1982. Reproductive behavior in *Varecia variegata.* Folia Primatol. 38:108–21.

Fogden, M. P. L. 1974. A preliminary field study of the western tarsier, *Tarsius bancanus* Horsfield. *In* Martin, Doyle, and Walker (1974), 151–65.

Fontaine, R., and F. V. Du Mond. 1977. The red ouakari in a seminatural environment: potentials for propagation and study. *In* Rainier III and Bourne (1977), 167–236.

Fooden, J. 1963. A revision of the woolly monkeys (genus *Lagothrix*). J. Mamm. 44:213–47.

———. 1969. Taxonomy and evolution of the monkeys of Celebes (Primates: Cercopithecidae). Bibl. Primatol. 10:1–148.

———. 1971. Report on primates collected in western Thailand January–April, 1967. Fieldiana Zool. 59:1–62.

———. 1975. Taxonomy and evolution of liontail and pigtail macaques (Primates: Cercopithecidae). Fieldiana Zool. 67:1–169.

———. 1976a. Primates obtained in peninsular Thailand June–July, 1973, with notes on the distribution of continental Southeast Asian leaf-monkeys *(Presbytis).* Primates 17:95–118.

———. 1976b. Provisional classification and key to living species of macaques (Primates: *Macaca*). Folia Primatol. 25:225–36.

———. 1979. Taxonomy and evolution of the *sinica* group of macaques: I. Species and subspecies accounts of *Macaca sinica.* Primates 10:109–40.

———. 1982. Ecogeographic segregation of macaque species. Primates 23:574–79.

———. 1988. Taxonomy and evolution of the *sinica* group of macaques: 6. Interspecific comparisons and synthesis. Fieldiana Zool., no. 45, vi + 44 pp.

———. 1989. Classification, distribution, and ecology of Indian macaques. *In* Seth, P. K., and S. Seth, eds., Perspectives in primate biology, Today and Tomorrow's Printers and Publishers, New Delhi, vol. 2, 33–46.

———. 1990. The bear macaque, *Macaca arctoides:* a systematic review. J. Human Evol. 19:607–86.

———. 1991a. Eastern limit of distribution of the slow loris, *Nycticebus coucang.* Internatl. J. Primatol. 12:287–90.

———. 1991b. Systematic review of Philippine macaques (Primates, Cercopithecidae: *Macaca fascicularis* subspp.). Fieldiana Zool., n.s., no. 64, iv + 44 pp.

Fooden, J., and R. J. Izor. 1983. Growth curves, dental emergence norms, and supplementary morphological observations in known-age captive orangutans. Amer. J. Primatol. 5:285–301.

Fooden, J., A. Mahabal, and S. S. Saha. 1981. Redefinition of rhesus macaque—bonnet macaque boundary in peninsular India (Primates: *Macaca mulatta, M. radiata*). J. Bombay Nat. Hist. Soc. 78:463–74.

Fooden, J., Quan Guoqiang, and Luo Yining. 1987. Gibbon distribution in China. Acta Theriol. Sinica 7:161–67.

Fooden, J., Quan Guoqiang, Wang Zongren, and Wang Yingxiang. 1985. The stumptail macaques of China. Amer. J. Primatol. 8:11–30.

Fooden, J., Quan Guoqiang, Zhang Yongzu, Wu Mingchuan, and Liang Monyuan. 1994. Southward extension of the range of *Macaca thibetana.* Internatl. J. Primatol. 15:623–27.

Ford, S. M. 1986a. Subfossil platyrrhine tibia (Primates: Callitrichidae) from Hispaniola: a possible further example of island gigantism. Amer. J. Phys. Anthropol. 70:47–62.

———. 1986b. Systematics of the New World monkeys. *In* Swindler and Erwin (1986), 73–135.

———. 1990. Platyrrhine evolution in the West Indies. J. Human Evol. 19:237–54.

———. 1994. Taxonomy and distribution of the owl monkey. *In* Baer, Weller, and Kakoma (1994), 1–57.

Fossey, D. 1972. Vocalizations of the mountain gorilla *(Gorilla gorilla beringei).* Anim. Behav. 20:36–53.

Fossey, D., and A. H. Harcourt. 1977. Feeding ecology of free-ranging mountain gorilla *(Gorilla gorilla beringei). In* Clutton-Brock (1977), 415–47.

Freese, C. 1976. Censusing *Alouatta palliata, Ateles geoffroyi,* and *Cebus capucinus* in the Costa Rican dry forest. *In* Thorington and Heltne (1976), 4–19.

Freese, C., M. A. Freese, and N. Castro R. 1977. The status of callitrichids in Peru. *In* Kleiman (1977a), 121–30.

Froehlich, J. W., and P. H. Froehlich. 1987. The status of Panama's endemic howling monkeys. Primate Conserv. 8:58–62.

Froehlich, J. W., R. W. Thorington, Jr., and J. S. Otis. 1981. The demography of howler monkeys *(Alouatta palliata)* on Barro Colorado Island, Panama. Internatl. J. Primatol. 2:207–36.

G

Gabow, S. A. 1975. Behavioral stabilization of a baboon hybrid zone. Amer. Nat. 109:701–12.

Galat, G., and A. Galat-Luong. 1976. La colonisation de la mangrove par *Cercopithecus aethiops sabaeus* au Senegal. Terre Vie 30:3–30.

———. 1977. Démographie et alimentaire d'une troupe de *Cercopithecus aethiops sabaeus* en habitat marginal au nord Senegal. Terre Vie 31:557–77.

———. 1978. Diet of green monkeys in Senegal. *In* Chivers and Herbert (1978), 257–58.

Galat-Luong, A. 1975. Notes préliminaires sur l'écologie de *Cercopithecus ayscanius schmidti* dans les environs de Bangui (R. C. A.). Terre Vie 29:288–97.

Galdikas, B. M. F. 1981. Orangutan reproduction in the wild. *In* Graham, C. E., ed., Reproductive biology of the great apes, Academic Press, New York, 281–300.

———. 1982. Orangutan tool-use at Tanjung Puting Reserve, central Indonesian Borneo (Kalimantan Tengah). J. Human Evol. 10:19–33.

———. 1985a. Orangutan sociality at Tanjung Puting. Amer. J. Primatol. 9:101–19.

———. 1985b. Sub-adult male orangutan sociality and reproductive behavior at Tanjung Puting. Amer. J. Primatol. 8:87–99.

———. 1988. Orangutan diet, range, and activity at Tanjung Puting, central Borneo. Internatl. J. Primatol. 9:1–35.

Ganzhorn, J. U., J. P. Abraham, and M. Razanahoera-Rakotomalala. 1985. Some aspects of the natural history and food selection of *Avahi laniger.* Primates 26:452–63.

Ganzhorn, J. U., and J. Rabesoa. 1986. The aye-aye *(Daubentonia madagascariensis)* found in the eastern rainforest of Madagascar. Folia Primatol. 46:125–26.

Garber, P. A. 1993. Feeding ecology and behaviour of the genus *Saguinus. In* Rylands (1993b), 273–95.

Garber, P. A., L. Moya, and C. Malaga. 1984. A preliminary field study of moustached tamarin monkey *(Saguinus mystax)* in northeastern Peru: questions concerned with the evolution of a communal breeding system. Folia Primatol. 42:17–32.

Garber, P. A., and M. F. Teaford. 1986. Body weights in mixed species troops of *Saguinus mystax mystax* and *Saguinus fuscicollis nigrifrons* in Amazonian Peru. Amer. J. Phys. Anthropol. 71:331–36.

Gartlan, J. S. 1970. Preliminary notes on the ecology and behavior of the drill, *Mandrillus leucophaeus* Ritgen, 1824. *In* Napier and Napier (1970), 445–80.

Gartlan, J. S., and C. K. Brain. 1968. Ecology and social variability in *Cercopithecus aethiops* and *C. mitis. In* Jay (1968), 253–92.

Gartlan, J. S., and T. T. Struhsaker. 1972. Polyspecific associations and niche separation of rain-forest anthropoids in Cameroon, West Africa. J. Zool. 168:221–66.

Gautier, J.-P. 1985. Quelques caractéristiques écologiques du singe des marais: *Allenopithecus nigroviridis* Lang 1923. Rev. Ecol. 40:331–42.

———. 1988. Interspecific affinities among guenons as deduced from vocalizations. *In* Gautier-Hion et al. (1988), 194–226.

Gautier, J.-P., F. Moysan, A. T. C. Feistner, J.-N. Loireau, and R. W. Cooper. 1992. The distribution of *Cercopithecus (lhoesti) solatus,* an endemic guenon of Gabon. Rev. Ecol. (Terre Vie) 47:367–80.

Gautier-Hion, A. 1971. L'écologie du talapoin du Gabon. Terre Vie 25:427–90.

———. 1973. Social and ecological features of talapoin monkey—comparisons with sympatric cercopithecines. *In* Michael and Crook (1973), 147–70.

———. 1988. The diet and dietary habits of forest guenons. *In* Gautier-Hion et al. (1988), 257–83.

Geissmann, T. 1991a. Reassessment of age of sexual maturity in gibbons (*Hylobates* spp.). Amer. J. Primatol. 23:11–22.

———. 1991b. Sympatry between whitehanded gibbons *(Hylobates lar)* and pileated gibbons *(H. pileatus)* in southeastern Thailand. Primates 32:357–63.

———. 1994. Systematik der gibbons. Z. Kölner Zoo 37(2):65–77.

Gelvin, B. R. 1980. Morphometric affinities of *Gigantopithecus*. Amer. J. Phys. Anthropol. 53:541–68.

Gengozian, N., J. S. Batson, and T. A. Smith. 1977. Breeding of tamarins (*Saguinus* spp.) in the laboratory. *In* Kleiman (1977a), 207–13.

Gevaerts, H. 1992. Birth seasons of *Cercopithecus, Cercocebus,* and *Colobus* in Zaire. Folia Primatol. 59:105–13.

Gingerich, P. D. 1984. Paleobiology of tarsiiform primates. *In* Niemitz (1984a), 33–44.

Ginsburg, L., and P. Mein. 1987. *Tarsius thailandica* nov. sp., premier Tarsiidae (Primates, Mammalia) fossile d'Asie. Compt. Rend. Acad. Sci. Paris, ser. 2, 304:1213–15.

Glander, K. E. 1978. Howling monkey feeding behavior and plant secondary compounds: a study of strategies. *In* Montgomery (1978), 561–73.

Godfrey, L. R., and A. J. Petto. 1981. Clinal size variation of *Archaeolemur* spp. on Madagascar. *In* Chiarelli, A. B., and R. S. Corruccini, eds., Primate evolutionary biology, Springer-Verlag, Berlin, 14–34.

Godfrey, L. R., E. L. Simons, P. J. Chatrath, and B. Rakotosamimanana. 1990. A new fossil lemur (*Babakotia,* Primates) from northern Madagascar. Compt. Rend. Acad. Sci. Paris, ser. 2, 310:81–87.

Godfrey, L. R., M. R. Sutherland, A. J. Petto, and D. S. Boy. 1990. Size, space, and adaptation in some subfossil lemurs from Madagascar. Amer. J. Phys. Anthropol. 81:45–66.

Goodall, A. G. 1977. Feeding and ranging behaviour of a mountain gorilla group *(Gorilla gorilla beringei)* in the Tshibinda-Khuzi region (Zaire). *In* Clutton-Brock (1977), 449–79.

Goodall, A. G., and C. P. Groves. 1977. The conservation of eastern gorillas. *In* Rainier III and Bourne (1977), 599–637.

Goodall, J. 1977. Infant killing and cannibalism in free-living chimpanzees. Folia Primatol. 28:259–82.

———. 1986. The chimpanzees of Gombe: patterns of behavior. Harvard Univ. Press, Belknap Press, Cambridge, 673 pp.

———. 1994. Conservation and the future of chimpanzee and bonobo research in Africa. *In* Wrangham, R. W., W. C. McGrew, F. B. M. De Waal, P. G. Heltne, and L. A. Marquardt, eds., Chimpanzee cultures, Harvard Univ. Press, Cambridge, 397–404.

Goodman, M. 1975. Protein sequence and immunological specificity: their role in phylogenetic studies of primates. *In* Luckett and Szalay (1975), 219–48.

———. 1992. Hominoid evolution at the DNA level and the position of humans in a phylogenetic classification. *In* Nishida, T., W. C. McGrew, P. Marler, M. Pickford, and F. B. M. de Waal, eds., Topics in primatology, vol. 1, Human origins, Univ. Tokyo Press, 331–46.

Goodwin, G. G., and A. M. Greenhall. 1961. A review of the bats of Trinidad and Tobago. Bull. Amer. Mus. Nat. Hist. 122:187–302.

Goonan, P. M. 1993. Behaviour and reproduction of the slender loris *(Loris tardigradus)* in captivity. Folia Primatol. 60:146–57.

Goss, C. M., L. T. Popejoy, II, J. L. Fusiler, and T. M. Smith. 1968. Observations on the relationship between embryological development, time of conception, and gestation. *In* Rosenblum and Cooper (1968), 171–91.

Gozalo, A., and E. Montoya. 1990. Reproduction of the owl monkey *(Aotus nancymai)*

(Primates: Cebidae) in captivity. Amer. J. Primatol. 21:61–68.

Graham, C. E. 1988. Reproductive physiology. *In* Schwartz (1988b), 91–103.

Green, S., and K. Minkowski. 1977. The liontailed monkey and its South Indian forest habitat. *In* Rainier III and Bourne (1977), 289–337.

Groves, C. P. 1970a. The forgotten leafeaters, and the phylogeny of the Colobinae. *In* Napier and Napier (1970), 555–87.

———. 1970b. Population systematics of the gorilla. J. Zool. 161:287–300.

———. 1971b. Systematics of the genus *Nycticebus*. Proc. 3rd Internatl. Congr. Primatol. 1:44–53.

———. 1972b. Systematics and phylogeny of gibbons. *In* Rumbaugh (1972), 1–89.

———. 1974b. Taxonomy and phylogeny of prosimians. *In* Martin, Doyle, and Walker (1974), 449–73.

———. 1976. The origin of the mammalian fauna of Sulawesi (Celebes). Z. Saugetierk. 41:201–16.

———. 1978a. Phylogenetic and population systematics of the mangabeys (Primates: Cercopithecoidea). Primates 19:1–34.

———. 1980c. Speciation in *Macaca:* the view from Sulawesi. *In* Lindburg (1980), 84–124.

———. 1984. A new look at the taxonomy and phylogeny of gibbons. *In* Preuschoft et al. (1984), 542–61.

———. 1985a. The case of the pygmy gorilla: a cautionary tale for cryptozoology. Cryptozoology 4:37–44.

———. 1986a. Systematics of the great apes. *In* Swindler and Erwin (1986), 187–217.

———. 1989. A theory of human and primate evolution. Clarendon Press, Oxford, xii + 375 pp.

———. 1993. Speciation in living hominoid primates. *In* Kimbel and Martin (1993), 109–21.

Groves, C. P., R. Angst, and C. Westwood. 1993. The status of *Colobus polykomos dollmani* Schwarz. Internatl. J. Primatol. 14: 573–86.

Groves, C. P., and R. H. Eaglen. 1988. Systematics of the Lemuridae (Primates, Strepsirhini). J. Human Evol. 17:513–38.

Groves, C. P., and L. B. Holthius. 1985. The nomenclature of the orangutan. Zool. Meded. 59:411–17.

Groves, C. P., and I. Tattersall. 1991. Geographical variation in the fork-marked lemur, *Phaner furcifer* (Primates, Cheirogaleidae). Folia Primatol. 56:39–49.

Groves, C. P., and Wang Yingxiang. 1990. The gibbons of the subgenus *Nomascus* (Primates, Mammalia). Zool. Res. 11:147–54.

Groves, C. P., C. Westwood, and B. T. Shea. 1992. Unfinished business: Mahalanobis and a clockwork orang. J. Human Evol. 22:327–40.

Grubb, P. 1973. Distribution, divergence, and speciation of the drill and mandrill. Folia Primatol. 20:161–77.

Grzimek, B., ed. 1975. Grzimek's animal life encyclopedia: mammals, I–IV. Van Nostrand Reinhold, New York, vols. 10–13.

———, ed. 1990. Grzimek's encyclopedia of mammals. McGraw-Hill, New York, 5 vols.

Gust, D. A., C. D. Busse, and T. P. Gordon. 1990. Reproductive parameters in the sooty mangabey *(Cercocebus torquatus atys)*. Amer. J. Primatol. 22:241–50.

H

Hadidian, J., and I. S. Bernstein. 1979. Female reproductive cycles and birth data from an Old World monkey colony. Primates 20:429–42.

Haimoff, E. H. 1984. The organization of song in the Hainan black gibbon *(Hylobates concolor hainanus)*. Primates 25:225–35.

Haimoff, E. H., D. J. Chivers, S. P. Gittins, and T. Whitten. 1982. A phylogeny of gibbons (*Hylobates* spp.) based on morphological and behavioural characters. Folia Primatol. 39:213–37.

Hall, E. R. 1981. The mammals of North America. John Wiley & Sons, New York, 2 vols.

Hall, E. R., and W. W. Dalquest. 1963. The mammals of Veracruz. Univ. Kansas Publ. Mus. Nat. Hist. 14:165–362.

Hall, K. R. L. 1967. Social interactions of the adult male and adult females of a patas monkey group. *In* Altmann (1967), 261–80.

———. 1968. Behaviour and ecology of the wild patas monkey, *Erythrocebus patas*, in Uganda. *In* Jay (1968), 32–119 (reprinted from J. Zool. 148:15–87).

Hall, K. R. L., and I. De Vore. 1965. Baboon social behavior. *In* De Vore (1965), 53–110.

Hamada, Y., T. Oi, and T. Watanabe. 1994. *Macaca nigra* on Bacan Island, Indonesia: its morphology, distribution, and present habitat. Internatl. J. Primatol. 15:487–93.

Hamilton, W. J., III, R. E. Buskirk, and W. H. Buskirk. 1978. Omnivory and utilization of food resources by chacma baboons, *Papio ursinus*. Amer. Nat. 112:911–24.

Handley, C. O., Jr. 1976. Mammals of the Smithsonian Venezuelan Project. Brigham Young Univ. Sci. Bull., Biol. Ser., 20(5):1–89.

Hannah, A. C., and W. C. McGrew. 1987.

Chimpanzees using stones to crack open oil palm nuts in Liberia. Primates 28:31–46.

Happel, R. 1982. Ecology of *Pithecia hirsuta* in Peru. J. Human Evol. 11:581–90.

Happel, R., and T. Cheek. 1986. Evolutionary biology and ecology of *Rhinopithecus*. *In* Taub, D. M., and F. A. King, eds., Current perspectives in primate social dynamics, Van Nostrand Reinhold, New York, 305–24.

Happold, D. C. D. 1987. The mammals of Nigeria. Clarendon Press, Oxford, xvii + 402 pp.

Harako, R. 1981. The cultural ecology of hunting behaviour among Mbuti pygmies in the Ituri Forest, Zaire. *In* Harding and Teleki (1981), 499–555.

Harcourt, A. H. 1978. Strategies of emigration and transfer by primates, with particular reference to gorillas. Z. Tierpsychol. 48:401–20.

———. 1979. Social relationships between adult male and female mountain gorillas in the wild. Anim. Behav. 27:325–42.

———. 1981. Can Uganda's gorillas survive?—A survey of the Bwindi Forest Reserve. Biol. Conserv. 19:269–82.

———. 1986. Gorilla conservation: anatomy of a campaign. *In* Benirschke (1986), 31–46.

Harcourt, A. H., and K. Curry-Lindahl. 1978. The FPS mountain gorilla project—a report from Rwanda. Oryx 14:316–24.

Harcourt, A. H., and D. Fossey. 1981. The Virunga gorillas: decline of an "island" population. Afr. J. Ecol. 19:83–97.

Harcourt, A. H., D. Fossey, and J. Sabater Pi. 1981. Demography of *Gorilla gorilla*. J. Zool. 195:215–33.

Harcourt, A. H., and S. A. Harcourt. 1984. Insectivory by gorillas. Folia Primatol. 43:229–33.

Harcourt, A. H., and K. J. Stewart. 1984. Gorillas' time feeding: aspects of methodology, body size, competition, and diet. Afr. J. Ecol. 22:207–15.

Harcourt, A. H., K. J. Stewart, and D. Fossey. 1976. Male emigration and female transfer in wild mountain gorilla. Nature 263:226–27.

———. 1981. Gorilla reproduction in the wild. *In* Graham, C. E., ed., Reproductive biology of the great apes, Academic Press, New York, 265–79.

Harcourt, A. H., K. J. Stewart, and I. M. Inahoro. 1989. Gorilla quest in Nigeria. Oryx 23:7–13.

Harcourt, C. 1981. An examination of the function of urine washing in *Galago senegalensis*. Z. Tierpsychol. 55:119–28.

———. 1986. *Galago zanzibaricus*: birth seasonality, litter size, and perinatal behaviour of females. J. Zool. 210:451–57.

———. 1991. Diet and behaviour of a nocturnal lemur, *Avahi laniger*, in the wild. J. Zool. 223:667–74.

Harcourt, C., and S. K. Bearder. 1989. A comparison of *Galago moholi* in South Africa with *Galago zanzibaricus* in Kenya. Internatl. J. Primatol. 10:35–45.

Harcourt, C., and J. Thornback. 1990. Lemurs of Madagascar and the Comoros: the IUCN red data book. IUCN (World Conservation Union), Gland, Switzerland, 240 pp.

Harcourt, C., and L. T. Nash. 1986. Social organization of galagos in Kenyan coastal forests: I. *Galago zanzibaricus*. Amer. J. Primatol. 10:339–55.

Harding, R. S. O., and D. K. Olson. 1986. Patterns of mating among male patas monkeys *(Erythrocebus patas)*. Amer. J. Primatol. 11:343–58.

Harrington, J. E. 1975. Field observations of social behavior of *Lemur fulvus fulvus* E. Geoffroy 1812. *In* Tattersall and Sussman (1975), 259–79.

———. 1978. Diurnal behavior of *Lemur mongoz* at Ampijora, Madagascar. Folia Primatol. 29:291–302.

Harrison, D. L., and P. J. Bates. 1991. The mammals of Arabia. Harrison Zoological Museum, Sevenoaks, Kent, England, xvi + 354 pp.

Harrison, M. J. S. 1988. A new species of guenon (genus *Cercopithecus*) from Gabon. J. Zool. 215:561–75.

Harrison, T. 1993. Cladistic concepts and the species problem in hominoid evolution. *In* Kimbel and Martin (1993), 345–71.

Hasegawa, M. 1992. Evolution of hominoids as inferred from DNA sequences. *In* Nishida, T., W. C. McGrew, P. Marler, M. Pickford, and F. B. M. de Waal, eds., Topics in primatology, vol. 1, Human origins, Univ. Tokyo Press, 347–57.

Hausfater, G. 1974. History of three little-known New World populations of macaques. Lab. Primate Newsl. 13(1):16–18.

Hayden, B. 1981. Subsistence and ecological adaptations of modern hunter/gatherers. *In* Harding and Teleki (1981), 344–421.

Hayes, V. J., L. Freedman, and C. E. Oxnard. 1990. The taxonomy of savannah baboons: an odontomorphometric analysis. Amer. J. Primatol. 22:171–90.

Hayssen, V., A. Van Tienhoven, and A. Van Tienhoven. 1993. Asdell's patterns of mammalian reproduction: a compendium of species-specific data. Comstock/Cornell Univ. Press, Ithaca, viii + 1023 pp.

Heltne, P. G., D. C. Turner, and N. J. Scott, Jr.

1976. Comparison of census data on *Alouatta palliata* from Costa Rica and Panama. *In* Thorington and Heltne (1976), 10–19.

Hernandez-Camacho, J., and R. W. Cooper. 1976. The nonhuman primates of Colombia. *In* Thorington and Heltne (1976), 35–69.

Hershkovitz, P. 1972. Notes on New World monkeys. Internatl. Zoo Yearbook 12:3–12.

———. 1975a. Comments on the taxonomy of Brazilian marmosets (Callithrix, Callithricidae). Folia Primatol. 24:137–72.

———. 1977. Living New World monkeys (Platyrrhini). Vol. 1. Univ. Chicago Press, xiv + 1117 pp.

———. 1979. The species of sakis, genus *Pithecia* (Cebidae, Primates), with notes on sexual dichromatism. Folia Primatol. 31:1–22.

———. 1983. Two new species of night monkeys, genus *Aotus* (Cebidae, Platyrrhini): a preliminary report on *Aotus* taxonomy. Amer. J. Primatol. 4:209–43.

———. 1984a. On the validity of the family-group name Callitrichidae (Platyrrhini, Primates). Mammalia 48:153.

———. 1985. A preliminary taxonomic review of the South American bearded saki monkeys genus *Chiropotes* (Cebidae, Platyrrhini), with the description of a new subspecies. Fieldiana Zool., n.s., no. 27, 46 pp.

———. 1987b. The taxonomy of South American sakis, genus *Pithecia* (Cebidae, Platyrrhini): a preliminary report and critical review with the description of a new species and a new subspecies. Amer. J. Primatol. 12:387–468.

———. 1987c. Uacaries, New World monkeys of the genus *Cacajao* (Cebidae, Platyrrhini): a preliminary taxonomic review with the description of a new subspecies. Amer. J. Primatol. 12:1–53.

———. 1988a. Origin, speciation, and distribution of South American titi monkeys, genus *Callicebus* (family Cebidae, Platyrrhini). Proc. Acad. Nat. Sci. Philadelphia 140:240–72.

———. 1988b. The subfossil monkey femur and subfossil monkey tibia of the Antilles: a review. Internatl. J. Primatol. 9:365–84.

———. 1990c. Titis, New World monkeys of the genus *Callicebus* (Cebidae, Platyrrhini): a preliminary taxonomic review. Fieldiana Zool., n.s., no. 55, v + 109 pp.

Hill, A., S. Ward, A. Deino, G. Curtis, and R. Drake. 1992. Earliest *Homo*. Nature 355:719–22.

Hill, C. 1975. The longevity record for *Colobus*. Primates 16:235.

Hill, C. A. 1967. A note on the gestation period of the siamang *Hylobates syndactylus*. Internatl. Zoo Yearbook 7:93–94.

Hill, D., and D. Sprague. 1991. Update on the Yakushima macaque. Asian Primates 1(3):5–6.

Hill, W. C. O. 1970. Primates: comparative anatomy and taxonomy. VIII. Cynopithecinae. *Papio, Mandrillus, Theropithecus*. Wiley-Interscience, New York, xix + 680 pp.

Hladik, C. M. 1978. Adaptive strategies of primates in relation to leaf-eating. *In* Montgomery (1978), 373–95.

———. 1979. Diet and ecology of prosimians. *In* Doyle and Martin (1979), 307–57.

Hladik, C. M., and P. Charles-Dominique. 1974. The behaviour and ecology of the sportive lemur *(Lepilemur mustelinus)* in relation to its dietary peculiarities. *In* Martin, Doyle, and Walker (1974), 23–37.

Hladik, C. M., P. Charles-Dominique, and J.-J. Petter. 1980. Feeding strategies of five nocturnal prosimians in the dry forest of the west coast of Madagascar. *In* Charles-Dominique et al. (1980), 41–73.

Homewood, K. 1975. Can the Tana mangabey survive? Oryx 13:53–59.

Homewood, K., and W. A. Rodgers. 1981. A previously undescribed mangabey from southern Tanzania. Internatl. J. Primatol. 2:47–55.

Horn, A. D. 1979. The taxonomic status of the bonobo chimpanzee. Amer. J. Phys. Anthropol. 51:273–82.

———. 1987a. The socioecology of the black mangabey *(Cercocebus aterrimus)* near Lake Tumba, Zaire. Amer. J. Primatol. 12:165–80.

———. 1987b. Taxonomic assessment of the allopatric gray-cheeked mangabey *(Cercocebus albigena)* and black mangabey *(C. aterrimus)*: comparative socioecological data and the species concept. Amer. J. Primatol. 12:181–87.

Horr, D. A. 1975. The Borneo orang-utan: population structure and dynamics in relationship to ecology and reproductive strategy. Primate Behav. 4:307–23.

Horrocks, J. A. 1986. Life-history characteristics of a wild population of vervets *(Cercopithecus aethiops sabaeus)* in Barbados, West Indies. Internatl. J. Primatol. 7:31–47.

Horwich, R. H. 1983. Species status of the black howler monkey, *Alouatta pigra*, of Belize. Primates 24:288–89.

Horwich, R. H., and K. Gebhard. 1983. Roaring rhythms in black howler monkeys *(Alouatta pigra)* of Belize. Primates 24:290–96.

Horwich, R. H., and E. D. Johnson. 1986. Geographical distribution of the black howler *(Alouatta pigra)* in Central America. Primates 27:53–62.

Hoshino, J. 1985. Feeding ecology of man-

drills *(Mandrillus sphinx)* in Campo Animal Reserve, Cameroon. Primates 26:248–73.

Hoshino, J., A. Mori, H. Kudo, and M. Kawai. 1984. Preliminary report on the grouping of mandrills *(Mandrillus sphinx)* in Cameroon. Primates 25:295–307.

I

IUCN (World Conservation Union). 1972–78. Red data book. I. Mammalia. Morges, Switzerland.

Iwamoto, T. 1993. The ecology of *Theropithecus gelada. In* Jablonski (1993b), 441–52.

Iwano, T. 1975. Distribution of Japanese monkey *(Macaca fuscata)*. Proc. 5th Internatl. Congr. Primatol., 389–91.

Izard, M. K. 1987. Lactation length in three species of *Galago*. Amer. J. Primatol. 13:73–76.

Izard, M. K., and D. T. Rasmussen. 1985. Reproduction in the slender loris *(Loris tardigradus malabaricus)*. Amer. J. Primatol. 8:153–65.

Izard, M. K., K. A. Weisenseel, and R. L. Ange. 1988. Reproduction in the slow loris *(Nycticebus coucang)*. Amer. J. Primatol. 16:331–39.

Izard, M. K., P. C. Wright, and E. L. Simons. 1985. Gestation length in *Tarsius bancanus*. Amer. J. Primatol. 9:327–31.

Izawa, K. 1970. Unit groups of chimpanzees and their nomadism in the savanna woodland. Primates 11:1–46.

———. 1976. Group sizes and compositions of monkeys in the upper Amazon basin. Primates 17:367–99.

———. 1978. A field study of the ecology and behavior of the black-mantle tamarin *(Saguinus nigricollis)*. Primates 19:241–74.

———. 1980. Social behavior of the wild black-capped capuchin *(Cebus apella)*. Primates 21:443–67.

J

Jablonski, N. G. 1993a. The phylogeny of *Theropithecus. In* Jablonski (1993b), 209–24.

———. 1995. The phyletic position and systematics of the douc langurs of Southeast Asia. Amer. J. Primatol. 35:185–205.

Jablonski, N. G., and Peng Yan-zhang. 1993. The phylogenetic relationships and classification of the doucs and snub-nosed langurs of China and Vietnam. Folia Primatol. 60:36–55.

Jenkins, P. D. 1987. Catalogue of primates in the British Museum (Natural History) and elsewhere in the British Isles, Part IV: suborder Strepsirhini, including the subfossil Madagascan lemurs and family Tarsiidae. British Mus. (Nat. Hist.), London, x + 189 pp.

Johanson, D. C., and T. D. White. 1979. A systematic assessment of early African hominids. Science 203:321–30.

Johns, A. 1985. Current status of the southern bearded saki *(Chiropotes satanas satanas)*. Primate Conserv. 5:28.

———. 1986. Notes on the ecology and current status of the buffy saki, *Pithecia albicans*. Primate Conserv. 7:26–29.

Jolly, A. 1966. Lemur behavior: a Madagascar field study. Univ. Chicago Press, xiv + 187 pp.

———. 1972. Troop continuity and troop spacing in *Propithecus verreauxi* and *Lemur catta* at Berenty (Madagascar). Folia Primatol. 17:335–62.

Jolly, C. J. 1993. Species, subspecies, and baboon systematics. *In* Kimbel and Martin (1993), 67–107.

Jones, C., and S. Anderson. 1978. *Callicebus moloch*. Mammalian Species, no. 112, 5 pp.

Jones, C., and J. Sabater Pi. 1968. Comparative ecology of *Cercocebus albigena* (Gray) and *Cercocebus torquatus* (Kerr) in Rio Muni, West Africa. Folia Primatol. 9:99–113.

———. 1971. Comparative ecology of *Gorilla gorilla* (Savage and Wyman) and *Pan troglodytes* (Blumenbach) in Rio Muni, West Africa. Bibl. Primatol., no. 13, 96 pp.

Jones, M. L. 1982. Longevity of captive mammals. Zool. Garten 52:113–28.

Jordan, C. 1982. Object manipulation and tool-use in captive pygmy chimpanzee *(Pan paniscus)*. J. Human Evol. 11:35–39.

Jouventin, P. 1975a. Observations sur la socio-écologie du mandrill. Terre Vie 29:493–532.

———. 1975b. Les roles des colorations du mandrill *(Mandrillus sphinx)*. Z. Tierpsychol. 39:455–62.

Jungers, W. L., L. R. Godfrey, E. L. Simons, P. S. Chatrath, and B. Rakotosamimanana. 1991. Phylogenetic and functional affinities of *Babakotia* (Primates), a fossil lemur from northern Madagascar. Proc. Natl. Acad. Sci. 88:9082–86.

Jungers, W. L., and R. L. Susman. 1984. Body size and skeletal allometry in African apes. *In* Susman (1984b), 131–77.

K

Kahlke, H. D. 1973. A review of the Pleistocene history of the orang-utan *(Pongo* Lacépède 1799). Asian Perspectives 15:514.

Kano, T. 1992. The last ape: pygmy chimpanzee behavior and ecology. Stanford Univ. Press, Stanford, California, xxvi + 248 pp.

Kanâo, T., and M. Mulavwa. 1984. Feeding ecology of the pygmy chimpanzees *(Pan paniscus)* of Wamba. *In* Susman (1984b), 233–74.

Kavanagh, M., and E. Laursen. 1984. Breeding seasonality among long-tailed macaques, *Macaca fascicularis*, in peninsular Malaysia. Internatl. J. Primatol. 5:17–29.

Kawabe, M., and T. Mano. 1972. Ecology and behavior of the wild proboscis monkey, *Nasalis larvatus* (Wurmb), in Sabah, Malaysia. Primates 13:213–28.

Kawai, M., ed. 1979. Ecological and sociological studies of gelada baboons. Contrib. Primatol., no. 16, xxiv + 344 pp.

Kawai, M., R. Dunbar, H. Ohsawa, and U. Mori. 1983. Social organization of gelada baboons: social units and definitions. Primates 24:13–24.

Kay, R. F. 1990. The phyletic relationships of extant and fossil Pitheciinae (Platyrrhini, Anthropoidea). J. Human Evol. 19:175–208.

Kay, R. F., C. Ross, and B. A. Williams. 1997. Anthropoid origins. Science 275:797–804.

Kelley, J., and D. Pilbeam. 1986. The dryopithecines: taxonomy, comparative anatomy, and phylogeny of Miocene large hominoids. *In* Swindler and Erwin (1986), 361–411.

Kern, J. A. 1964. Observations on the habits of the proboscis monkey, *Nasalis larvatus* (Wurmb), made in the Brunei Bay area, Borneo. Zoologica 49:183–92.

Khan, Mohd Khan Bin Momin. 1978. Man's impact on the primates of peninsular Malaysia. *In* Chivers and Lane-Petter (1978), 41–46.

Kingdon, J. 1971. East African mammals: an atlas of evolution in Africa. I. Academic Press, London, ix + 446 pp.

———. 1974a. East African mammals: an atlas of evolution in Africa. II(A). Insectivores and bats. Academic Press, London, xi + 341 + l pp.

———. 1988. Comparative morphology of hands and feet in the genus *Cercopithecus*. *In* Gautier-Hion et al. (1988), 184–93.

Kinzey, W. G. 1977. Diet and feeding behaviour of *Callicebus torquatus*. *In* Clutton-Brock (1977), 127–51.

Kinzey, W. G., A. L. Rosenberger, P. S. Heisler, D. L. Prowse, and J. S. Trilling. 1977. A preliminary field investigation of the yellow-handed titi monkey, *Callicebus torquatus torquatus*, in northern Peru. Primates 18:159–81.

Kirkpatrick, R. C. 1995. The natural history and conservation of the snub-nosed monkeys (genus *Rhinopithecus*). Biol. Conserv. 72:363–69.

Kleiman, D. G. 1977b. Characteristics of reproduction and sociosexual interactions in pairs of lion tamarins *(Leontopithecus rosalia)* during the reproductive cycle. *In* Kleiman (1977a), 181–92.

———. 1981. *Leontopithecus rosalia*. Mammalian Species, no. 148, 7 pp.

Kleiman, D. G., B. B. Beck, J. M. Dietz, L. A. Dietz, J. D. Ballou, and A. F. Coimbra-Filho. 1986. Conservation program for the golden lion tamarin: captive research management, ecological studies, educational strategies, and reintroduction. *In* Benirschke (1986), 959–79.

Klein, L. L., and D. J. Klein. 1976. Neotropical primates: aspects of habitat usage, population density, and regional distribution in La Macarena, Colombia. *In* Thorington and Heltne (1976), 70–78.

———. 1977. Feeding behaviour of the Colombian spider monkey. *In* Clutton-Brock (1977), 153–81.

Klopfer, P. H., and K. J. Boskoff. 1979. Maternal behavior in prosimians. *In* Doyle and Martin (1979), 123–56.

Kool, K. M. 1992. The status of endangered primates in Gunung Halimun Reserve, Indonesia. Oryx 26:29–33.

Kortlandt, A. 1995. A survey of the geographical range, habitats, and conservation of the pygmy chimpanzee *(Pan paniscus):* an ecological perspective. Primate Conserv. 16:20–35.

Kortlandt, A., and E. Holzhaus. 1987. New data on the use of stone tools by chimpanzees in Guinea and Liberia. Primates 28:473–96.

Koyama, N., K. Norikoshi, and T. Mano. 1975. Population dynamics of Japanese monkeys at Arashiyama. Proc. 5th Internatl. Congr. Primatol., 411–17.

Kramer, A. 1993. Human taxonomic diversity in the Pleistocene: does *Homo erectus* represent multiple hominid species? Amer. J. Phys. Anthropol. 91:161–71.

Kudo, H., and M. Mitani. 1985. New record of predatory behavior by the mandrill in Cameroon. Primates 26:161–67.

Kummer, H. 1968. Social organization of hamadryas baboons: a field study. Univ. Chicago Press, vi + 189 pp.

Kummer, H., W. Goetz, and W. Angst. 1970. Cross-species modifications of social behavior in baboons. *In* Napier and Napier (1970), 351–63.

Kuntz, R. E., and B. J. Myers. 1969. A checklist of parasites and commensals reported for the Taiwan macaque. Primates 10:71–80.

Kurland, J. A. 1973. A natural history of Kra macaques *(Macaca fascicularis* Raffles, 1821) at the Kutai Reserve, Kalimantan Timur, Indonesia. Primates 14:245–62.

———. 1977. Kin selection in the Japanese monkey. Contrib. Primatol., no. 12, x + 145 pp.

Kuroda, S., T. Kano, and K. Muhindo. 1985. Further information on the new monkey species *Cercopithecus salongo* Thys van den Audenaerde, 1977. Primates 26:325–33.

Kurup, G. 1988. The present status of the

lion-tailed macaque. Primate Conserv. 9: 34–36.

L

Lawes, M. J. 1990. The distribution of the samango monkey (*Cercopithecus mitis erythrarchus* Peters, 1852 and *Cercopithecus mitis labiatus* I. Geoffroy, 1843) and forest history in southern Africa. J. Biogeogr. 17:669–80.

Leakey, L. S. B., P. V. Tobias, and J. R. Napier. 1964. A new species of the genus *Homo* from Olduvai Gorge. Nature 202:7–9.

Lee Ling-Ling and Lin Yao-Sung. 1990. Status of Formosan macaques in Taiwan. Primate Conserv. 11:18–20.

Lee, P. C., J. Thornback, and E. L. Bennett. 1988. Threatened primates of Africa: the IUCN red data book. IUCN (World Conservation Union), Gland, Switzerland, xx + 155 pp.

Le Gros Clark, W. E. 1955. The fossil evidence for human evolution. Univ. Chicago Press, 180 pp.

Lekagul, B., and J. A. McNeely. 1977. Mammals of Thailand. Sahakarnbhat, Bangkok, li + 758 pp.

Lemos de Sá, R. M., T. R. Pope, K. E. Glander, T. T. Struhsaker, and G. A. B. Da Fonseca. 1990. A pilot study of genetic and morphological variation in the muriqui (*Brachyteles arachnoides*). Primate Conserv. 11:26–30.

Leopold, A. S. 1959. Wildlife of Mexico. Univ. California Press, Berkeley, xiii + 568.

Lernold, J.-M. 1988. Classification and geographical distribution of guenons: a review. *In* Gautier-Hion et al. (1988), 54–77.

Lewin, R. 1993. Human evolution: an illustrated introduction. Blackwell, Boston, vii + 208 pp.

Li Zhixiang, Ma Shilai, Hua Chenghui, and Wang Yingxiang. 1982. The distribution and habits of the Yunnan golden monkey, *Rhinopithecus bieti*. J. Human Evol. 11:633–38.

Lilly, A. A., and P. T. Mehlman. 1993. Conservation update on the Barbary macaque: I. Declining distribution and population size in Morocco. Amer. J. Primatol. 30:327.

Lindburg, D. G. 1971. The rhesus monkey in north India: an ecological and behavioral study. Primate Behav. 2:1–106.

Lindburg, D. G., and L. Gledhill. 1992. Captive breeding and conservation of lion-tailed macaques. Endangered Species Update 10(1): 1–10.

Lindburg, D. G., A. M. Lyles, and N. M. Czekala. 1989. Status and reproductive potential of lion-tailed macaques in captivity. Zoo Biol. Suppl. 1:5–16.

Lippold, L. K. 1977. The duoc langur: a time

for conservation. *In* Rainier III and Bourne (1977), 513–38.

Long Yongcheng, C. R. Kirkpatrick, Zhongtai, and Xiaolin. 1994. Report on the distribution, population, and ecology of the Yunnan snub-nosed monkey (*Rhinopithecus bieti*). Primates 35:241–50.

Lorenz, R., C. O. Anderson, and W. A. Mason. 1973. Notes on reproduction in captive squirrel monkeys (*Saimiri sciureus*). Folia Primatol. 19:286–92.

Lorini, M. L., and V. G. Persson. 1990. Nova espécie de *Leontopithecus* Lesson, 1840, do sul do Brasil (Primates, Callitrichidae). Bol. Mus. Nac. Rio de Janeiro (Zool.), n.s., no. 338, 14 pp.

———. 1994. Status of field research on *Leontopithecus caissara*: the black-faced lion tamarin project. Neotropical Primates 2(suppl.):52–55.

Loy, J. 1981. The reproductive and heterosexual behaviours of adult patas monkeys in captivity. Anim. Behav. 29:714–26.

Luoma, J. R. 1989. The chimp connection. Anim. Kingdom 92(1):38–51.

M

Ma Shilai and Wang Yingxiang. 1986. The taxonomy and distribution of the gibbons in southern China and its adjacent region— with description of three new subspecies. Zool. Res. 7:393–410.

Macdonald, D. W. 1982. Notes on the size and composition of groups of proboscis monkey, *Nasalis larvatus*. Folia Primatol. 37:95–98.

MacKinnon, J. 1971. The orang-utan in Sabah today. Oryx 11:141–91.

———. 1973. Orang-utans in Sumatra. Oryx 12:234–42.

———. 1974. The behaviour and ecology of wild orang-utans (*Pongo pygmaeus*). Anim. Behav. 22:3–74.

MacKinnon, J., and K. MacKinnon. 1980. The behavior of wild spectral tarsiers. Internatl. J. Primatol. 1:361–79.

———. 1987. Conservation status of the primates of the Indo-Chinese subregion. Primate Conserv. 8:187–95.

MacKinnon, K. 1986. The conservation status of nonhuman primates in Indonesia. *In* Benirschke (1986), 99–126.

MacPhee, R. D. E. 1984. Quaternary mammal localities and heptaxodontid rodents of Jamaica. Amer. Mus. Novit., no. 2803, 34 pp.

———. 1993a. From Cuba: a mandible of *Paralouatta*. Evol. Anthropol. 2(2):42.

———. 1993b. Summary. *In* Macphee, R. D. E., ed., Primates and their relatives in phylo-

genetic perspective, Plenum Press, New York, 363–73.

MacPhee, R. D. E., and J. G. Fleagle. 1991. Postcranial remains of *Xenothrix mcgregori* (Primates, Xenotrichidae) and other late Quaternary mammals from Long Mile Cave, Jamaica. Bull. Amer. Mus. Nat. Hist. 206: 287–321.

MacPhee, R. D. E., I. Horovitz, O. Arredondo, and O. Jiménez Vasquez. 1995. A new genus for the extinct Hispaniolan monkey *Saimiri bernensis* Rímoli, 1977, with notes on its systematic position. Amer. Mus. Novit., no. 3134, 21 pp.

MacPhee, R. D. E., and M. A. Iturralde-Vinent. 1995a. Earliest monkey from Greater Antilles. J. Human Evol. 28:197–200.

MacPhee, R. D. E., and E. M. Raholimavo. 1988. Modified subfossil aye-aye incisors from southwestern Madagascar: species allocation and paleoecological significance. Folia Primatol. 51:126–42.

MacPhee, R. D. E., and M. Rivero. 1996. Accelerator mass spectrometry C14 age determination for the alleged "Cuban spider monkey," *Ateles* (= Montaneia) anthropomorphus. J. Human Evol. 30:89–94.

MacPhee, R. D. E., and C. A. Woods. 1982. A new fossil cebine from Hispaniola. Amer. J. Phys. Anthropol. 58:419–36.

Mallinson, J. J. C. 1974. Establishing mammal gestation at the Jersey Zoological Park. Internatl. Zoo Yearbook 14:184–87.

———. 1977. Maintenance of marmosets and tamarins at Jersey Zoological Park with special reference to the design of the new marmoset complex. *In* Kleiman (1977a), 323–29.

———. 1987. International efforts to secure a viable population of the golden-headed lion tamarin. Primate Conserv. 8:124–25.

———. 1994. Saving the world's richest rainforest. Biologist 41:57–60.

Manson, J. H., and R. W. Wrangham. 1991. Intergroup aggression in chimpanzees and humans. Current Anthropol. 32:369–90.

Maples, W. R. 1972. Systematic reconsideration and a revision of the nomenclature of Kenya baboons. Amer. J. Phys. Anthropol. 36:9–19.

Marks, J. 1988. The phylogenetic status of orang-utans from a genetic perspective. *In* Schwartz (1988b), 53–67.

Marshall, J., and J. Sugardjito. 1986. Gibbon systematics. *In* Swindler and Erwin (1986), 137–85.

Martin, R. D. 1973. A review of the behaviour and ecology of the lesser mouse lemur (*Microcebus murinus* J. F. Miller 1777). *In* Michael and Crook (1973), 1–68.

———. 1975a. The bearing of reproductive behavior and ontogeny on strepsirhine phylogeny. *In* Luckett and Szalay (1975), 265–97.

———. 1990. Primate origins and evolution: a phylogenetic reconstruction. Princeton Univ. Press, xiv + 804 pp.

———. 1992. Goeldi and the dwarfs: the evolutionary biology of the small New World monkeys. J. Human Evol. 22:367–93.

———. 1993. Primate origins: plugging the gaps. Nature 363:223–34.

Martuscelli, P., L. M. Petroni, and F. Olmos. 1994. Fourteen new localities for the muriqui *Brachyteles arachnoides*. Neotropical Primates 2(2):12–15.

Maruhashi, T., D. S. Sprague, and K. Matsubayashi. 1992. What future for Japanese monkeys? Asian Primates 2(1):1–4.

Masataka, N. 1981a. A field study of the social behavior of Goeldi's monkeys *(Callimico goeldii)* in north Bolivia. I. Group composition, breeding cycle, and infant development. *In* Kyoto Univ. Overseas Res. Rept. of New World Monkeys, 23–32.

———. 1981b. A field study of the social behavior of Goeldi's monkeys *(Callimico goeldii)* in north Bolivia. II. Grouping pattern and intragroup relationship. *In* Kyoto Univ. Overseas Res. Rept. of New World Monkeys, 33–41.

———. 1982. A field study on the vocalizations of Goeldi's monkeys *(Callimico goeldii)*. Primates 23:206–19.

Mason, W. A. 1968. Use of space by *Callicebus* groups. *In* Jay (1968), 200–216.

———. 1971. Field and laboratory studies of social organization in *Saimiri* and *Callicebus*. Primate Behav. 2:107–37.

Masters, J. 1986. Geographic distributions of karyotypes and morphotypes within the greater galagines. Folia Primatol. 46:127–41.

———. 1988. Speciation in the greater galagos (Prosimii: Galaginae): review and synthesis. Biol. J. Linnean Soc. 34:149–74.

Masui, K., Y. Sugiyama, A. Nishimura, and H. Ohsawa. 1975. The life table of Japanese monkeys at Takasakiyama: a preliminary report. Proc. 5th Internatl. Congr. Primatol., 401–6.

Mayr, E. 1951. Taxonomic categories in fossil hominids. Cold Spring Harbor Symp. Quant. Biol. 15:109–18.

McCarthy, T. J. 1982b. *Chironectes, Cyclopes, Cabassous*, and probably *Cebus* in southern Belize. Mammalia 46:397–400.

McCrossin, M. L., and B. R. Benefit. 1993. Recently discovered *Kenyapithecus* mandible and its implications for great ape and human origins. Proc. Natl. Acad. Sci. 90:1962–66.

McGrew, W. C. 1992. Tool-use by free-ranging chimpanzees: the extent of diversity. J. Zool. 228:689–94.

McGrew, W. C., and D. A. Collins. 1985. Tool use by wild chimpanzees *(Pan troglodytes)* to obtain termites *(Macrotermes hersus)* in the Mahale Mountains, Tanzania. Amer. J. Primatol. 9:47–62.

McGuire, M. T. 1974. The St. Kitts vervet. Contrib. Primatol., no. 1, x + 199 pp.

McKenna, M. C. 1975. Toward a phylogenetic classification of the Mammalia. *In* Luckett and Szalay (1975), 21–46.

McKey, D., and P. G. Waterman. 1982. Ranging behaviour of a group of black colobus *(Colobus satanas)* in the Doula-Edea Reserve, Cameroon. Folia Primatol. 39:264–304.

McLanahan, E. B., and K. M. Green. 1977. The vocal repertoire and an analysis of the contexts of vocalizations in *Leontopithecus rosalia*. *In* Kleiman (1977a), 251–69.

Medley, K. E. 1993. Primate conservation along the Tana River, Kenya: an examination of the forest habitat. Conserv. Biol. 7:109–21.

Medway, Lord. 1970. The monkeys of Sundaland. *In* Napier and Napier (1970), 513–53.

———. 1977. Mammals of Borneo. Monogr. Malaysian Branch Roy. Asiatic Soc., no. 7, xii + 172 pp.

———. 1978. The wild mammals of Malaya (peninsular Malaysia) and Singapore. Oxford Univ. Press, Kuala Lumpur, xxii + 128 pp.

Meester, J., I. L. Rautenbach, N. J. Dippenaar, and C. M. Baker. 1986. Classification of southern African mammals. Transvaal Mus. Monogr., no. 5, x + 359 pp.

Meester, J., and H. W. Setzer. 1977. The mammals of Africa: an identification manual. Smithson. Inst. Press, Washington, D.C.

Meier, B., and R. Albignac. 1991. Rediscovery of *Allocebus trichotis* Günther 1875 (Primates) in northeast Madagascar. Folia Primatol. 56:57–63.

Meier, B., R. Albignac, A. Peyrieras, Y. Rumpler, and P. Wright. 1987. A new species of *Hapalemur* from south east Madagascar. Folia Primatol. 48:211–15.

Meier, B., and Y. Rumpler. 1987. Preliminary survey of *Hapalemur simus* and of a new species of *Hapalemur* in eastern Betsileo, Madagascar. Primate Conserv. 8:40–43.

Ménard, N., and D. Vallet. 1993. Dynamics of fission in a wild Barbary macaque group *(Macaca sylvanus)*. Internatl. J. Primatol. 14:479–500.

Meritt, D. A., Jr. 1977. Second-generation owl monkey birth. Amer. Assoc. Zool. Parks Aquar. Newsl. 18(3):12.

———. 1980. Captive reproduction and husbandry of the douroucouli *Aotus trivirgatus* and the titi monkey *Callicebus* spp. Internatl. Zoo Yearbook 20:52–59.

Milton, K. 1984. Habitat, diet, and activity patterns of free-ranging woolly spider monkeys *(Brachyteles arachnoides* E. Geoffroy 1806). Internatl. J. Primatol. 5:491–514.

———. 1985. Mating patterns of woolly spider monkeys, *Brachyteles arachnoides:* implications for female choice. Behav. Ecol. Sociobiol. 17:53–59.

———. 1986. Ecological background and conservation priorities for woolly spider monkeys *(Brachyteles arachnoides)*. *In* Benirschke (1986), 241–50.

Mitani, J. C., and T. Nishida. 1993. Contexts and social correlates of long-distance calling by male chimpanzees. Anim. Behav. 45:735–46.

Mittermeier, R. A. 1973. Group activity and population dynamics of the howler monkey on Barro Colorado Island. Primates 14:1–19.

———. 1987. Rescuing Brazil's muriqui: monkey in peril. Natl. Geogr. 171:386–95.

Mittermeier, R. A., and A. F. Coimbra-Filho. 1977. Primate conservation in Brazilian Amazonia. *In* Rainier III and Bourne (1977), 117–66.

Mittermeier, R. A., W. R. Konstant, H. Ginsberg, M. G. M. Van Roosmalen, and E. C. Da Silva, Jr. 1983. Further evidence of insect consumption in the bearded saki monkey, *Chiropotes satanas chiropotes*. Primates 24:602–5.

Mittermeier, R. A., W. R. Konstant, M. E. Nicoll, and O. Langrand. 1992. Lemurs of Madagascar: an action plan for their conservation. IUCN (World Conservation Union) Species Survival Comm., Gland, Switzerland, iv + 58 pp.

Mittermeier, R. A., H. de Macedo-Ruiz, B. A. Luscombe, and J. Cassidy. 1977. Rediscovery and conservation of the Peruvian yellow-tailed woolly monkey *(Lagothrix flavicauda)*. *In* Rainier III and Bourne (1977), 95–115.

Mittermeier, R. A., J. F. Oates, A. E. Eudey, and J. Thornback. 1986. Primate conservation. *In* Mitchell, G., and J. Erwin, eds., Comparative primate biology, vol. 2, Behavior, conservation, and ecology, John Wiley & Sons, New York, 3–72.

Mittermeier, R. A., C. V. Padua, C. Valle, and A. F. Coimbra-Filho. 1985. Major program underway to save the black lion tamarin in Sao Paulo, Brazil. Primate Conserv. 6:19–21.

Mittermeier, R. A., M. Schwarz, and J. M. Ayres. 1992. A new species of marmoset, genus *Callithrix* Erxleben, 1777 (Callitrichidae, Primates) from the Rio Maués region, state of Amazonas, central Brazilian Amazonia. Goeldiana Zool., no. 14, 17 pp.

Mittermeier, R. A., I. Tattersall, W. R. Konstant, D. M. Meyers, R. B. Mast, and S. D. Nash. 1994. Lemurs of Madagascar. Conservation International, Washington, D.C., 356 pp.

Mohnot, S. M. 1978. The conservation of non-human Primates in India. *In* Chivers and Lane-Petter (1978), 47–53.

Montagnon, D., S. Crovella, and Y. Rumpler. 1993. Confirmation of the taxonomic position of *Callimico goeldi* (Primates, Platyrrhini) on the basis of its highly repeated DNA patterns. Compt. Rend. Acad. Sci. Paris, ser. 3, 316:219–23.

Moore, A. J., and J. M. Cheverud. 1992. Systematics of the *Saguinus oedipus* group of the bare-face tamarins: evidence from facial morphology. Amer. J. Phys. Anthropol. 89: 73–84.

Morell, V. 1994a. Decoding chimp genes and lives. Science 265:1172–73.

———. 1994b. Will primate genetics split one gorilla into two? Science 265:1661.

Mori, A., and G. Belay. 1990. The distribution of baboon species and a new population of gelada baboons along the Wabi-Shebeli River, Ethiopia. Primates 31:495–508.

Morin, P. A., J. J. Moore, R. Chakraborty, Li Jin, J. Goodall, and D. S. Woodruff. 1994. Kin selection, social structure, gene flow, and the evolution of chimpanzees. Science 265: 1193–1201.

Morland, H. S. 1991. Preliminary report on the social organization of ruffed lemurs *(Varecia variegata variegata)* in a northeast Madagascar rain forest. Folia Primatol. 56: 157–61.

Morris, J. 1995. Gorilla conservation programme suffers setback. Oryx 29:219.

Moynihan, M. 1976. Notes on the ecology and behavior of the pygmy marmoset *(Cebuella pygmaea)* in Amazonian Colombia. *In* Thorington and Heltne (1976), 79–84.

Mu Wenwei and Yang Dehua. 1982. A primary observation on the group figures, moving lines, and food of *Rhinopithecus bieti* at the east side of Baima-Snow Mountain. Acta Theriol. Sinica 2:125–31.

Mukherjee, R. P. 1982. Phayre's leaf monkey *(Presbytis phayrei* Blyth, 1847) of Tripura. J. Bombay Nat. Hist. Soc. 79:47–56.

Muriuki, J. W., and M. H. Tsingalia. 1990. A new population of de Brazza's monkey in Kenya. Oryx 24:157–62.

Muskin, A. 1984. Field notes and geographic distribution of *Callithrix aurita* in eastern Brazil. Amer. J. Primatol. 7:377–80.

Musser, G. G., and M. Dagosto. 1987. The identity of *Tarsius pumilus*, a pygmy species endemic to the montane mossy forests of central Sulawesi. Amer. Mus. Novit., no. 2867, 53 pp.

Mutschler, T., and A. T. C. Feistner. 1995. Conservation status and distribution of the Alaotran gentle lemur *Hapalemur griseus alaotrensis*. Oryx 29:267–74.

Myers, N. 1987b. Trends in the destruction of rain forests. *In* Marsh and Mittermeier (1987), 3–22.

N

Nadler, R. D. 1988. Sexual and reproductive behavior. *In* Schwartz (1988b), 105–16.

Nagel, U. 1971. Social organization in a baboon hybrid zone. Proc. 3rd Internatl. Congr. Primatol. 3:48–57.

———. 1973. A comparison of anubis baboons, hamadryas baboons, and their hybrids at a species border in Ethiopia. Folia Primatol. 19:104–65.

Napier, J. R., and P. H. Napier. 1967. A handbook of living primates. Academic Press, New York, xiv + 456 pp.

———. 1985. The natural history of primates. MIT Press, Cambridge, 200 pp.

Nash, L. T., S. K. Bearder, and T. R. Olson. 1989. Synopsis of galago characteristics. Internatl. J. Primatol. 10:57–80.

Nash, L. T., and C. Harcourt. 1986. Social organization of galagos in Kenyan coastal forests: II. *Galago garnettii*. Amer. J. Primatol. 10:357–69.

Natori, M. 1988. A cladistic analysis of interspecific relationships of *Saguinus*. Primates 29:263–76.

———. 1990. Numerical analysis of the taxonomical status of *Callithrix kuhli* based on measurements of the postcanine dentition. Primates 31:555–62.

———. 1994. Craniometrical variations among eastern Brazilian marmosets and their systematic relationships. Primates 35:167–76.

Natori, M., and T. Hanihara. 1988. An analysis of interspecific relationships of *Saguinus* based on cranial measurements. Primates 29:255–62.

———. 1992. Variations in dental measurements between *Saguinus* species and their systematic relationships. Folia Primatol. 58:84–92.

Neugebauer, W. 1980. The status and management of the pygmy chimpanzee *Pan paniscus* in European zoos. Internatl. Zoo Yearbook 20:64–70.

Neville, M., N. Castro, A. Marmol, and J. Revilla. 1976. Censusing primate populations in the reserved area of the Pacaya and Samiria rivers, Department Loreto, Peru. Primates 17:151–81.

Neyman, P. F. 1977. Aspects of the ecology and social organization of free-ranging cotton-top tamarins *(Saguinus oedipus)* and the conservation status of the species. *In* Kleiman (1977a), 39–71.

Niemitz, C. 1984b. An investigation and review of the territorial behaviour and social organisation of the genus *Tarsius*. *In* Niemitz (1984a), 117–27.

Niemitz, C., A. Nietsch, S. Warter, and Y. Rumpler. 1991. *Tarsius dianae:* a new primate species from central Sulawesi (Indonesia). Folia Primatol. 56:105–16.

Nieuwenhuijsen, K., Ad J. J. C. Lammers, K. J. de Neef, and A. K. Slob. 1985. Reproduction and social rank in female stumptail macaques *(Macaca arctoides)*. Internatl. J. Primatol. 6:77–99.

Nisbett, R. A., and R. L. Ciochon. 1993. Primates in northern Viet Nam: a review of the ecology and conservation status of extant species, with notes on Pleistocene localities. Internatl. J. Primatol. 14:765–95.

Nishida, T. 1972a. A note on the ecology of the red-colobus monkeys *(Colobus badius tephrosceles)* living in the Mahali Mountains. Primates 13:57–64.

———. 1972b. Preliminary information of the pygmy chimpanzees *(Pan paniscus)* of the Congo Basin. Primates 13:415–25.

Nishida, T., S. Uehara, and R. Nyundo. 1979. Predatory behavior among wild chimpanzees of the Mahale Mountains. Primates 20:1–20.

Nishimura, A. 1990. A sociological and behavioral study of woolly monkeys, *Lagothrix lagotricha*, in the upper Amazon. Sci. Engineering Rev. Doshisha Univ. (Kyoto) 31(2):1–121.

Nishimura, A., and K. Izawa. 1975. The group characteristics of woolly monkeys *(Lagothrix lagotricha)* in the upper Amazonian Basin. Proc. 5th Internatl. Congr. Primatol., 351–57.

O

Oates, J. F. 1977a. The guereza and its food. *In* Clutton-Brock (1977), 276–321.

———. 1977b. The guereza and man. *In* Rainier III and Bourne (1977), 419–67.

———. 1977c. The social life of a black-and-white colobus monkey, *Colobus guereza*. Z. Tierpsychol. 45:1–60.

———. 1982. In search of rare forest primates in Nigeria. Oryx 16:431–36.

———. 1984. The niche of the potto, *Perodicticus potto*. Internatl. J. Primatol. 5:51–61.

———. 1985. Action plan for African primate conservation: 1986–1990. IUCN (World Conservation Union), Gland, Switzerland, 41 pp.

———. 1988. The distribution of *Cercopithecus* monkeys in West African forests. *In* Gautier-Hion et al. (1988), 78–103.

———. 1994. The natural history of African colobines. *In* Davies and Oates (1994), 75–128.

———. 1996. African primates: status survey and conservation action plan. IUCN (World Conservation Union), Gland, Switzerland, viii + 80 pp.

Oates, J. F., and P. A. Anadu. 1989. A field observation of Sclater's guenon (*Cercopithecus sclateri* Pocock, 1904). Folia Primatol. 52: 93–96.

Oates, J. F., P. A. Anadu, E. L. Gadsby, and J. L. Werre. 1992. Sclater's guenon. Natl. Geogr. Res. Explor. 8:476–91.

Oates, J. F., A. G. Davies, and E. Delson. 1994. The diversity of living colobines. *In* Davies and Oates (1994), 45–73.

Oates, J. F., and T. F. Trocco. 1983. Taxonomy and phylogeny of black-and-white colobus monkeys: inferences from an analysis of loud call variation. Folia Primatol. 40:83–113.

Ohsawa, H., and M. Kawai. 1975. Social structure of gelada baboons: studies of the gelada society (I). Proc. 5th Internatl. Congr. Primatol., 464–69.

O'Leary, H. 1993. Monkey business in Gibraltar. Oryx 27:55–57.

Oliveira, J. M. S., M. G. Lima, C. Bonvincino, J. M. Ayres, and J. G. Fleagle. 1985. Preliminary notes on the ecology and behavior of the Guianan saki. Acta Amazonica 15:249–63.

Oliver, W. L. R., and I. B. Santos. 1991. Threatened endemic mammals of the Atlantic forest region of southeast Brazil. Wildlife Preservation Trust Spec. Sci. Rept., no. 4, 126 pp.

Olney, P. J. S., P. Ellis, and F. A. Fisken. 1994. Census of rare animals in captivity. Internatl. Zoo Yearbook 33:408–53.

Olupot, W., C. A. Chapman, C. H. Brown, and P. M. Waser. 1994. Mangabey (*Cercocebus albigena*) population density, group size, and ranging: a twenty-year comparison. Amer. J. Primatol. 32:197–205.

Omar, A., and A. DeVos. 1971. The annual reproductive cycle of an African monkey (*Cercopithecus mitis kolbi* Neuman). Folia Primatol. 16:206–15.

Oppenheimer, J. R. 1969. Changes in forehead patterns and group composition of the white-faced monkey (*Cebus capucinus*). Proc. 2nd Internatl. Congr. Primatol. 1:36–42.

———. 1977. *Presbytis entellus,* the hanuman langur. *In* Rainier III and Bourne (1977), 469–512.

Oppenheimer, J. R., and E. C. Oppenheimer. 1973. Preliminary observations of *Cebus nigrivittatus* (Primates: Cebidae) on the Venezuelan llanos. Folia Primatol. 19:409–36.

Ottocento, R., L. Marquez, R. Bodini, and G. A. Cordero R. 1989a. On the presence of *Cebus apella margaritae* on Margarita Island, northeastern Venezuela. Primate Conserv. 10:19–21.

Owen, R. C. 1984. The Americas: the case against an Ice-Age human population. *In* Smith, F. H., and F. Spencer, eds., The origins of modern humans: a world survey of the fossil evidence, Alan R. Liss, New York, 517–63.

Oxnard, C. E. 1981. The uniqueness of *Daubentonia.* Amer. J. Phys. Anthropol. 54:1–21.

———. 1985. Humans, apes, and Chinese fossils: new implication for human evolution. Hong Kong Univ. Press Occas. Pap. Ser., no. 4, 46 pp.

P

Pages, E. 1978. Home range, behaviour, and tactile communication in a nocturnal Malagasy lemur *Microcebus coquereli. In* Chivers and Joysey (1978), 171–77.

———. 1980. Ethoecology of *Microcebus coquereli* during the dry season. *In* Charles-Dominique et al. (1980), 97–116.

Patterson, F. 1986. The mind of the gorilla: conversation and conservation. *In* Benirschke (1986), 933–47.

Payne, K., and R. Payne. 1985. Large scale changes over nineteen years in songs of humpback whales in Bermuda. Z. Tierpsychol. 68:89–114.

Peng Yan-Zhang, Pan Ru-Liang, and N. G. Jablonski. 1993. Classification and evolution of Asian colobines. Folia Primatol. 60:106–17.

Pereira, M. E., M. L. Seeligson, and J. M. Macedonia. 1988. The behavioral repertoire of the black-and-white ruffed lemur, *Varecia variegata variegata* (Primates: Lemuridae). Folia Primatol. 51:1–32.

Peres, C. A. 1989. Costs and benefits of territorial defense in wild golden lion tamarins, *Leontopithecus rosalia.* Behav. Ecol. Sociobiol. 25:227–33.

———. 1991a. Humboldt's woolly monkeys decimated by hunting in Amazonia. Oryx 25:89–95.

———. 1994. Which are the largest New World monkeys? J. Human Evol. 26:245–49.

Perkins, L. 1993. Orangutan GASP Report. CBSG News, IUCN (World Conservation Union) Captive Breeding Specialist Group Newsl. 4(3):16–17.

Petter, J.-J. 1965. The lemurs of Madagascar. *In* De Vore (1965), 292–319.

———. 1975. Breeding of Malagasy lemurs in captivity. *In* Martin (1975b), 187–202.

———. 1977. The aye-aye. *In* Rainier III and Bourne (1977), 37–57.

———. 1978. Ecological and physiological adaptations of five sympatric nocturnal lemurs to seasonal variations in food production. *In* Chivers and Herbert (1978), 211–23.

Petter, J.-J., and P. Charles-Dominique. 1979. Vocal communication in prosimians. *In* Doyle and Martin (1979), 247–305.

Petter, J.-J., and A. Petter. 1967. The aye-aye of Madagascar. *In* Altmann (1967), 195–205.

Petter, J.-J., and A. Petter-Rousseaux. 1979. Classification of the prosimians. *In* Doyle and Martin (1979), 1–44.

Petter, J.-J., and A. Peyrieras. 1975. Preliminary notes on the behavior and ecology of *Hapalemur griseus. In* Tattersall and Sussman (1975), 281–86.

Petter, J.-J., A. Schilling, and G. Pariente. 1975. Observations on behavior and ecology of *Phaner furcifer. In* Tattersall and Sussman (1975), 209–18.

Petter-Rousseaux, A., and J.-J. Petter. 1967. Contribution á la systématique des Cheirogaleinae (lémuriens malagaches): *Allocebus,* gen. nov. pour *Cheirogaleus trichotis* Günther 1875. Mammalia 31:574–82.

Phillips-Conroy, J. E., C. J. Jolly, and F. L. Brett. 1991. Characteristics of hamadryas-like male baboons living in anubis baboon troops in the Awash hybrid zone, Ethiopia. Amer. J. Phys. Anthropol. 86:353–68.

Pickford, M. 1991. Paradise lost: mitochondrial Eve refuted. Human Evol. 6:263–68.

———. 1993. Climatic change, biogeography, and *Theropithecus. In* Jablonski (1993b), 227–43.

Pieczarka, J. C., R. M. de Souza Barros, F. M. de Faria, Jr., and C. Y. Nagamachi. 1993. *Aotus* from the southwestern Amazon region is geographically and chromosomally intermediate between *A. azarae boliviensis* and *A. infulatus.* Primates 34:197–204.

Pilbeam, D. 1984. The descent of hominoids and hominids. Sci. Amer. 250(3):84–96.

Pinto, L. P. S., and L. I. Tavares. 1994. Inventory and conservation status of wild populations of golden-headed lion tamarins, *Leontopithecus chrysomelas.* Neotropical Primates 2(suppl.):24–27.

Poirier, F. E. 1970. The Nilgiri langur (*Presbytis johnii*) of south India. Primate Behav. 1:254–383.

———. 1972. The St. Kitts green monkey (*Cercopithecus aethiops sabaeus*): ecology, population dynamics, and selected behavior traits. Folia Primatol. 17:20–55.

———. 1982. Taiwan macaques: ecology and conservation needs. *In* Chiarelli and Corruccini (1982), 138–42.

Pola, Y., and C. T. Snowdon. 1975. The vocalizations of pygmy marmosets *(Cebuella pygmaea)*. Anim. Behav. 23:826–42.

Pollock, J. I. 1975. Field observations on *Indri indri*: a preliminary report. *In* Tattersall and Sussman (1975), 287–311.

———. 1977. The ecology and sociobiology of feeding in *Indri indri*. *In* Clutton-Brock (1977), 37–69.

———. 1979. Spatial distribution and ranging behavior in lemurs. *In* Doyle and Martin (1979), 359–409.

———. 1986. The song of the indris *(Indri indri*; Primates: Lemuroidea): natural history, form, and function. Internatl. J. Primatol. 7:225–64.

Pollock, J. I., I. D. Constable, R. A. Mittermeier, J. Ratsirarson, and H. Simons. 1985. A note on the diet and feeding behavior of the aye-aye *Daubentonia madagascariensis*. Internatl. J. Primatol. 6:435–47.

Pook, A. G., and G. Pook. 1981. A field study of the socio-ecology of the Goeldi's monkey *(Callimico goeldii)* in northern Bolivia. Folia Primatol. 35:288–312.

Preuschoft, H. 1971. Mode of locomotion in subfossil giant lemuroids from Madagascar. Proc. 3rd Internatl. Congr. Primatol. 1:79–90.

Prouty, L. A., P. D. Buchanan, W. S. Politzer, and A. R. Mootnick. 1983. *Bunopithecus*: a genus-level taxon for the hoolock gibbon *(Hylobates hoolock)*. Amer. J. Primatol. 5: 83–87.

Q

Qu Wenyuan, Zhang Yongzu, D. Manry, and C. H. Southwick. 1993. Rhesus monkeys *(Macaca mulatta)* in the Taihang Mountains, Jiyuan County, Henan, China. Internatl. J. Primatol. 14:607–21.

Queiroz, H. L. 1992. A new species of capuchin monkey, genus *Cebus* Erxleben, 1777 (Cebidae: Primates), from eastern Brazilian Amazonia. Goeldiana Zool., no. 15, 13 pp.

Quris, R. 1975. Ecologie et organisation sociale de *Cercocebus galeritus agilis* dans le nord-est du Gabon. Terre Vie 20:337–98.

R

Raemaekers, J. 1984. Large versus small gibbons: relative roles of bioenergetics and competition in their ecological segregation in sympatry. *In* Preuschoft et al. (1984), 209–18.

Rahm, U. 1970a. Ecology, zoogeography, and systematics of some African forest monkeys. *In* Napier and Napier (1970), 589–626.

Rakotoarison, N., T. Mutschler, and U. Thalmann. 1993. Lemurs in Bemaraha (World Heritage Landscape, western Madagascar). Oryx 27:35–40.

Ramirez, M. F., C. Freese, and J. Revilla C. 1977. Feeding ecology of the pygmy marmoset, *Cebuella pygmaea*, in northeastern Peru. *In* Kleiman (1977a), 91–104.

Rathbun, G. B., and M. Gache. 1980. Ecological survey of the night monkey, *Aotus trivirgatus*, in Formosa Province, Argentina. Primates 21:211–19.

Ratomponirina, C., J. Andrianivo, and Y. Rumpler. 1982. Spermatogenesis in several intra- and interspecific hybrids of the lemur *(Lemur)*. J. Reprod. Fert. 66:717–21.

Ravosa, M. J. 1992. Allometry and heterochrony in extant and extinct Malagasy primates. J. Human Evol. 23:197–217.

Redford, K. H., and J. F. Eisenberg, eds. 1992. Mammals of the neotropics: the southern cone. Univ. Chicago Press, ix + 430 pp.

Rendall, D. 1993. Does female social precedence characterize captive aye-ayes *(Daubentonia madagascariensis)?* Internatl. J. Primatol. 14:125–30.

Reynolds, V., and F. Reynolds. 1965. Chimpanzees of the Budongo Forest. *In* De Vore (1965), 368–424.

Richard, A. 1974. Patterns of mating in *Propithecus verreauxi verreauxi*. *In* Martin, Doyle, and Walker (1974), 49–74.

———. 1977. The feeding behaviour of *Propithecus verreauxi*. *In* Clutton-Brock (1977), 72–96.

———. 1978. Variability in the feeding behavior of a Malagasy prosimian, *Propithecus verreauxi*: Lemuriformes. *In* Montgomery (1978), 519–33.

———. 1985. Social boundaries in a Malagasy prosimian, the sifaka *(Propithecus verreauxi)*. Internatl. J. Primatol. 6:553–68.

Richard, A., and R. W. Sussman. 1975. Future of the Malagasy lemurs: conservation or extinction? *In* Tattersall and Sussman (1975), 335–50.

Rigamonti, M. M. 1993. Home range and diet in red ruffed lemurs *(Varecia variegata rubra)* on the Masoala Peninsula, Madagascar. *In* Kappeler and Ganzhorn (1993), 25–39.

Rijksen, H. D. 1978. A field study on Sumatran orangutans *(Pongo pygmaeus abelli* Lesson 1827). Meded. Landbouwhogeschool, Wageningen, Netherlands, 420 pp.

———. 1986. Conservation of orangutans: a status report, 1985. *In* Benirschke (1986), 153–59.

Rivero, M., and O. Arredondo. 1991. *Paralouatta varonai*, a new Quaternary platyrrhine from Cuba. J. Human Evol. 21:1–11.

Rodman, P. S. 1973. Population composition and adaptive organisation among orang-utans of the Kutai Reserve. *In* Michael and Crook (1973), 171–209.

———. 1977. Feeding behaviour of orangutans of the Kutai Nature Reserve, East Kalimantan. *In* Clutton-Brock (1977), 384–413.

———. 1978. Diets, densities, and distributions of Bornean primates. *In* Montgomery (1978), 465–78.

———. 1988. Diversity and consistency in ecology and behavior. *In* Schwartz (1988b), 31–51.

Rogers, J. 1993. The phylogenetic relationships among *Homo, Pan*, and *Gorilla*: a population genetics perspective. J. Human Evol. 25:201–15.

Roonwal, M. L., and S. M. Mohnot. 1977. Primates of south Asia. Harvard Univ. Press, Cambridge, xviii + 421 pp.

Rosenberger, A. L. 1977. *Xenothrix* and ceboid phylogeny. J. Human Evol. 6:461–81.

Rosenberger, A. L., and A. F. Coimbra-Filho. 1984. Morphology, taxonomic status, and affinities of the lion tamarins, *Leontopithecus* (Callitrichinae, Cebidae). Folia Primatol. 42:149–79.

Rosenberger, A. L., T. Setoguchi, and N. Shigehara. 1990. The fossil record of callitrichine primates. J. Human Evol. 19:209–36.

Rosenblum, L. A. 1968. Some aspects of female reproductive physiology in the squirrel monkey. *In* Rosenblum and Cooper (1968), 147–69.

Ross, C. 1991. Life history patterns of New World monkeys. Internatl. J. Primatol. 12: 481–502.

———. 1992. Life history patterns and ecology of macaque species. Primates 33: 207–15.

Rowell, T. E. 1971. Organization of caged groups of *Cercopithecus* monkeys. Anim. Behav. 19:625–45.

———. 1977a. Reproductive cycles of the talapoin monkey *(Miopithecus talapoin)*. Folia Primatol. 28:188–202.

———. 1977b. Variation in age at puberty in monkeys. Folia Primatol. 27:284–96.

Rudran, R. 1973. The reproductive cycles of two subspecies of purple-faced langurs *(Presbytis senex)* with relation to environmental factors. Folia Primatol. 19:41–60.

———. 1978. Socioecology of the blue monkeys *(Cercopithecus mitis stuhlmani)* of the Kibale Forest, Uganda. Smithson. Contrib. Zool., no. 249, iv + 88 pp.

Ruhiyat, Y. 1983. Socio-ecological study of *Presbytis aygula* in west Java. Primates 24: 344–59.

Rumbaugh, D. M. 1970. Learning skills of anthropoids. Primate Behav. 1:1–70.

———, ed. 1972. Gibbon and siamang. I. Evolution, ecology, behavior, and captive maintenance. S. Karger, Basel, x + 263 pp.

Rumpler, Y. 1974. Cytogenetic contributions to a new classification of lemurs. In Martin, Doyle, and Walker (1974), 865–69.

———. 1975. The significance of chromosomal studies in the systematics of the Malagasy lemurs. In Tattersall and Sussman (1975), 25–40.

Rumpler, Y., and R. Albignac. 1975. Intraspecific chromosome variability in a lemur from the north of Madagascar: Lepilemur septentrionalis, species nova. Amer. J. Phys. Anthropol. 42:425–30.

———. 1977. Chromosome studies of the Lepilemur, an endemic Malagasy genus of lemurs: contribution of the cytogenetics to their taxonomy. J. Human Evol. 7:191–96.

Rumpler, Y., and B. Dutrillaux. 1978. Chromosomal evolution in Malagasy lemurs. III. Chromosome banding studies in the genus Hapalemur and the species Lemur catta. Cytogenet. Cell Genet. 21:201–11.

Rumpler, Y., S. Warter, J.-J. Petter, R. Albignac, and B. Dutrillaux. 1988. Chromosomal evolution of Malagasy lemurs. XI. Phylogenetic position of Daubentonia madagascariensis. Folia Primatol. 50:124–29.

Rumpler, Y., S. Warter, C. Rabarivola, J. J. Petter, and B. Dutrillaux. 1990. Chromosomal evolution in Malagasy lemurs. XII. Chromosomal banding study of Avahi laniger occidentalis (syn: Lichanotus laniger occidentalis) and cytogenetic data in favour of its classification in a species apart—Avahi occidentalis. Amer. J. Primatol. 21:307–16.

Rylands, A. B. 1993a. The ecology of the lion tamarins, Leontopithecus: some intrageneric differences and comparisons with other callitrichids. In Rylands (1993b), 296–313.

———. 1994. Population viability analyses and the conservation of the lion tamarins, Leontopithecus, of south-east Brazil. Primate Conserv. 14:34–42.

Rylands, A. B., A. F. Coimbra-Filho, and R. A. Mittermeier. 1993. Systematics, geographic distribution, and some notes on the conservation status of the Callitrichidae. In Rylands (1993b), 11–77.

Rylands, A. B., and D. S. de Faria. 1993. Habitats, feeding ecology, and home range size in the genus Callithrix. In Rylands (1993b), 262–72.

Rylands, A. B., and E. R. Luna. 1993. A new species of untufted capuchin from the Brazilian Amazon. Neotropical Primates 1(2):5–7.

Rylands, A. B., R. A. Mittermeier, and E. R. Luna. 1995. A species list for the New World Primates (Platyrrhini): distribution by coun-

try, endemism, and conservation status according to the Mace-Land system. Neotropical Primates 3(suppl.):113–60.

S

Sabater Pi, J. 1972. Contribution to the ecology of Mandrillus sphinx Linnaeus 1758 of Rio Muni (Republic of Equatorial Guinea). Folia Primatol. 17:304–19.

———. 1973. Contribution to the ecology of Colobus polykomos satanas (Waterhouse 1838) of Rio Muni, Republic of Equatorial Guinea. Folia Primatol. 19:193–207.

Sade, D. S. 1967. Determinants of dominance in a group of free-ranging rhesus monkeys. In Altmann (1967), 99–114.

Salgado, E. J., D. G. Calvache, R. D. E. MacPhee, and G. C. Gould. 1992. The monkey caves of Cuba. Cave Sci. 19(1):25–28.

Salter, R. E., N. A. MacKenzie, N. Nightingale, K. M. Aken, and P. P. K. Chai. 1985. Habitat use, ranging behaviour, and food habits of the proboscis monkey, Nasalis larvatus (van Wurmb), in Sarawak. Primates 26:436–51.

Sarmiento, E. E., T. M. Butynski, and J. Kalina. 1995. Taxonomic status of the gorillas of the Bwindi-Impenetrable Forest, Uganda. Primate Conserv. 16:39–42.

Sauther, M. L., and R. W. Sussman. 1993. A new interpretation of the social organization and mating system of the ringtailed lemur (Lemur catta). In Kappeler and Ganzhorn (1993), 111–21.

Schaller, G. B. 1963. The mountain gorilla. Univ. Chicago Press, xvii + 431 pp.

———. 1965. The behavior of the mountain gorilla. In De Vore (1965), 324–67.

Schlitter, D. A., J. Phillips, and G. E. Kemp. 1973. The distribution of the white-collared mangabey, Cercocebus torquatus, in Nigeria. Folia Primatol. 19:380–83.

Schmid, J., and P. M. Kappeler. 1994. Sympatric mouse lemurs (Microcebus spp.) in western Madagascar. Folia Primatol. 63:162–70.

Schneider, H., M. P. C. Schneider, I. Sampaio, M. L. Harada, M. Stanhope, J. Czelusniak, and M. Goodman. 1993. Molecular phylogeny of the New World monkeys (Platyrrhini, Primates). Molecular Phylogenet. Evol. 2: 225–42.

Schrenk, F., T. G. Bromage, C. G. Betzler, U. Ring, and Y. M. Juwayeyi. 1993. Oldest Homo and Pliocene biogeography of the Malawi Rift. Nature 365:833–36.

Schultz, A. H. 1942. Growth and development of the proboscis monkey. Bull. Mus. Comp. Zool. 89:279–314.

Schulze, H., and B. Meier. 1994. The subspecies of Loris tardigradus and their con-

servation status: a review. In Alterman, L., G. A. Doyle, and M. K. Izard, eds., Creatures of the dark: the nocturnal prosimians, Plenum Press, New York.

Schwartz, J. H. 1986. Primate systematics and a classification of the order. In Swindler and Erwin (1986), 1–41.

———. 1987. The red ape: orang-utans and human origins. Houghton Mifflin, Boston, 352 pp.

———. 1988a. History, morphology, paleontology, and evolution. In Schwartz (1988b), 69–85.

———. 1996. Pseudopotto martini: a new genus and species of extant lorisiform primate. Anthropol. Pap. Amer. Mus. Nat. Hist. 78:1–14.

Schwartz, J. H., and I. Tattersall. 1985. Evolutionary relationships of living lemurs and lorises (Mammalia, Primates) and their potential affinities with European Eocene Adapidae. Anthropol. Pap. Amer. Mus. Nat. Hist. 60:1–100.

———. 1987. Tarsiers, adapids, and the integrity of Strepsirhini. J. Human Evol. 16: 23–40.

Sekulic, R. 1982a. Daily and seasonal patterns of roaring and spacing in four red howler Alouatta seniculus troops. Folia Primatol. 39:22–48.

———. 1982b. The function of howling in red howler monkeys (Alouatta seniculus). Behaviour 81:38–54.

Setoguchi, T., and A. L. Rosenberger. 1985. Miocene marmosets: first fossil evidence. Internatl. J. Primatol. 6:615–25.

Shea, B. T. 1984. Between the gorilla and the chimpanzee: a history of debate concerning the existence of the kooloo-kamba or gorilla-like chimpanzee. J. Ethnobiol. 4:1–13.

Shea, B. T., S. R. Leigh, and C. P. Groves. 1993. Multivariate craniometric variation in chimpanzees: implications for species identification in paleoanthropology. In Kimbel and Martin (1993), 265–95.

Shedd, D. H., and J. M. Macedonia. 1991. Metachromism and its phylogenetic implications for the genus Eulemur (Prosimii: Lemuridae). Folia Primatol. 57:221–31.

Shoemaker, A. H. 1982a. Fecundity in the captive howler monkey, Alouatta caraya. Zoo Biol. 1:149–56.

———. 1982b. Notes on the reproductive biology of the white-faced saki Pithecia pithecia in captivity. Internatl. Zoo Yearbook 22: 124–27.

Sigg, H., A. Stolba, J.-J. Abegglen, and V. Dasser. 1982. Life history of hamadryas baboons: physical development, infant mortality, reproductive parameters, and family relationships. Primates 23:473–87.

Silk, J., J. Short, J. Roberts, and J. Kusnitz. 1993. Gestation length in rhesus macaques *(Macaca mulatta)*. Internatl. J. Primatol. 14: 95–104.

Silva, B. T. F., M. I. C. Sampaio, H. Schneider, M. P. C. Schneider, E. Montoya, F. Encarnación, and F. M. Salzano. 1992. Natural hybridization between *Saimiri* taxa in the Peruvian Amazonia. Primates 33:107–13.

Simonds, P. E. 1965. The bonnet macaque in south India. *In* De Vore (1965), 175–96.

Simons, E. L. 1988. A new species of *Propithecus* (Primates) from northeast Madagascar. Folia Primatol. 50:143–51.

———. 1989. Human origins. Science 245: 1343–50.

———. 1993. Discovery of the western aye-aye. Lemur News 1(1):6.

Simons, E. L., and T. M. Bown. 1985. *Afrotarsius chatrathi*, first tarsiform primate (? Tarsiidae) from Africa. Nature 313:475–77.

Simons, E. L., L. R. Godfrey, W. L. Jungers, P. S. Chatrath, and B. Rakotosamimanana. 1992. A new giant subfossil lemur, *Babakotia*, and the evolution of the sloth lemurs. Folia Primatol. 58:197–203.

Simons, E. L., L. R. Godfrey, M. Vuillaume-Randriamanantena, P. S. Chatrath, and M. Gagnon. 1990. Discovery of new giant subfossil lemurs in the Ankarana Mountains of northern Madagascar. J. Human Evol. 19: 311–19.

Simons, E. L., and Y. Rumpler. 1988. *Eulemur:* new generic name for species of *Lemur* other than *Lemur catta*. Compt. Rend. Acad. Sci. Paris, ser. 3, 307:547–51.

Simpson, G. G. 1945. The principles of classification and a classification of the mammals. Bull. Amer. Mus. Nat. Hist. 85:i–xvi + 1–350.

Sineo, L., R. Stanyon, and B. Chiarelli. 1986. Chromosomes of the *Cercopithecus aethiops* species group: *C. aethiops* (Linnaeus, 1758), *C. cynosurus* (Scopoli, 1786), *C. pygerythrus* (Cuvier, 1821), and *C. sabaeus* (Linnaeus, 1766). Internatl. J. Primatol. 7:569–82.

Skelton, R. R., and H. M. McHenry. 1992. Evolutionary relationships among early hominids. J. Human Evol. 23:309–49.

Skinner, C. 1991. Justification for reclassifying Geoffroy's tamarin from *Saguinus oedipus geoffroyi* to *Saguinus geoffroyi*. Primate Rept. 31:77–84.

Smith, C. C. 1977. Feeding behaviour and social organization in howling monkeys. *In* Clutton-Brock (1977), 96–126.

Smithers, R. H. N. 1983. The mammals of the southern African subregion. Univ. Pretoria, xxii + 736 pp.

Snyder, P. A. 1974. Behavior of *Leontopithecus rosalia* (golden-lion marmoset) and related species: a review. J. Human Evol. 3:109–22.

Soini, P. 1982. Ecology and population dynamics of the pygmy marmoset, *Cebuella pygmaea*. Folia Primatol. 39:1–21.

———. 1993. The ecology of the pygmy marmoset, *Cebuella pygmaea*: some comparisons with two sympatric tamarins. *In* Rylands (1993b), 257–72.

Southwick, C. H., M. A. Beg, and M. F. Siddiqi. 1965. Rhesus monkeys in north India. *In* De Vore (1965), 111–59.

Southwick, C. H., and D. G. Lindburg. 1986. The primates of India: status, trends, and conservation. *In* Benirschke (1986), 171–87.

Southwick, C. H., and M. F. Siddiqi. 1977. Population dynamics of rhesus monkeys in northern India. *In* Rainier III and Bourne (1977), 339–62.

———. 1988. Partial recovery and a new population estimate of rhesus populations in India. Amer. J. Primatol. 16:187–97.

———. 1994a. Population status of nonhuman primates in Asia, with emphasis on rhesus macaques in India. Amer. J. Primatol. 34:51–59.

———. 1994b. Primate commensalism: the rhesus monkey in India. Rev. Ecol. (Terre Vie) 49:223–31.

Southwick, C. H., M. F. Siddiqi, J. A. Cohen, J. R. Oppenheimer, J. Khan, and S. W. Ashraf. 1982. Further declines in rhesus populations of India. *In* Chiarelli and Corruccini (1982), 128–37.

Southwick, C. H., M. F. Siddiqi, M. Y. Farooqui, and B. C. Pal. 1974. Xenophobia among free-ranging rhesus groups in India. *In* Holloway (1974), 185–209.

Southwick, C. H., M. F. Siddiqi, and J. R. Oppenheimer. 1983. Twenty-year changes in rhesus monkey populations in agricultural areas of northern India. Ecology 64:434–39.

Srikosamatara, S. 1984. Ecology of pileated gibbons in south-east Thailand. *In* Preuschoft et al. (1984), 242–57.

Stanford, C. B. 1992. Comparative ecology of the capped langur *Presbytis pileata* Blyth in two forest types in Bangladesh. J. Bombay Nat. Hist. Soc. 89:187–93.

Stanger, K. F., B. S. Coffman, and M. K. Izard. 1995. Reproduction in Coquerel's dwarf lemur *(Mirza coquereli)*. Amer. J. Primatol. 36:223–37.

Stanyon, R., J. Wienberg, E. L. Simons, and M. K. Izard. 1992. A third karyotype for *Galago demidovii* suggests the existence of multiple species. Folia Primatol. 59:33–38.

Starin, E. D. 1981. Monkey moves. Nat. Hist. 90(9):36–43.

Sterling, E. I. 1993. Patterns of range use and social organization in aye-ayes *(Daubentonia madagascariensis)* on Nosy Mangabe. *In* Kappeler and Ganzhorn (1993), 1–10.

Stevenson, M. F. 1973. Notes on pregnancy in the sooty mangabey *Cercocebus atys*. Internatl. Zoo Yearbook 13:134–35.

Steyn, H., and A. T. C. Feistner. 1994. Development of a captive-bred infant Alaotran gentle lemur *Hapalemur griseus alaotrensis*. Dodo 30:47–57.

Strahan, R., ed. 1983. The Australian Museum complete book of Australian mammals. Angus & Robertson, London, xxi + 530 pp.

Strier, K. B. 1990. New World primates, new frontiers: insights from the woolly spider monkey, or muriqui *(Brachyteles arachnoides)*. Internatl. J. Primatol. 11:7–19.

———. 1992. Faces in the forest: the endangered muriqui monkeys of Brazil. Oxford Univ. Press, New York, 138 pp.

———. 1995. Viability analyses of an isolated population of muriqui monkeys *(Brachyteles arachnoides):* implications for primate conservation and demography. Primate Conserv. 15:43–52.

Strier, K. B., F. D. C. Mendes, J. Rímoli, and A. O. Rímoli. 1993. Demography and social structure of one group of muriquis *(Brachyteles arachnoides)*. Internatl. J. Primatol. 14:513–26.

Struhsaker, T. T. 1967a. Auditory communication among vervet monkeys *(Cercopithecus aethiops)*. *In* Altmann (1967), 281–324.

———. 1967b. Ecology of vervet monkeys *(Cercopithecus aethiops)* in the Masai-Amboseli Game Reserve, Kenya. Ecology 48: 891–904.

———. 1971. Notes on *Cercocebus a. atys* in Senegal, West Africa. Mammalia 35:343–44.

———. 1975. The red colobus monkey. Univ. Chicago Press. xiv + 311 pp.

———. 1981. Vocalizations, phylogeny, and palaeogeography of red colobus monkeys *(Colobus badius)*. Afr. J. Ecol. 19:265–83.

Struhsaker, T. T., and J. S. Gartlan. 1970. Observations on the behaviour and ecology of the patas monkey *(Erythrocebus patas)* in the Waza Reserve. Cameroon. J. Zool. 161: 49–63.

Struhsaker, T. T., and L. Leland. 1988. Group fission in redtail monkeys *(Cercopithecus ascanius)* in the Kibale Forest, Uganda. *In* Gautier-Hion et al. (1988), 364–88.

Sugardjito, J., and N. Nurhuda. 1981. Meat-eating behaviour in wild orang utans, *Pongo pygmaeus*. Primates 22:414–16.

Sugiyama, Y. 1973. The social structure of wild chimpanzees: a review of field studies. *In* Michael and Crook (1973), 375–41.

Supriatna, J., J. W. Froehlich, J. M. Erwin, and C. H. Southwick. 1992. Population, habitat, and conservation status of *Macaca maurus, Macaca tonkeana,* and their putative hybrids. Trop. Diversity 1:31–48.

Susman, R. L. 1984a. The locomotor behavior of *Pan paniscus* in the Lomako Forest. *In* Susman (1984b), 369–93.

———, ed. 1984b. The pygmy chimpanzee: evolutionary biology and behavior. Plenum Press, New York, xxviii + 435 pp.

Susman, R. L. 1995. The only way to determine the conservation status of the pygmy chimpanzee is to conduct a survey in the Zaire Basin: a reply to Dr. Kortlandt. Primate Conserv. 16:36–38.

Sussman, R. W. 1975. A preliminary study of the behavior and ecology of *Lemur fulvus rufus* Audebert 1800. *In* Tattersall and Sussman (1975), 237–58.

———. 1977. Feeding behaviour of *Lemur catta* and *Lemur fulvus. In* Clutton-Brock (1977), 1–36.

———. 1993. A current controversy in human evolution. Amer. Anthropol. 95:9–13.

Sussman, R. W., and W. G. Kinzey. 1984. The ecological role of the Callitrichidae: a review. Amer. J. Phys. Anthropol. 64:419–49.

Sussman, R. W., and A. Richard. 1974. The role of aggression among diurnal prosimians. *In* Holloway (1974), 49–76.

Sussman, R. W., and I. Tattersall. 1986. Distribution, abundance, and putative ecological strategy of *Macaca fascicularis* on the island of Mauritius, southwestern Indian Ocean. Folia Primatol. 46:28–43.

Suzuki, A. 1971. Carnivority and cannibalism observed among forest-living chimpanzees. J. Anthropol. Soc. Nippon 79:30–48.

Symington, M. M. 1988. Demography, ranging patterns, and activity budgets of black spider monkeys *(Ateles paniscus chamek)* in the Manu National Park, Peru. Amer. J. Primatol. 15:45–67.

Szalay, F. S., A. L. Rosenberger, and M. Dagosto. 1987. Diagnosis and differentiation of the order Primates. Yearbook Phys. Anthropol. 30:75–105.

T

Tan Bangjie. 1985. The status of primates in China. Primate Conserv. 5:63–81.

Tardiff, S. D., M. L. Harrison, and M. A. Simek. 1993. Communal infant care in marmosets and tamarins: relation to energetics, ecology, and social organization. *In* Rylands (1993b), 220–34.

Tattersall, I. 1971. Revision of the subfossil Indriinae. Folia Primatol. 16:257–69.

———. 1977a. Ecology and behavior of *Lemur fulvus mayottensis* (Primates, Lemuriformes). Anthropol. Pap. Amer. Mus. Nat. Hist. 54:421–82.

———. 1977b. The lemurs of the Comoro Islands. Oryx 13:445–48.

———. 1978a. Behavioural variation in *Lemur mongoz* (= L. m. mongoz). In Chivers and Joysey (1978), 127–32.

———. 1978b. Functional cranial anatomy of the subfossil Malagasy lemurs. Natl. Geogr. Soc. Res. Rept., 1969 Proj., 559–68.

———. 1982. The primates of Madagascar. Columbia Univ. Press, New York, xiv + 382 pp.

———. 1986. Systematics of the Malagasy strepsirhine primates. *In* Swindler and Erwin (1986), 43–72.

———. 1992. Human origins and the origins of humanity. Human Evol. 7:17–24.

———. 1993. Speciation and morphological differentiation in the genus *Lemur. In* Kimbel and Martin (1993), 163–76.

Tattersall, I., and J. H. Schwartz. 1975. Relationships among the Malagasy lemurs: the craniodental evidence. *In* Luckett and Szalay (1975), 299–312.

———. 1991. Phylogeny and nomenclature in the *Lemur*-group of Malagasy strepsirhine primates. Anthropol. Pap. Amer. Mus. Nat. Hist. 69:1–18.

Taub, D. M. 1977. Geographic distribution and habitat diversity of the Barbary macaque *Macaca sylvanus* L. Folia Primatol. 27:108–33.

———. 1984. A brief historical account of the recent decline in geographic distribution of the Barbary macaque in North Africa. *In* Fa (1984), 71–78.

Thalmann, U., T. Geissmann, A. Simona, and T. Mutschler. 1993. The indris of Anjanaharibe-Sud, northeastern Madagascar. Internatl. J. Primatol. 14:357–81.

Thompson-Handler, N., R. K. Malenky, and N. Badrian. 1984. Sexual behavior of *Pan paniscus* under natural conditions in the Lomako Forest, Equateur, Zaire. *In* Susman (1984b), 347–68.

Thompson-Handler, N., R. K. Malenky, and G. Reinartz. 1995. Action plan for *Pan paniscus*: report on free-ranging populations and proposals for their preservation. Zoological Society of Milwaukee County, Wis., x + 105 pp.

Thorington, R. W., Jr. 1968. Observations of squirrel monkeys in a Colombian forest. *In* Rosenblum and Cooper (1968), 69–85.

———. 1985. The taxonomy and distribution of squirrel monkeys *(Saimiri). In* Rosenblum, L. A., and C. L. Coe, eds., Handbook of squirrel monkey research, Plenum Press, New York, 1–33.

———. 1988. Taxonomic status of *Saguinus tripartitus* (Milne-Edwards, 1878). Amer. J. Primatol. 15:367–71.

Thorington, R. W., Jr., and S. Anderson. 1984. Primates. *In* Anderson and Jones (1984), 187–217.

Thorington, R. W., Jr., and C. P. Groves. 1970. An annotated classification of the Cercopithecoidea. *In* Napier and Napier (1970), 629–47.

Thorington, R. W., Jr., N. A. Muckenhirn, and G. G. Montgomery. 1976. Movements of a wild night monkey *(Aotus trivirgatus). In* Thorington and Heltne (1976), 32–34.

Thorington, R. W., Jr., J. C. Ruiz, and J. F. Eisenberg. 1984. A study of a black howling monkey *(Alouatta caraya)* population in northern Argentina. Amer. J. Primatol. 6:357–66.

Thorington, R. W., Jr., and R. E. Vorek. 1976. Observations on the geographic variation and skeletal development of *Aotus.* Lab. Anim. Sci. 26:1006–21.

Thornback, J., and M. Jenkins. 1982. The IUCN mammal red data book. Part 1: Threatened mammal taxa of the Americas and the Australasian zoogeographic region (excluding Cetacea). Internatl. Union Conserv. Nat., Gland, Switzerland, xl + 516 pp.

Tilson, R. L. 1977. Social organization of simakobu monkeys *(Nasalis concolor)* in Siberut Island, Indonesia. J. Mamm. 58: 202–12.

———. 1981. Family formation strategies of Kloss's gibbons. Folia Primatol. 35:259–87.

Tilson, R. L., and A. Eudey. 1993. IUCN/SSC CBSG Orangutan PHVA Report. CBSG News, IUCN (World Conservation Union) Captive Breeding Specialist Group Newsl. 4(3):14–15.

Timm, R. M., and E. C. Birney. 1992. Systematic notes on the Philippine slow loris, *Nycticebus coucang menagensis* (Lydekker, 1893) (Primates: Lorisidae). Internatl. J. Primatol. 13:679–86.

Torres de Assumpcao, C. 1983. Ecological and behavioural information on *Brachyteles arachnoides.* Primates 24:584–93.

Tsingalia, H. M., and T. E. Rowell. 1984. The behaviour of adult male blue monkeys. Z. Tierpsychol. 64:253–68.

Turnbull, C. M. 1976. Man in Africa. Anchor Press/Doubleday, Garden City, New York, xx + 313 pp.

Turner, A., and A. Chamberlain. 1989. Speciation, morphological change, and the status of African *Homo erectus.* J. Human Evol. 18:115–30.

Tutin, C. E. G., and M. Fernandez. 1983. Gorillas feeding on termites in Gabon, West Africa. J. Mamm. 64:530–31.

———. 1984. Nationwide census of gorilla (*Gorilla g. gorilla*) and chimpanzee (*Pan t. troglodytes*) populations in Gabon. Amer. J. Primatol. 6:313–36.

Tutin, C. E. G., and P. R. McGinnis. 1981. Chimpanzee reproduction in the wild. *In* Graham, C. E., ed., Reproductive biology of the great apes, Academic Press, New York, 239–64.

Tutin, C. E. G., W. C. McGrew, and P. J. Baldwin. 1983. Social organization of savanna-dwelling chimpanzees, *Pan troglodytes verus*, at Mt. Assirik, Senegal. Primates 24:154–73.

Tuttle, R. H. 1986. Apes of the world. Noyes, Park Ridge, New Jersey, xix + 421 pp.

Tyler, D. E. 1991*a*. The evolutionary relationships of *Aotus*. Folia Primatol. 56:50–52.

———. 1991*b*. The problems of the Pliopithecidae as a hylobatid ancestor. Human Evol. 6:73–80.

U

United States Department of Health, Education and Welfare. 1975. Restrictions on importation of nonhuman primates. Federal Register 40:33659.

USDI (United States Department of the Interior). 1976. Proposal to list twenty-seven species of primates as endangered or threatened species. Federal Register 41:1646–49.

V

Valladares-Padua, C., S. M. Padua, and L. Cullen, Jr. 1994. The conservation biology of the black lion tamarin, *Leontopithecus chrysopygus:* first ten years' report. Neotropical Primates 2(suppl.):36–39.

Van Horn, R. N., and G. G. Eaton. 1979. Reproductive physiology and behavior in prosimians. *In* Doyle and Martin (1979), 79–122.

Van Lawick–Goodall, J. 1968. The behaviour of free-living chimpanzees in the Gombe Stream Reserve. Anim. Behav. Monogr. 1: 165–311.

———. 1973. Cultural elements in a chimpanzee community. Symp. 4th Internatl. Congr. Primatol. 1:144–84.

Van Roosmaien, M. G. M. 1985. Habitat preferences, diet, feeding strategy, and social organization of the black spider monkey (*Ateles paniscus paniscus* Linnaeus 1758) in Surinam. Acta Amazonica 15(3–4):1–238.

Vedder, A. 1987. Report from the Gorilla Advisory Committee on the status of *Gorilla gorilla*. Primate Conserv. 8:75–81.

Vessey, S. H., B. K. Mortenson, and N. Muckenhirn. 1978. Size and characteristics of primate groups in Guyana. *In* Chivers and Herbert (1978), 187–88.

Visalberghi, E. 1990. Tool use in *Cebus*. Folia Primatol. 54:146–54.

Vitousek, P. M., H. A. Mooney, J. Lubchenco, and J. M. Melillo. 1997. Human domination of earth's ecosystem. Science 277:494–99.

Vuillaume-Randriamanantena, M. 1988. The taxonomic attributions of giant subfossil lemur bones from Ampasambazimba: *Archaeoindris* and *Lemuridotherium*. J. Human Evol. 17:379–91.

Vuillaume-Randriamanantena, M., L. R. Godfrey, W. L. Jungers, Jr., and E. L. Simons. 1992. Morphology, taxonomy, and distribution of *Megaladapis*—giant subfossil lemur from Madagascar. Compt. Rend. Acad. Sci. Paris, ser. 2, 315:1835–42.

Vuillaume-Randriamanantena, M., L. R. Godfrey, and M. R. Sutherland. 1985. Revision of *Hapalemur (Prohapalemur) gallieni* (Standing 1905). Folia Primatol. 45:89–116.

W

Wahome, J. M., T. E. Rowell, and H. M. Tsingalia. 1993. The natural history of de Brazza's monkey in Kenya. Internatl. J. Primatol. 14:445–66.

Walker, A. 1970. Nuchal adaptations in *Perodicticus potto*. Primates 11:134–44.

———. 1979. Prosimian locomotor behavior. *In* Doyle and Martin (1979), 543–65.

Wang Sung and Quan Guoqiang. 1986. Primate status and conservation in China. *In* Benirschke (1986), 213–20.

Ward, S. C. 1991. Taxonomy, paleobiology, and adaptations of the "robust" australopithecines. J. Human Evol. 21:469–83.

Waser, P. M. 1977. Feeding, ranging, and group size in the mangabey, *Cercocebus albigena*. *In* Clutton-Brock (1977), 183–222.

Waser, P. M., and O. Floody. 1974. Ranging patterns of the mangabey, *Cercocebus albigena*, in the Kibale Forest, Uganda. Z. Tierpsychol. 35:85–101.

Watanabe, K., H. Lapasere, and R. Tantu. 1991. External characteristics and associated developmental changes in two species of Sulawesi macaques, *Macaca tonkeana* and *M. hecki*, with special reference to hybrids and the borderland between the species. Primates 32:61–76.

Watanabe, K., and S. Matsumura. 1991. The borderlands and possible hybrids between three species of macaques, *M. nigra*, *M. nigrescens*, and *M. hecki*, in the northern peninsula of Sulawesi. Primates 32:365–69.

Watanabe, K., S. Matsumura, T. Watanabe, and Y. Hamada. 1991. Distribution and possible intergradation between *Macaca tonkeana* and *M. ochreata* at the borderland of the species in Sulawesi. Primates 32:385–89.

Watts, D. P. 1991. Mountain gorilla reproduction and sexual behavior. Amer. J. Primatol. 24:211–25.

———. 1992. Social relationships of immi-

grant and resident female mountain gorillas. I. Male-female relationships. Amer. J. Primatol. 28:159–81.

———. 1994. Social relationships of immigrant and resident female mountain gorillas. II. Relatedness, residence, and relationships between females. Amer. J. Primatol. 32:13–30.

Weber, A. W., and A. Vedder. 1983. Population dynamics of the Virunga gorillas: 1959–1978. Biol. Conserv. 26:341–66.

Weitzel, V. 1992. A review of the taxonomy of *Trachypithecus francoisi*. Austral. Primatol. 7(2):2–4.

Weitzel, V., and C. P. Groves. 1985. The nomenclature and taxonomy of the colobine monkeys of Java. Internatl. J. Primatol. 6: 399–409.

Westergaard, G. C. 1988. Lion-tailed macaques (*Macaca silenus*) manufacture and use tools. J. Compar. Psychol. 102:152–59.

Wheatley, B. P. 1980. Feeding and ranging of east Bornean *Macaca fascicularis*. *In* Lindburg (1980), 215–46.

White, T. D., G. Suwa, and B. Asfaw. 1994. *Australopithecus ramidus*, a new species of early hominid from Aramis, Ethiopia. Nature 371:306–12.

White, T. D., G. Suwa, and B. Asfaw. 1995. *Australopithecus ramidus*, a new species of early hominid from Aramis, Ethiopia. Nature 375:88.

Whitten, A. J. 1982. Home range use by Kloss gibbons (*Hylobates klossii*) on Siberut Island, Indonesia. Anim. Behav. 30:182–98.

Williams, E. E., and K. F. Koopman. 1952. West Indian fossil monkeys. Amer. Mus. Novit., no. 1546, 16 pp.

Wilson, D(on). E., and D. M. Reeder, eds. 1993. Mammal species of the world: a taxonomic and geographic reference. Smithsonian Inst. Press, Washington, D.C., xviii + 1206 pp.

Wilson, J. M., P. D. Stewart, G.-S. Ramangason, A. M. Denning, and M. S. Hutchings. 1989. Ecology and conservation of the crowned lemur, *Lemur coronatus*, at Ankarana, n. Madagascar. Folia Primatol. 52:1–26.

Winkler, P., H. Loch, and C. Vogel. 1984. Life history of hanuman langurs (*Presbytis entellus*). Folia Primatol. 43:1–23.

Winn, R. M. 1989. The aye-ayes, *Daubentonia madagascariensis*, at the Paris Zoological Garden: maintenance and preliminary behavioural observations. Folia Primatol. 52: 109–23.

Winter, P. 1968. Social communication in the squirrel monkey. *In* Rosenblum and Cooper (1968), 235–53.

Wirth, R., H. J. Adler, and N. Q. Thang. 1991. Douc langurs: how many species are there? Zoonooz 64(6):12–13.

Wolfe, L. D., and E. H. Peters. 1987. History of the free-ranging rhesus monkeys *(Macaca mulatta)* of Silver Springs. Florida Sci. 50:234–45.

Wolfheim, J. H. 1983. Primates of the world: distribution, abundance, and conservation. Univ. Washington Press, Seattle, xxiii + 831 pp.

Wood, B. 1992. Early hominid species and speciation. J. Human Evol. 22:351–65.

———. 1994. The oldest hominid yet. Nature 371:280–81.

Wrangham, R. W. 1977. Feeding behavior of chimpanzees in Gombe National Park, Tanzania. *In* Clutton-Brock (1977), 503–37.

Wright, P. C. 1994. The behavior and ecology of the owl monkey. *In* Baer, Weller, and Kakoma (1994), 97–112.

Wright, P. C., D. Haring, E. L. Simons, and P. Andau. 1987. Tarsiers: a conservation perspective. Primate Conserv. 8:51–54.

Wright, P. C., M. K. Izard, and E. L. Simons. 1986. Reproductive cycles in *Tarsius bancanus*. Amer. J. Primatol. 11:207–15.

Wu Hai-yin and Lin Yao-sung. 1992. Life history variables of wild troops of Formosan macaques *(Macaca cyclopis)* in Kenting, Taiwan. Primates 33:85–97.

Y

Yalden, D. W., M. J. Largen, and D. Kock. 1977. Catalogue of the mammals of Ethiopia. 3. Primates. Italian J. Zool., Suppl., n.s., 9:1–52.

Yamagiwa, J., N. Mwanza, A. Spangenberg, T. Maruhashi, T. Yumoto, A. Fischer, and B. Steinhauer-Burkart. 1993. A census of the eastern lowland gorillas *Gorilla gorilla graueri* in Kahuzi-Biega National Park with reference to mountain gorillas *G. g. beringei* in the Virunga region, Zaire. Biol. Conserv. 64:83–89.

Yeager, C. P. 1989. Feeding ecology of the proboscis monkey *(Nasalis larvatus)*. Internatl J. Primatol. 10:497–530.

———. 1990. Proboscis monkey *(Nasalis larvatus)* social organization: group structure. Amer. J. Primatol. 20:95–106.

———. 1991. Proboscis monkey *(Nasalis larvatus)* social organization: intergroup patterns of association. Amer. J. Primatol. 23:73–86.

———. 1992. Proboscis monkey *(Nasalis larvatus)* social organization: nature and possible functions of intergroup patterns of association. Amer. J. Primatol. 26:133–37.

Yoder, A. D. 1989. Survey of cheirogaleid primates in Madagascar. Rept. to World Wildl. Fund, mimeographed, 7 pp.

———. 1994. Relative position of the Cheirogaleidae in strepsirhine phylogeny: a comparison of morphological and molecular methods and results. Amer. J. Phys. Anthropol. 94:25–46.

Z

Zhang Ya-ping, Chen Zhi-ping, and Shi Li-ming. 1993. Phylogeny of the slow lorises (genus *Nycticebus*): an approach using mitochondrial DNA restriction enzyme analysis. Internatl. J. Primatol. 14:167–75.

Zhang Yongzu, Quan Guogiang, Lin Yonglei, and C. H. Southwick. 1989. Extinction of rhesus monkeys *(Macaca mulatta)* in Xinglung, north China. Internatl. J. Primatol. 10:375–81.

Zhang Yongzu, Quan Guogiang, Zhao Tigong, and C. H. Southwick. 1991. Distribution of macaques *(Macaca)* in China. Acta Theriol. Sinica 11:171–85.

Zhang Yongzu and L. Sheeran. 1994. Current status of the Hainan black gibbon *(Hylobates concolor hainanus)*. Asian Primates 3(3–4):3.

Zhang Ya-ping and Shi Li-ming. 1993. Phylogeny of rhesus monkeys *(Macaca mulatta)* as revealed by mitochondrial DNA restriction enzyme analysis. Internatl. J. Primatol. 14:587–605.

Zihlman, A. 1984. Body build and tissue composition in *Pan paniscus* and *Pan troglodytes*, with comparisons to other hominoids. *In* Susman (1984b), 179–200.

Zihlman, A., and J. M. Lowenstein. 1979. False start of the human parade. Nat. Hist. 88(7):86–91.

Zimmermann, E. 1985. Vocalizations and associated behaviours in adult slow loris *(Nycticebus coucang)*. Folia Primatol. 44:52–64.

———. 1989. Reproductive, physical growth, and behavioral development in slow loris *(Nycticebus coucang, Lorisidae)*. Human Evol. 4:171–79.

Index

The scientific names of orders, families, and genera, which have titled accounts in the text, are in boldfaced type, as are the page numbers on which such accounts begin. Other scientific names, and vernacular names, appear in roman.